Methods in Molecular Biology™

Series Editor
John M. Walker
School of Life Sciences
University of Hertfordshire
Hatfield, Hertfordshire, AL10 9AB, UK

For other titles published in this series, go to
www.springer.com/series/7651

Muscle Gene Therapy

Methods and Protocols

Edited by

Dongsheng Duan

Department of Molecular Microbiology and Immunology
University of Missouri, Columbia, MO, USA

Editor
Dongsheng Duan, Ph.D.
Department of Molecular Microbiology
and Immunology
University of Missouri
Columbia, MO
USA
duand@missouri.edu

ISSN 1064-3745 e-ISSN 1940-6029
ISBN 978-1-61737-981-9 e-ISBN 978-1-61737-982-6
DOI 10.1007/978-1-61737-982-6
Springer New York Dordrecht Heidelberg London

Printed on acid-free paper

Humana Press is part of Springer Science+Business Media (www.springer.com)

Preface

The mechanic that would perfect his work must first sharpen his tools.
Confucius (c. 551 BC–479 BC), a Chinese philosopher

Give us the tools, and we will finish the job.
Winston Churchill (1874–1965)

Gene therapy offers many conceptual advantages to treat muscle diseases, especially various forms of muscular dystrophies. Many of these diseases are caused by a single gene mutation. While the traditional approaches may ameliorate some symptoms, the ultimate cure will depend on molecular correction of the genetic defect. The clinical feasibility of gene therapy has been recently demonstrated in treatment of a type of inherited blindness. By delivering a therapeutic gene to the retina, investigators were able to partially recover the vision in a disease once thought incurable. Compared to retinal gene therapy, muscle gene therapy faces a number of unique challenges. Muscle is one of the most abundant tissues in the body. An effective therapy will require systemic infusion and targeted muscle delivery of huge amounts of therapeutic vectors. Severe inflammation associated with muscle degeneration and necrosis may further complicate immune reactions to the viral vectors and the therapeutic gene products. Furthermore, the vast majority of our current knowledge on muscle gene therapy is obtained from rodent models. Although these proof-of-concept studies have provided the critical foundation, the results are not easily translatable to human patients. With this in mind, we compiled this collection of muscle gene therapy methods and protocols with the intention of bridging the translational gap in muscle gene therapy.

The book is divided into three sections. The first section includes basic protocols for optimizing the muscle gene expression cassette and for evaluating the therapeutic outcomes. The chapters on the muscle-specific promoters and codon optimization outline strategies to generate powerful cassettes for muscle expression. Four chapters are devoted to end-point analysis. These include the use of epitope-specific antibodies, noninvasive monitoring of myofiber survival, and physiology assays of skeletal muscle and heart function.

Technology breakthroughs are the driving force in muscle gene therapy. Early muscle gene transfer studies were largely performed using vectors based on retrovirus, adenovirus, or plasmid DNA. Inherent limitations of these vectors (such as low transduction efficiency, transient expression, and a strong immune response) suggest that they are unlikely to meet the clinical need. These traditional gene delivery vehicles have now been replaced with the robust adeno-associated viral vector (AAV), oligonucleotide-mediated exon-skipping, and novel RNA-based strategies such as microRNA and RNA interference. The second section of this book is dedicated to the new developments in muscle gene therapy technology. Two chapters describe new strategies to generate muscle-specific AAV vectors by *in vivo* evolution and capsid reengineering. Two chapters provide methods for optimizing exon-skipping, and three chapters detail different applications of RNA-based approaches in muscle gene therapy.

Considering the importance of large animal studies, it is not surprising that the bulk of the protocols are devoted to muscle gene transfer in large animals models. In the last section, ten chapters provide step-by-step guidance on muscle gene delivery in swine, ovine, canine, and nonhuman primates. Methods include local delivery, isolated limb perfusion, myocardial gene transfer, and whole body systemic delivery. Ages range from fetal and neonatal to adult subjects.

In summary, this book presents a comprehensive collection of state-of-the-art muscle gene therapy protocols from leaders in the field. I would also like to mention that this collection of muscle gene therapy techniques complements the recently published book entitled "Muscle Gene Therapy" (Duan D eds., Springer, 2010, ISBN 978-1-4419-1205-3). Together, they will serve as a valuable resources for graduate students, postdoctoral fellows, and principle investigators who are interested in muscle gene therapy.

I would like to thank the contributors of each chapter for their excellent contributions. There is no doubt that these hard-to-find techniques, tricks, and the hands-on experience from the leading investigators will play an important role in bench-side to bedside translation of muscle gene therapy. I would like to thank Dr. John Walker, the series editor, for his guidance in the development of this book. I would like to thank Ms. Karen Ehlert for her administrative assistance in the final stage of preparation.

I am also very grateful to the National Institutes of Health and the Muscular Dystrophy Association for the funding of muscle gene therapy studies in my laboratory. I also thank the Parent Project Muscular Dystrophy and Jesse's Journey, The Foundation for Gene and Cell Therapy for their recent support in expanding our research in developing muscular dystrophy gene therapy. I am also much indebted to the patients and their families. I truly believe their dream will one day come true.

Finally, I'd like to dedicate this book to boys like Mark McDonald, they are our driving force.

Columbia, MO *Dongsheng Duan*

Contents

Contributors

KHALIL N. ABI-NADER • *Fetal Medicine Unit and Prenatal Cell and Gene Therapy Group, EGA Institute for Women's Health, University College London Hospitals, London, UK*

ARAVIND ASOKAN • *Gene Therapy Center, University of North Carolina at Chapel Hill, Chapel Hill, NC, USA; Department of Genetics, University of North Carolina at Chapel Hill, Chapel Hill, NC, USA*

TAKIS ATHANASOPOULOS • *Institute of Biomedical and Life Sciences, South West London Academic Network, St. George's University of London, London, UK; Centre for Biomedical Sciences, School of Biological Sciences, Royal Holloway, University of London, UK*

JÉRÔME AUSSEIL • *Généthon – CNRS-UMR8587 LAMBE, 1 bis rue de l'Internationale, France*

LAWRENCE T. BISH • *Department of Physiology, University of Pennsylvania School of Medicine, B400 Richards Building, 3700 Hamilton Walk, Philadelphia, USA*

ANGELA M. BODLES-BRAKHOP • *Inovio Biomedical Corporation, 2700 Research Forest Drive, The Woodlands, TX, USA*

DANIEL J. BOGAN • *Department of Pathology and Laboratory Medicine and The Gene Therapy Center, School of Medicine, University of North Carolina-Chapel Hill, Chapel Hill, NC, USA*

JANET R. BOGAN • *Department of Pathology and Laboratory Medicine and The Gene Therapy Center, School of Medicine, University of North Carolina-Chapel Hill, Chapel Hill, NC, USA*

BRIAN BOSTICK • *Department of Molecular Microbiology and Immunology, School of Medicine, The University of Missouri, One Hospital Drive, Columbia, MO, USA*

CHARLES R. BRIDGES • *Department of Surgery, Division of Cardiovascular Surgery, University of Pennsylvania School of Medicine, BRB II/III Building, 421 Currie Boulevard, Philadelphia, PA, USA*

KATE BRODERICK • *Inovio Biomedical Corporation, 2700 Research Forest Drive, The Woodlands, TX, USA*

CATHERINE CAILLAUD • *Département Génétique et Développement, Institut Cochin, Université Paris Descartes, CNRS (UMR 8104), Paris, France*

XIAOLAN CHEN • *Department of Biochemistry, University of Washington, Seattle, USA*

LOUIS G. CHICOINE • *Center for Gene Therapy, The Research Institute at Nationwide Children's Hospital and Department of Pediatrics, The Ohio State University, Columbus, OH, USA*

MARTIN K. CHILDERS • *Department of Neurology and Wake Forest Institute for Regenerative Medicine, School of Medicine, Wake Forest University, Winston-Salem, NC, USA*

ANNA L. DAVID • *Fetal Medicine Unit and Prenatal Cell and Gene Therapy Group, EGA Institute for Women's Health, University College London Hospitals, London, UK*
KAY E. DAVIES • *MRC Functional Genomics Unit, Department of Physiology, Anatomy, and Genetics, University of Oxford, South Parks Road, UK*
GEORGE DICKSON • *Institute of Biomedical and Life Sciences, South West London Academic Network, St. George's University of London, London, UK; Centre for Biomedical Sciences, School of Biological Sciences, Royal Holloway, University of London, Egham, UK; School of Biological Sciences, Royal Holloway, University of London, Egham, UK*
GAELLE DOUILLARD-GUILLOUX • *Department of Pathology, University of Pittsburgh, Pittsburgh, PA, USA*
RUXANDRA DRAGHIA-AKLI • *Inovio Biomedical Corporation, 2700 Research Forest Drive, The Woodlands, TX, USA*
DONGSHENG DUAN • *Department of Molecular Microbiology and Immunology, School of Medicine, The University of Missouri, 1 Hospital Drive, M610, Columbia, MO, USA*
HELEN FOSTER • *Institute of Biomedical and Life Sciences, South West London Academic Network, St. George's University of London, London, UK; Centre for Biomedical Sciences, School of Biological Sciences, Royal Holloway, University of London, Egham, UK*
KEITH FOSTER • *Institute of Biomedical and Life Sciences, South West London Academic Network, St. George's University of London, London, UK; Centre for Biomedical Sciences, School of Biological Sciences, Royal Holloway, University of London, Egham, UK*
AURÉLIE GOYENVALLE • *MRC Functional Genomics Unit, Department of Physiology, Anatomy, and Genetics, University of Oxford, South Parks Road, UK*
IAN R. GRAHAM • *School of Biological Sciences, Royal Holloway, University of London, Egham, UK*
ROBERT W. GRANGE • *Department of Human Nutrition, Foods, and Exercise, College of Agriculture and Life Sciences, Virginia Tech University, Blacksburg, USA*
ROGER J. HAJJAR • *The Cardiovascular Research Center, Mount Sinai School of Medicine, Atran Berg Laboratory Building, Floor 05, 1428 Madison Avenue New York, NY, USA*
CHADY H. HAKIM • *Department of Molecular Microbiology and Immunology, School of Medicine, University of Missouri, 1 Hospital Drive, M610, Columbia, MO, USA*
STEPHEN D. HAUSCHKA • *Department of Biochemistry, University of Washington, Seattle, WA, USA*
JULIA HEGGE • *Mirus BioCorporation, Madison WI, USA*
ARMEN HENDERSON • *Department of Surgery, Division of Cardiovascular Surgery, University of Pennsylvania School of Medicine, Philadelphia, PA, USA*
CHARIS L. HIMEDA • *Department of Biochemistry, University of Washington, Seattle, WA, USA*
ERIC HOFFMAN • *Research Center for Genetic Medicine, Children's National Medical Center, 111 Michigan Avenue, NW ,Washington, DC, USA*

MEI HUANG • *Department of Medicine, Stanford University School of Medicine, Stanford, CA, USA; Department of Radiology, Stanford University School of Medicine, Stanford, CA, USA*
ZHAN-PENG HUANG • *Cardiovascular Research Division, Department of Cardiology, Children's Hospital Boston, Harvard Medical School, 320 Longwood Avenue, Boston, MA, USA*
MICHAEL G. KATZ • *Department of Surgery, Division of Cardiovascular Surgery, University of Pennsylvania School of Medicine, Philadelphia, PA, USA*
YOSHIAKI KAWASE • *The Cardiovascular Research Center, Mount Sinai School of Medicine, Atran Berg Laboratory Building, Floor 05, 1428 Madison Avenue, New York, NY, USA*
AMIR S. KHAN • *Inovio Biomedical Corporation, 2700 Research Forest Drive, The Woodlands, TX, USA*
JOE N. KORNEGAY • *Department of Pathology and Laboratory Medicine and The Gene Therapy Center, School of Medicine, University of North Carolina-Chapel Hill, Chapel Hill, NC, USA*
DENNIS LADAGE • *The Cardiovascular Research Center, Mount Sinai School of Medicine, Atran Berg Laboratory Building, Floor 05, 1428 Madison Avenue, New York, NY, USA*
DEJIA LI • *Department of Molecular Microbiology and Immunology, School of Medicine, The University of Missouri, 1 Hospital Drive, M610, Columbia, MO, USA*
JUAN LI • *Division of Molecular Pharmaceutics, Eshelman School of Pharmacy, University of North Carolina at Chapel Hill, Chapel Hill, NC, USA*
ALBERTO MALERBA • *School of Biological Sciences, Royal Holloway, University of London, Egham, UK*
ALOCK MALIK • *Department of Surgery, University of Pennsylvania School of Medicine, BRB II/III Building, 421 Currie Boulevard, Philadelphia, PA, USA*
NGUYEN THI MAN • *Wolfson Centre for Inherited Neuromuscular Disease, RJAH Orthopaedic Hospital, Oswestry and Institute for Science and Technology in Medicine, Keele University, UK; Institute for Science and Technology in Medicine, Keele University, Keele, UK*
ANDREW MEAD • *Department of Surgery, Division of Gastrointestinal Surgery, University of Pennsylvania School of Medicine, BRB II/III Building, 421 Currie Boulevard, Philadelphia, PA, USA*
JERRY R. MENDELL • *Center for Gene Therapy, The Research Institute at Nationwide Children's Hospital and Department of Pediatrics, The Ohio State University, Columbus, OH, USA*
CHRYSTAL L. MONTGOMERY • *Center for Gene Therapy, The Research Institute at Nationwide Children's Hospital and Department of Pediatrics, The Ohio State University, Columbus, OH, USA*
GLENN MORRIS • *Wolfson Centre for Inherited Neuromuscular Disease, RJAH Orthopaedic Hospital, Oswestry and Institute for Science and Technology in Medicine, Keele University, UK; Institute for Science and Technology in Medicine, Keele University, Keele, UK*

RONALD L. NEPPL JR. • *Cardiovascular Research Division, Department of Cardiology, Children's Hospital Boston, Harvard Medical School, 320 Longwood Avenue, Boston, MA, USA*
MIHAIL PETROV • *Department of Surgery, Division of Gastrointestinal Surgery, University of Pennsylvania School of Medicine, BRB II/III Building, 421 Currie Boulevard, Philadelphia, PA, USA*
JANA L. PHILLIPS • *Gene Therapy Center, University of North Carolina at Chapel Hill, Chapel Hill, NC, USA*
LINDA J. POPPLEWELL • *School of Biological Sciences, Royal Holloway, University of London, Egham, UK*
JÉRÔME POUPIOT • *Généthon – CNRS-UMR8587 LAMBE, 1 bis rue de l'Internationale, France*
EMMANUEL RICHARD • *INSERM U876, IFR 66, Université Bordeaux 2, 146 rue Léo Saignat, Bordeaux, France*
ISABELLE RICHARD • *Généthon – CNRS-UMR8587 LAMBE, 1 bis rue de l'Internationale, France*
LOUISE R. RODINO-KLAPAC • *Center for Gene Therapy, The Research Institute at Nationwide Children's Hospital and Department of Pediatrics, The Ohio State University, Columbus, OH, USA*
R. JUDE SAMULSKI • *Gene Therapy Center, University of North Carolina at Chapel Hill, Chapel Hill, NC, USA*
CAROLINE A. SEWRY • *Wolfson Centre for Inherited Neuromuscular Disease, RJAH Orthopaedic Hospital, Oswestry and Institute for Science and Technology in Medicine, Keele University, UK; Institute for Science and Technology in Medicine, Keele University, Keele, UK*
JIN-HONG SHIN • *Department of Molecular Microbiology and Immunology, School of Medicine, University of Missouri, Columbia, MO, USA*
MEG M. SLEEPER • *Section of Cardiology, Department of Clinical Studies, Veterinary Hospital of the University of Pennsylvania, Philadelphia, PA, USA*
HANSELL H. STEDMAN • *Department of Surgery, Division of Gastrointestinal Surgery, University of Pennsylvania School of Medicine, BRB II/III Building, 421 Currie Boulevard, Philadelphia, PA, USA*
RAINER STORB • *Program in Transplantation Biology, Division of Clinical Research, Fred Hutchinson Cancer Research Center, Seattle, WA, USA; Department of Medicine, University of Washington, Seattle, WA, USA*
JABARIS D. SWAIN • *Department of Surgery, Division of Cardiovascular Surgery, University of Pennsylvania School of Medicine, Philadelphia, USA*
H. LEE SWEENEY • *Department of Physiology, University of Pennsylvania School of Medicine, B400 Richards Building, 3700 Hamilton Walk, Philadelphia, PA, USA*
SHIN'ICHI TAKEDA • *Department of Molecular Therapy, National Institute of Neuroscience, National Center of Neurology and Psychiatry (NCNP), Tokyo, Japan*
STEPHEN J. TAPSCOTT • *Division of Human Biology, Fred Hutchinson Cancer Research Center, Seattle, WA, USA; Department of Neurology, University of Washington, Seattle WA, USA*

DANIELLE M. THESIER • *Department of Surgery, Division of Cardiovascular Surgery, University of Pennsylvania, School of Medicine, Philadelphia, PA, USA*
DA-ZHI WANG • *Cardiovascular Research Division, Department of Cardiology, Children's Hospital Boston, Harvard Medical School, 320 Longwood Avenue, Boston, MA, USA*
ZEJING WANG • *Program in Transplantation Biology, Division of Clinical Research, Fred Hutchinson Cancer Research Center, Fairview Av N, D1-100 Seattle, WA, USA*
JENNIFER D. WHITE • *Department of Surgery, Division of Cardiovascular Surgery, University of Pennsylvania School of Medicine, Philadelphia, PA, USA*
JON A. WOLFF • *Mirus BioCorporation, Madison, WI, USA*
JOSEPH C. WU • *Department of Medicine, Stanford University School of Medicine, Stanford, CA, USA; Department of Radiology, Stanford University School of Medicine, Stanford, CA, USA*
XIAO XIAO • *Division of Molecular Pharmaceutics, Eshelman School of Pharmacy, University of North Carolina at Chapel Hill, Chapel Hill, NC, USA*
LIN YANG • *Division of Molecular Pharmaceutics, Eshelman School of Pharmacy, University of North Carolina at Chapel Hill, Chapel Hill, NC, USA*
TOSHIFUMI YOKOTA • *Research Center for Genetic Medicine, Children's National Medical Center, 111 Michigan Avenue, NW, Washington, DC, USA*
YONGPING YUE • *Department of Molecular Microbiology and Immunology, School of Medicine, The University of Missouri, One Hospital Drive, Columbia, USA*

Part I

Basic Methodology Related to Muscle Gene Therapy

Chapter 1

Design and Testing of Regulatory Cassettes for Optimal Activity in Skeletal and Cardiac Muscles

Charis L. Himeda, Xiaolan Chen, and Stephen D. Hauschka

Abstract

Gene therapy for muscular dystrophies requires efficient gene delivery to the striated musculature and specific, high-level expression of the therapeutic gene in a physiologically diverse array of muscles. This can be achieved by the use of recombinant adeno-associated virus vectors in conjunction with muscle-specific regulatory cassettes. We have constructed several generations of regulatory cassettes based on the enhancer and promoter of the muscle creatine kinase gene, some of which include heterologous enhancers and individual elements from other muscle genes. Since the relative importance of many control elements varies among different anatomical muscles, we are aiming to tailor these cassettes for high-level expression in cardiac muscle, and in fast and slow skeletal muscles. With the achievement of efficient intravascular gene delivery to isolated limbs, selected muscle groups, and heart in large animal models, the design of cassettes optimized for activity in different muscle types is now a practical goal. In this protocol, we outline the key steps involved in the design of regulatory cassettes for optimal activity in skeletal and cardiac muscle, and testing in mature muscle fiber cultures. The basic principles described here can also be applied to engineering tissue-specific regulatory cassettes for other cell types.

Key words: Skeletal muscle, Cardiac muscle, Regulatory cassette, Muscular dystrophy, Gene therapy, Transcriptional regulation, Muscle creatine kinase

1. Introduction

Duchenne muscular dystrophy (DMD) is caused by a lack of functional dystrophin, resulting in patient death due to cardiac and/or respiratory failure. Gene therapy for muscle diseases such as DMD requires efficient gene delivery to the striated musculature and specific, high-level expression of the therapeutic gene in a physiologically diverse array of muscles. This can be achieved by the use of regulatory cassettes composed of enhancers and promoters that contain combinations of muscle-specific and ubiquitous

Dongsheng Duan (ed.), *Muscle Gene Therapy: Methods and Protocols*, Methods in Molecular Biology, vol. 709,
DOI 10.1007/978-1-61737-982-6_1,

control elements. Recombinant adeno-associated virus (rAAV) vectors (particularly serotypes 1, 6, 7, 8, and 9, which exhibit preferential transduction of striated muscle) are well-suited for this challenge, since when combined with appropriate regulatory cassettes, they mediate high-level, long-term transgene expression in striated muscle. Despite the preferential targeting of striated muscle by certain AAV serotypes, the widespread dissemination of vector following systemic delivery still raises concerns about inadvertent transduction of nonmuscle tissues, which may give rise to an unwanted immune response. Thus, even though viral promoters (such as the CMV and RSV promoters) mediate high-level expression in skeletal and cardiac muscle (1, 2), these promoters also exhibit high activity in dendritic cells and relatively high activity in spleen and testes (1). Thus, the construction of high-activity, muscle-specific regulatory cassettes is a necessary goal. Since the effective genome size limit for AAV packaging is limited, miniaturization of the transcription regulatory cassettes is another important priority for the expression of cDNAs larger than ~3.5 kb.

Muscle-specific gene expression is determined by combinatorial interactions between muscle-specific and ubiquitous trans-acting factors bound to the enhancers and promoters, and the interactions of these factors with associated protein complexes. The exact nature of these combinatorial interactions is unknown, but many of the *cis* elements and *trans* components regulating muscle-specific genes have been identified and partially characterized.

The muscle creatine kinase (MCK) gene has served as a useful model of muscle-specific gene transcription since its protein product is specifically and abundantly expressed in striated muscle, and its regulatory regions have been extensively characterized. MCK is also expressed at different levels in different anatomical skeletal muscles, and in skeletal vs. cardiac muscle (3, 4). This allows for the identification of control elements and binding factors important for expression in different muscle types, as well as those specific to each myogenic lineage. Importantly, the MCK enhancer and promoter synergize in all striated muscle types in vivo to give 100-fold elevated activity over that of either alone (3, 4), making the composite enhancer-promoter a useful cassette for muscle gene therapy.

To date, cassettes built from regulatory regions of the MCK, skeletal α-actin, myosin heavy chain, and myosin light chain (MLC) 1/3 genes, and randomized muscle gene control elements fulfill the criteria for muscle-specific expression (1, 2, 5–9). However, these cassettes either lack high-level activity in certain muscle types, have not been studied quantitatively in different tissues, or their size exceeds the limitation for packaging into an AAV vector in conjunction with the smallest microdystrophin cDNAs (~3.5 kb) that provide reasonable function in the mdx mouse model of DMD.

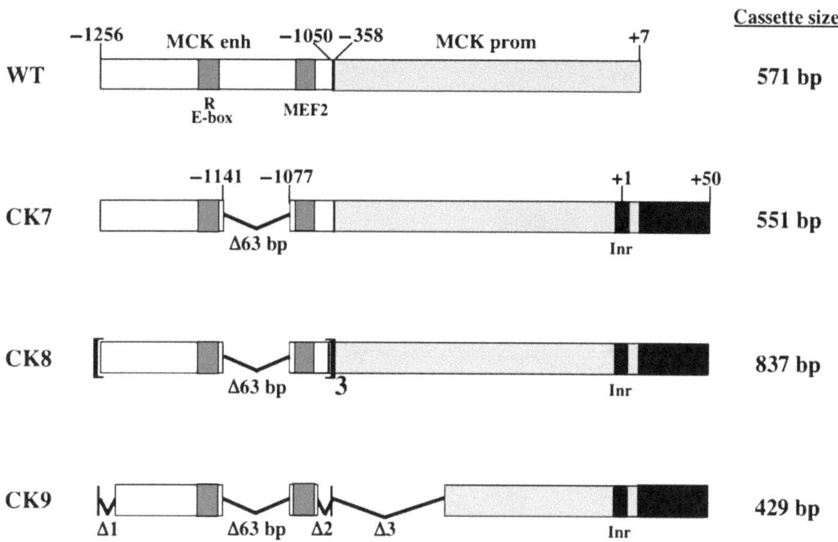

Fig. 1. MCK-based regulatory cassettes for expression in skeletal and cardiac muscle. The original cassette (WT) includes the MCK enhancer (–1256 to –1050) linked to the MCK proximal promoter (–358 to +7). The Right E-box and MEF2 site within the MCK enhancer are shown. CK7 incorporates three changes from WT: (1) deletion of 63 bp between the R E-box and MEF2 site (–1140 to –1078), (2) mutation of sequence overlapping +1 to a consensus Initiator element (Inr), and (3) insertion of 43 bp 3′ of +7. CK8 is identical to CK7 except that it contains three copies of the modified enhancer. CK9 is identical to CK7 except that it contains three additional deletions: (Δ1) –1256 to –1240, (Δ2) –1063 to –1050, and (Δ3) –358 to –268. Refer to text for more details.

Recently, our lab has constructed several new generations of regulatory cassettes based on the MCK enhancer and promoter, with the addition of enhancers and individual elements from other muscle genes. In designing these cassettes, we have tried to utilize existing information regarding the function of control elements and their cognate binding factors. For example, we theorized that deleting nonconserved sequences between the Right E-box and MEF2 site in the MCK enhancer (Figs. 1 and 2) might better facilitate interactions between the myogenic regulatory factors (MyoD, Myogenin, Myf5, and MRF4) and MEF2, which are known to synergize. In addition to increasing MCK enhancer activity in skeletal myocytes, this change also significantly reduced the size of the cassette, allowing other useful elements to be incorporated.

Since the relative importance of many control elements appears to vary among different anatomical muscles, we are aiming to tailor regulatory cassettes for high-level expression in cardiac muscle, and in fast and slow skeletal muscle. With the achievement of efficient intravascular gene delivery to isolated limbs, selected muscle groups, and heart in a large animal model (10–12), the design of cassettes optimized for activity in different muscle types is now a practical goal. Such cassettes would be useful, not only for DMD, but for many other neuromuscular diseases as well.

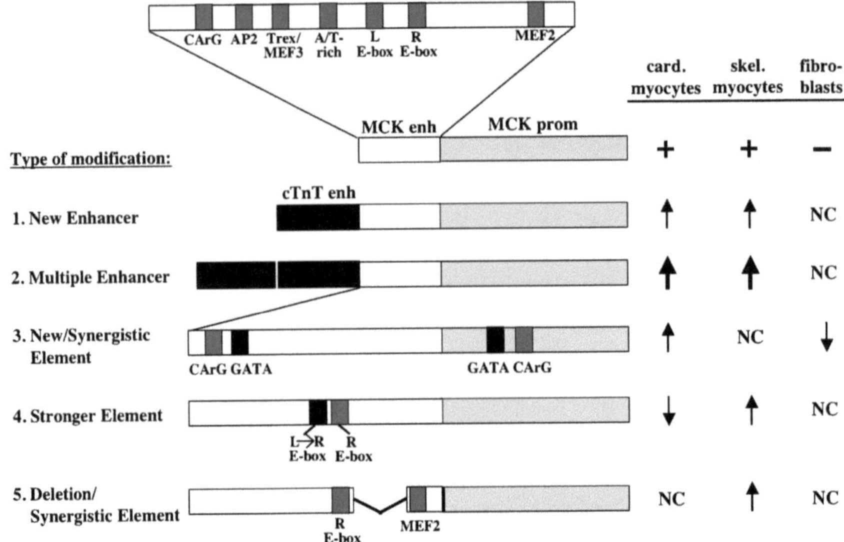

Fig. 2. Activity-enhancing modifications to MCK-based regulatory cassettes. The wild-type MCK enhancer-promoter was modified in a number of ways, including the examples shown here: (1) insertion of the cTnT enhancer 5′ of the MCK enhancer; (2) insertion of two copies of the cTnT enhancer 5′ of the MCK enhancer; (3) insertion of two GATA sites near CArG sites in the MCK enhancer and promoter; (4) mutation of the Left E-box in the MCK enhancer to the sequence of the Right E-box; (5) deletion of 63 bp between the Right E-box and MEF2 site in the MCK enhancer. Effects on cassette activity in cultured cardiomyocytes, skeletal myocytes, and fibroblasts are shown. Although these modifications are shown for simplicity in the context of the wild-type MCK enhancer, they were actually performed in the context of the CK6 (25), CK7, or CK9 cassettes. Refer to text for more details.

In this protocol, we outline the key steps involved in the design of regulatory cassettes for optimal activity in skeletal and cardiac muscle. The basic principles described here can also be applied to the engineering of tissue-specific regulatory cassettes for optimal expression in other cell types.

Since the vast majority of our in vitro assays for cassette activity are performed in the MM14 permanent line of mouse myoblasts (similar to C2C12 cells) and in neonatal cardiomyocytes, and since detailed protocols for the culture and transfection of these cell types have been described (13–16), the testing section of this protocol will focus on our ongoing efforts to isolate and transfect mature skeletal muscle fibers in culture, which have the potential to provide a more physiologically relevant model for cassette activity.

2. Materials

2.1. Designing Regulatory Cassettes for Expression in Striated Muscle

1. Access to the transcription factor binding site databases such as TRANSFAC.
2. DNA sequence analysis software such as Vector NTI.

2.2. Testing Regulatory Cassettes in Isolated Muscle Fiber Cultures

As a general rule, only sterile materials and supplies are used. All glassware and dissection tools are autoclaved, and all tissue culture (TC) steps are performed with sterile technique.

1. 35 × 100-mm cell culture dishes (Corning, Corning, NY, USA).
2. 100 × 20-mm Petri dishes (Corning).
3. Penicillin-streptomycin solution (Pen/strep): 10,000 U/mL penicillin and 10 mg/mL streptomycin (Sigma-Aldrich, St. Louis, MO, USA).
4. Pretested horse serum (HS) from any suppliers. Filter through a 0.45-μm filter just prior to use.
5. Cell culture medium: DMEM, high glucose with l-glutamine, sodium pyruvate, and pyridoxine hydrochloride (Gibco-Invitrogen, Grand Island, NY, USA) supplemented with 50 U/mL penicillin and 50 μg/mL streptomycin.
6. Collagenase type I (Sigma-Aldrich).
7. Matrigel (BD Biosciences, San Jose, CA, USA).
8. Five-inch sterile glass Pasteur pipettes (VWR, West Chester, PA). Use a file or a diamond knife to prepare pipettes with a bore diameter of ~3 and ~2 mm. Shake the pipettes to remove any glass fragments and fire-polish the sharp ends.
9. Hyaluronidase type I-S (Sigma-Aldrich).
10. Lipofectamine LTX reagent (Invitrogen, Carlsbad, CA, USA).
11. Optimem I reduced serum medium 1× (Invitrogen).

3. Methods

3.1. Designing Regulatory Cassettes for Expression in Striated Muscle

1. Define the boundaries of the starting enhancer/promoter. The starting enhancer/promoter should provide strong muscle-specific activity to a reporter gene in both cell culture and animal models (see Note 1). If the starting regions are uncharacterized, multispecies sequence alignments can provide strong clues to the presence of functional elements (see Note 2). Search candidate sequences against the TRANSFAC transcription factor binding site database. Since the DNA-binding motifs for many transcription factors have not been fully characterized, the absence of a TRANSFAC-identified motif does not rule out the presence of a functional element (see Note 3). If no information is available regarding function, perform in silico sequence analysis. This provides a starting point for functional analysis of candidate regulatory regions, usually by testing the effects of systematic deletions and putative control element mutations on the expression of a reporter gene (13) (see Note 4).

2. Multimerize enhancers. Multimerizing a single enhancer such as that from the MCK gene can provide significant increases in cassette activity (see Note 5). Tissue- or cell-type-specific enhancers from other striated muscle genes (such as the α-myosin heavy chain gene and the cardiac troponin T [cTnT] gene) can also be added to the MCK-based cassettes to further enhance expression in a particular muscle type (see Note 6) (Fig. 2). Introduce additional enhancers to the starting construct using standard cloning procedures. Use restriction sites that allow bi-directional insertion during cloning since enhancers may be more active in one orientation than the other (see Note 7). Screen for orientation by restriction mapping or sequencing.
3. Introduce additional control elements. Striated muscle gene enhancer/promoter sequences contain only a subset of the total control elements known to be important for expression in particular muscle types. In order to boost expression, new elements can be introduced. Introduce additional positive control elements by site-directed mutagenesis (nucleotide mutation or insertion). These elements include general muscle elements (A/T-rich/MEF2, CArG/SRF, MCAT/TEF-1, MEF3/Six, NFAT), skeletal muscle elements (E-box), slow/fast skeletal muscle elements (SURE/FIRE regions containing multiple elements), cardiac muscle elements (GATA-4, Nkx2.5, Tbx5), and ubiquitous elements (Sp, KLF, MAZ, AP2) (see Note 8). Sequence the resulting enhancer/promoter to verify the proper change and integrity of the remaining sequence (see Note 9). Since regulatory cassette improvement is an ongoing process, we typically test alterations in the context of the "best" current cassette. This shortcut is faster and less costly, but it could result in missing some beneficial effects that would have been detected by testing in the native enhancer/promoter context (see Note 10).

   Consider the following when deciding where to place new elements. (1) If no information is available regarding the optimal spacing of two elements, ~10 bp of intervening sequence should be left to avoid steric hindrance of binding factors. This amounts to ~1 turn of the DNA helix; therefore, factors binding adjacent sites will be on roughly the same side of the helix, potentially enhancing their ability to interact. (2) Mutation of native "nonfunctional" sequences to new elements is more space-efficient than insertion of elements. (3) Mutation of negative elements to positive elements is another space-efficient way to increase activity. (4) New elements can be placed close to preexisting elements to allow for potential synergy (see Note 11) (Fig. 2) (5). Many elements that are important for muscle gene regulation are recognized by factors whose expression is ubiquitous or widespread (e.g., SRF, TEF-1). Since sequences flanking a core element often play a role in tissue-specific binding, care should be taken to introduce the appropriate flanking sequences (see Note 12).

In some cases, control elements in the native enhancer/promoter will not be optimal motifs for binding/trans-activation by their cognate transcription factors. If data on stronger motifs is available, the weaker elements can be altered to increase activity of the cassette (see Note 13) (Fig. 2). Likewise, when data on synergy between two transcription factors or binding sites is available, elements can be incorporated to allow for potential synergy between their cognate factors (see Notes 8, 11, and 14) (Table 1, Fig. 2).

4. Delete negative elements and nonconserved sequences. Deletion of negative elements and poorly conserved sequences reduces the size of the cassette, allowing room for additional positive elements (see Notes 14 and 15) (Fig. 2). Perform sequence deletions using site-directed mutagenesis.

5. Consider species-specific issues. The effects of species-specific differences in control elements have not yet been fully explored.

Table 1
Examples of Synergy between muscle transcription factors

	SRF	GATA factors	Nkx2.5	MEF2	MRFs	HAND factors	Tbx factors	TEF-1	NFAT	YY1	KLF3
SRF	(46)	(32)	(47)[a]		(48)			(49)		(50)	[X][a]
GATA factors	(32)		(51–53)	(54)[a]	(55)	(56)[a]	(57)		(58)	(59)[a]	
Nkx2.5	(45)[a]	(51–53)					(60)[a]				
MEF2		(53)[a]			(61)[a]	(62)[a]		(63)			
MRFs	(48)	(55)		(61)[a]					(64)		
HAND factors		(56)[a]		(62)[a]							
Tbx factors		(57)	(60)[a]								
TEF-1	(49)			(63)							
NFAT		(58)			(64)						
YY1	(50)	(59)[a]									
KLF3	(29)[a]										

(#) Synergistic interaction between two factors. Reference numbers are shown.
[a]Synergistic interaction requiring the binding site of only one factor

Many, but not all sequence motifs are fully conserved between mammalian homologues of the same gene, and virtually all recognized control elements exhibit minor sequence differences between striated muscle genes. For example, the transcriptionally important CArG site, and Left and Right E-boxes differ between the mouse and primate versions of the MCK enhancer, whereas the MEF2, A/T-rich, and Six4/5 control elements are identical. The conservation of these differences among primates could imply that the optimal striated muscle regulatory cassettes based on mouse studies should be "humanized" for their eventual use in patients. This could be achieved by simply replacing the optimal elements in a cassette developed via mouse studies with what are thought to be the optimal human versions (see Note 16).

3.2. Testing Regulatory Cassettes in Isolated Muscle Fiber Cultures

Cultures derived from single isolated muscle fibers are a powerful in vitro model to examine satellite cell gene expression and activation, myogenesis, and regenerative capacity (17–21). This system enables researchers to monitor satellite cells as well as mature fibers in the more physiological microenvironment found in living muscle. Since MCK-based cassettes display different activities in fast vs. slow fiber types in vivo (1), mature fibers isolated from fast and slow skeletal muscles may serve as a useful and relatively inexpensive model prior to testing in animals. A logical order would be to test new cassettes first in cultured skeletal myocytes, followed by testing in isolated fiber cultures, followed by testing in vivo. Results should be evaluated and prioritized at each successive step so that progressively fewer gene constructs are tested at each level.

1. Coat TC dishes with Matrigel. Thaw Matrigel on ice for ~20 min (see Note 17). Dilute Matrigel with ice-cold DMEM to a final concentration of 1 mg/mL. Add ~500 μL diluted Matrigel to the required number of 35-mm dishes and spread to evenly coat (see Note 18). Allow coated dishes to sit at room temperature for 5–10 min. Transfer excess Matrigel solution from the dishes back to the original tube with diluted Matrigel that is kept on ice. Use this solution to coat additional dishes within the next 2 h. Incubate Matrigel-coated dishes at 37°C in 5% CO_2 in the humidified TC incubator for at least 30 min. Add 1.6 mL DMEM containing 10% HS and hyaluronidase to each dish and return dishes to the incubator until fibers are ready to be transferred (see Notes 19 and 20).
2. Coat glassware and Petri dishes to prevent fibers from sticking. Coat 10-cm Petri dishes by adding 1 mL HS to each dish and swirling to coat evenly. Use one dish for each individual muscle. Allow dishes to sit at room temperature for 5 min, then aspirate HS and add 10 mL DMEM to each dish. Place Petri dishes in the incubator until needed. Coat fire-polished

Pasteur pipettes with DMEM + 10% HS by passing solution through pipettes several times.

3. Prepare muscle digestion solution. Prepare 0.2% collagenase in DMEM (6 mg collagenase in 3 mL DMEM for each 35-mm dish) (see Note 21). Filter the solution through a 0.22-μm syringe filter.

4. Dissect muscle tissue. Sacrifice mice according to institute regulations. Remove muscles immediately, taking care to handle them only by their tendons to minimize damage to fibers. Our studies have concentrated on the extensor digitorum longus (EDL, primarily fast fibers) and soleus (primarily slow fibers) muscles. Rinse muscles with DMEM (see Note 22). Transfer to the collagenase solution and place in the incubator for 2 h (see Note 21) without shaking (see Note 23).

5. Isolate single fibers from bulk muscle tissue. Inspect each muscle under dissecting microscope to make sure myofibers are loosened. If not, continue enzymatic digestion for another 10–15 min and check again. Transfer one muscle from the collagenase solution to a 60-mm dish containing ~5 mL DMEM using the wide-bore Pasteur pipette. Swirl to rinse muscle. Remove one 10-cm Petri dish containing 10 mL media from incubator. Transfer muscle to Petri dish and return dish to incubator for ~15 min prior to trituration. Remove 10-cm Petri dish from incubator and use the wide-bore pipette to gently triturate the muscle. Liberate single fibers by repeated trituration (see Note 24). Remove one 35-mm Matrigel-coated dish from incubator once 20–30 single myofibers have been isolated. Pick up each fiber, one at a time, using a P10 pipette (tip preflushed with DMEM + 10% HS), minimizing uptake of media. Gently release each myofiber onto dish. Check under dissecting microscope to verify release (occasionally myofibers adhere to the tip and are not dispensed). Alternate between several 10-cm dishes so that muscles are not outside the incubator for longer than 15 min at a time. Repeat the procedure until the required number of single myofibers has been isolated and plated (see Note 25).

6. Transfect isolated muscle fibers. Transfection of muscle fibers can be performed on the day of isolation (2–3 h after fibers have been plated) or the following day. Follow the standard Lipofectamine LTX protocol with the following modifications. (1) Before adding DNA/LTX complexes, gently remove 0.8 mL media from the 35-mm culture dish. (2) Add DNA/LTX complexes carefully, with a minimum of disturbance. (3) Incubate for 4–6 h in TC incubator. (4) Add 0.6 mL fresh DMEM + 10% HS + Pen/strep (see Note 26). (5) Incubate for 48 h prior to assaying reporter expression (see Note 27).

4. Notes

1. We have used the MCK upstream enhancer (−1256 to −1050) linked to the proximal promoter (−358 to +7) as the foundation for our muscle-specific regulatory cassettes (see Figs. 1 and 2). This seemed strategically appealing since our initial goal was to develop a single cassette that would be optimal for expression in both skeletal and cardiac muscle in conjunction with systemic vector delivery. This original cassette drives striated muscle-specific expression of reporter genes in cell culture, transgenic mice, and virus-mediated gene transfer (1, 3, 4, 13, 22–25).
2. The sequences and relative positions of the seven known control elements in the MCK enhancer are highly conserved among mammalian species, whereas sequences between these elements are poorly conserved (26).
3. We have found that several transcription factors binding the MCK enhancer/promoter recognize sequences that diverge from the established binding motif. For example, the Trex site in the MCK enhancer is bound by Six4, a homeodomain protein of the Six/sine oculis family, in skeletal muscle, and Six5 in cardiac muscle (27). Six proteins recognize MEF3 motifs in the regulatory regions of their target genes; however, because the MCK Trex site deviates from the previously established MEF3 sequence in 2 out of 7 bp, this relationship was not identifiable by in silico screening against the TRANSFAC database.

   In more recent studies, we showed that MAZ and KLF3, two zinc-finger transcription factors that regulate the MCK gene and other muscle genes, also recognize a divergent spectrum of sequences (28, 29). Many of these alternate motifs are not present in the TRANSFAC database, but are found in the regulatory regions of many striated muscle genes.
4. While most enhancer/promoter sequences lie upstream of the transcription start site, many active regulatory regions (such as intronic or 3′ enhancers) lie downstream of +1. For example, the well-studied MLC 1/3 enhancer occurs >24 kb downstream of the MLC1 promoter and >14 kb downstream of the MLC3 promoter (30), and MCK contains a highly-conserved enhancer within the first intron (Tai and Hauschka, unpublished data). We have also found that adding 50 bp of highly conserved sequence from the noncoding MCK exon-1 (see Fig. 1) significantly increased activity of the MCK enhancer-promoter in both skeletal and cardiac myocytes (1).
5. We have found increasing transcriptional activity with up to three copies of the MCK enhancer in both skeletal and cardiac myocytes, but four or more copies provide little to no additional

benefit (Nguyen and Hauschka, unpublished data). In skeletal myocytes, two copies of the MCK CK7 enhancer (Fig. 1) are ~2-fold more active than a single copy, and three copies (CK8) (Fig. 1) are ~4-fold more active than a single copy (Nguyen and Hauschka, unpublished data). In cardiac myocytes, two copies of the CK7 enhancer are only 20% more active than a single copy, whereas three copies are ~3-fold more active than a single copy (Nguyen and Hauschka, unpublished data).

6. Adding the cTnT enhancer upstream of the MCK CK7 enhancer-promoter (Fig. 2) increased activity ~13-fold over CK7 in cardiac myocytes (Chen, Nguyen, and Hauschka, unpublished data). Unexpectedly, activity in skeletal myocytes was also increased by ~4-fold. However, extrapolation of the latter in vitro data to the in vivo steady-state expression levels in skeletal muscle fibers may not be warranted, because while cTnT is activated at the onset of skeletal muscle differentiation, its expression decreases upon maturation. Up to two copies of the cTnT enhancer upstream of the MCK CK7 enhancer-promoter (Fig. 2) further increased activity (~17-fold over CK7 in cardiomyocytes and ~5-fold in skeletal myocytes; Chen, Nguyen, and Hauschka, unpublished data). Thus, three total enhancers appear to be the maximum for increased activity in our test system (see Note 5).

7. One copy of the MCK enhancer is ~50% more active in the *negative* vs. *positive* orientation in skeletal myocytes, and, strikingly, ~7.5-fold more active in cardiomyocytes (13). However, two copies of the MCK CK7 enhancer are ~35% more active in the *positive* vs. *negative* orientation in skeletal myocytes and ~70% more active in cardiomyocytes (Nguyen and Hauschka, unpublished data). Since the cause(s) of these differences are not understood, potentially beneficial orientation effects should be established empirically.

8. Initiator elements have been shown to cooperate with TATA boxes in driving transcription of the downstream genes (31). Addition of a consensus Initiator site at +1 of the MCK enhancer-promoter (Fig. 1) resulted in a ~30% increase in activity in skeletal myocytes, but a ~5-fold increase in fibroblasts (Nguyen, Himeda, and Hauschka, unpublished data). Even though expression in fibroblasts is still very low compared to skeletal myocytes, this result underscores a potential pitfall (decrease in tissue specificity) of introducing strong ubiquitous elements into regulatory cassettes designed for cell type-specific expression.

9. Following sequencing, it is advisable to re clone the altered enhancer/promoter into the original parent vector. This eliminates the possibility of mutations introduced into the vector during site-directed mutagenesis.

10. When successive changes are made to a cassette, later changes are likely to have a dampened positive effect compared to earlier ones. There is also no formula for predicting changes in activity (i.e., adding an extra element whose deletion from the native enhancer gives a twofold decrease in activity will not necessarily result in a twofold increase). Additionally, there is always a trade-off between enhancing activity of the cassette and staying within the vector packaging size constraints (e.g., introducing a large insertion into a cassette may not be justified by only a slight increase in activity).
11. GATA sites have been shown to be important mediators of cardiac gene expression, and are capable of synergizing with CArG sites (32). Introduction of two GATA sites in CK9 (one just 3′ of the CArG site in the enhancer and one just 5′ of the CArG site in the promoter) (Fig. 2) increased activity by ~2-fold over CK9 in cardiomyocytes, but had no effect in skeletal myocytes (Nguyen and Hauschka, unpublished data). Interestingly, this alteration also decreased activity in fibroblasts, thus further increasing the muscle-specificity of the cassette.
12. Flanking sequences have been shown to modulate the muscle-specificity of several control elements, including E-boxes (33), MCAT binding motifs (34), and SRF binding sites (35). For ex-ample, while E-box sequences are often simplified to CANNTG, the E-box consensus for skeletal muscle genes is $(C/G)N(A/G)_2CA(C/G)_2TG(C/T)_2N(C/G)$ (core sequence underlined) (36). Interestingly, MyoD/E12 heterodimers preferentially bind the E-box consensus [A/G/C]CACCTGT [T/C] (37) while purified Myogenin preferentially binds AACAG[T/C]TGTT (38).
13. PCR-mediated random site selection was used to determine a set of overlapping, but distinct sequence preferences for GATA family members (39). In this study, high-affinity binding sites also mediated high levels of transactivation by GATA factors. Binding-site preferences for many other factors, including TEF-1 and MAZ, have also been explored using gel-shift studies to compare variant sequences (28, 40). However, binding-site affinity does not always correlate with in vivo transactivation potential, as demonstrated for MyoD (41). Using a functional random sequence selection approach, different flanking and core sequences of the MyoD binding motif (CANNTG) were shown to be required for in vitro binding vs. in vivo transactivation (41). Interestingly, sequences that mediated high-level transactivation were found to be identical or very similar to native E-boxes in the promoters of endogenous muscle genes, suggesting that muscle E-boxes have already been "optimized" by nature for high activity.

Since MyoD has been shown to activate reporter genes more strongly through paired E-boxes than through a single E-box (42), the Left E-box in the MCK enhancer (a low-affinity binding site for MyoD) was altered to conform to the sequence of the Right E-box (a high-affinity MyoD binding site) (Fig. 2). This change significantly increased activity in skeletal myocytes as found in our CK6 regulatory cassette (25). However, two Right E-boxes are not optimal for expression in cardiac and slow skeletal muscles. Thus, the Left E-box was restored in our CK7 cassette.

14. Deletion of 63 bp of sequence between the Right E-box and the MEF2 site in the MCK CK7 enhancer (Figs. 1 and 2) resulted in a ~30% increase in activity in skeletal myocytes, possibly due to enhanced synergy between myogenic regulatory factors and MEF2 (1).
15. We have made several deletions in nonconserved sequences within the MCK enhancer and promoter (Fig. 1), with varying effects. When deletions in the CK7 enhancer (–1256 to –1240 and –1063 to –1050) were combined with a large promoter deletion (–358 to –268) to construct CK9, activity increased ~4-fold over CK7 in skeletal myocytes, and ~5-fold in cardiomyocytes (Nguyen and Hauschka, unpublished data).
16. With respect to species optimization of control elements, it is of interest that a mouse version of the CK7 cassette is ~12-fold more active in rat cardiomyocytes (rat and mouse MCK enhancer sequences are essentially identical) and ~3-fold more active in human cardiomyocytes than the human version of CK7 (Welikson and Hauschka, unpublished data). Thus, it is possible that the optimal cassettes for expression in a particular species will contain control elements from a different species.
17. Matrigel is a solubilized basement membrane matrix composed of laminin, collagen IV, heparin sulfate proteoglycan, and entacin/nidogen. To ensure its stability, we thaw the stock solution on ice and aliquot 200 μL each to precooled tubes. Aliquots are stored at –20°C.
18. We usually use 35-mm dishes or a 2-well chamber slide (not a 24-well plate) for plating muscle fibers. Since fibers cannot be kept out of the incubator for longer than 15 min (43), this is much easier to accomplish working with a single dish at a time.
19. Several different media types have been used by investigators interested in propagating satellite cells from isolated muscle fibers (21). These media typically contain 20% FBS + 10% HS + 1% chick embryo extract (CEE). Since our goal is to test regulatory cassettes in mature muscle fibers rather than proliferating/differentiating satellite cells, and since the higher concentrations of serum or CEE, which promotes satellite

cell growth, cause most of the plated fibers to hypercontract and die (43), we opted to omit the FBS and CEE.

20. Hyaluronidase helps dissociate the extracellular matrix and has been shown to significantly enhance transfection efficiency in rat skeletal muscle (44). Therefore, we include hyaluronidase in the media prior to fiber plating. Different concentrations are used for plating EDL and soleus (12.5 U/mL for EDL and 25 U/mL for soleus). Our initial results using a GFP reporter showed that ~10–30% of viable fibers are transfected using this approach.
21. Digestion with 0.2% collagenase for 1.5–2 h is generally appropriate for the isolation of EDL fibers from a 2-month-old C57Bl/6 mouse. The concentration of collagenase and time of digestion may need to be adjusted depending on age, gender, strain, or tissue type. For example, we have found that both a higher concentration of collagenase and a longer digestion time is required for soleus vs. EDL tissue.
22. Be sure to keep each muscle fully immersed in media to prevent it from drying out.
23. In order to maintain quiescence of satellite cells, the muscle tissue is not agitated during collagenase digestion (45).
24. During trituration, we recommend using the wide-bore pipette until fibers are no longer liberated before switching to the narrow-bore pipette. (EDL and soleus contain relatively long fibers, and premature use of a small-bore pipette will cause dramatic damage to fibers.)
25. It usually takes a full day to isolate fibers from eight muscles (EDL and soleus from two mice). We usually obtain >200 fibers/EDL and >100 fibers/soleus and plate ~100 EDL fibers/dish and ~50 soleus fibers/dish (with ~70% plating efficiency).
26. In order to minimize disturbance to the plated fibers, we only replace a portion of the media during the transfection step.
27. The day after transfection, ~30–60% of EDL fibers and ~80–90% of soleus fibers are viable (based on a phase bright vs. dark appearance).

References

1. Salva, M. Z., Himeda, C. L., Tai, P. W., Nishiuchi, E., Gregorevic, P., Allen, J. M., Finn, E. E., Nguyen, Q. G., Blankinship, M. J., Meuse, L., Chamberlain, J. S., and Hauschka, S. D. (2007) Design of tissue-specific regulatory cassettes for high-level rAAV-mediated expression in skeletal and cardiac muscle. *Mol Ther 15*, 320–329.
2. Gregorevic, P., Blankinship, M. J., Allen, J. M., Crawford, R. W., Meuse, L., Miller, D. G., Russell, D. W., and Chamberlain, J. S. (2004) Systemic delivery of genes to striated muscles using adeno-associated viral vectors. *Nat Med 10*, 828–834.
3. Donoviel, D. B., Shield, M. A., Buskin, J. N., Haugen, H. S., Clegg, C. H., and Hauschka,

S. D. (1996) Analysis of muscle creatine kinase gene regulatory elements in skeletal and cardiac muscles of transgenic mice. *Mol Cell Biol 16*, 1649–1658.
4. Shield, M. A., Haugen, H. S., Clegg, C. H., and Hauschka, S. D. (1996) E-box sites and a proximal regulatory region of the muscle creatine kinase gene differentially regulate expression in diverse skeletal muscles and cardiac muscle of transgenic mice. *Mol Cell Biol 16*, 5058–5068.
5. Hagstrom, J. N., Couto, L. B., Scallan, C., Burton, M., McCleland, M. L., Fields, P. A., Arruda, V. R., Herzog, R. W., and High, K. A. (2000) Improved muscle-derived expression of human coagulation factor IX from a skeletal actin/CMV hybrid enhancer/promoter. *Blood 95*, 2536–2542.
6. Jerkovic, R., Vitadello, M., Kelly, R., Buckingham, M., and Schiaffino, S. (1997) Fibre type-specific and nerve-dependent regulation of myosin light chain 1 slow promoter in regenerating muscle. *J Muscle Res Cell Motil 18*, 369–373.
7. Li, X., Eastman, E. M., Schwartz, R. J., and Draghia-Akli, R. (1999) Synthetic muscle promoters: activities exceeding naturally occurring regulatory sequences. *Nat Biotechnol 17*, 241–245.
8. Kelly, R. G., and Buckingham, M. E. (2000) Modular regulation of the MLC1F/3F gene and striated muscle diversity. *Microsc Res Tech 50*, 510–521.
9. Skarli, M., Kiri, A., Vrbova, G., Lee, C. A., and Goldspink, G. (1998) Myosin regulatory elements as vectors for gene transfer by intramuscular injection. *Gene Ther 5*, 514–520.
10. Arruda, V. R., Stedman, H. H., Nichols, T. C., Haskins, M. E., Nicholson, M., Herzog, R. W., Couto, L. B., and High, K. A. (2005) Regional intravascular delivery of AAV-2-F.IX to skeletal muscle achieves long-term correction of hemophilia B in a large animal model. *Blood 105*, 3458–3464.
11. Su, L. T., Gopal, K., Wang, Z., Yin, X., Nelson, A., Kozyak, B. W., Burkman, J. M., Mitchell, M. A., Low, D. W., Bridges, C. R., and Stedman, H. H. (2005) Uniform scale-independent gene transfer to striated muscle after transvenular extravasation of vector. *Circulation 112*, 1780–1788.
12. Bridges, C. R., Gopal, K., Holt, D. E., Yarnall, C., Cole, S., Anderson, R. B., Yin, X., Nelson, A., Kozyak, B. W., Wang, Z., Lesniewski, J., Su, L. T., Thesier, D. M., Sundar, H., and Stedman, H. H. (2005) Efficient myocyte gene delivery with complete cardiac surgical isolation in situ. *J Thorac Cardiovasc Surg 130*, 1364.
13. Amacher, S. L., Buskin, J. N., and Hauschka, S. D. (1993) Multiple regulatory elements contribute differentially to muscle creatine kinase enhancer activity in skeletal and cardiac muscle. *Mol Cell Biol 13*, 2753–2764.
14. Clegg, C. H., Linkhart, T. A., Olwin, B. B., and Hauschka, S. D. (1987) Growth factor control of skeletal muscle differentiation: commitment to terminal differentiation occurs in G1 phase and is repressed by fibroblast growth factor. *J Cell Biol 105*, 949–956.
15. Neville, C., Rosenthal, N., McGrew, M., Bogdanova, N., and Hauschka, S. (1997) Skeletal muscle cultures. *Methods Cell Biol 52*, 85–116.
16. Nguyen, Q. G., Buskin, J. N., Himeda, C. L., Fabre-Suver, C., and Hauschka, S. D. (2003) Transgenic and tissue culture analyses of the muscle creatine kinase enhancer Trex control element in skeletal and cardiac muscle indicate differences in gene expression between muscle types. *Transgenic Res 12*, 337–349.
17. Bischoff, R. (1989) Analysis of muscle regeneration using single myofibers in culture. *Med Sci Sports Exerc 21*, S164–S172.
18. Bischoff, R. (1990) Interaction between satellite cells and skeletal muscle fibers. *Development 109*, 943–952.
19. Bischoff, R. (1990) Cell cycle commitment of rat muscle satellite cells. *J Cell Biol 111*, 201–207.
20. Konigsberg, U. R., Lipton, B. H., and Konigsberg, I. R. (1975) The regenerative response of single mature muscle fibers isolated in vitro. *Dev Biol 45*, 260–275.
21. Shefer, G., and Yablonka-Reuveni, Z. (2005) Isolation and culture of skeletal muscle myofibers as a means to analyze satellite cells. *Methods Mol Biol 290*, 281–304.
22. Sternberg, E. A., Spizz, G., Perry, W. M., Vizard, D., Weil, T., and Olson, E. N. (1988) Identification of upstream and intragenic regulatory elements that confer cell-type-restricted and differentiation-specific expression on the muscle creatine kinase gene. *Mol Cell Biol 8*, 2896–2909.
23. Johnson, J. E., Wold, B. J., and Hauschka, S. D. (1989) Muscle creatine kinase sequence elements regulating skeletal and cardiac muscle expression in transgenic mice. *Mol Cell Biol 9*, 3393–3399.
24. Jaynes, J. B., Johnson, J. E., Buskin, J. N., Gartside, C. L., and Hauschka, S. D. (1988) The muscle creatine kinase gene is regulated by multiple upstream elements, including a muscle-specific enhancer. *Mol Cell Biol 8*, 62–70.

25. Hauser, M. A., Robinson, A., Hartigan-O'Connor, D., Williams-Gregory, D. A., Buskin, J. N., Apone, S., Kirk, C. J., Hardy, S., Hauschka, S. D., and Chamberlain, J. S. (2000) Analysis of muscle creatine kinase regulatory elements in recombinant adenoviral vectors. *Mol Ther 2*, 16–25.
26. Himeda, C. L. (2003) Identification and Characterization of the Trex-Binding Factor in the Muscle Creatine Kinase Enhancer. University of Washington, Seattle, Washington.
27. Himeda, C. L., Ranish, J. A., Angello, J. C., Maire, P., Aebersold, R., and Hauschka, S. D. (2004) Quantitative proteomic identification of six4 as the trex-binding factor in the muscle creatine kinase enhancer. *Mol Cell Biol 24*, 2132–2143.
28. Himeda, C. L., Ranish, J. A., and Hauschka, S. D. (2008) Quantitative proteomic identification of MAZ as a transcriptional regulator of muscle-specific genes in skeletal and cardiac myocytes. Mol Cell Biol 28, 6521–6535.
29. Himeda, C. L., Ranish, J. A., Pearson, R. C., Crossley, M., and Hauschka, S. D. (2010) KLF3 regulates muscle-specific gene expression and synergizes with serum response factor on KLF binding sites. Mol Cell Biol 30, 3430–3443.
30. Donoghue, M., Ernst, H., Wentworth, B., Nadal-Ginard, B., and Rosenthal, N. (1988) A muscle-specific enhancer is located at the 3′ end of the myosin light-chain 1/3 gene locus. *Genes Dev 2*, 1779–1790.
31. Emami, K. H., Jain, A., and Smale, S. T. (1997) Mechanism of synergy between TATA and initiator: synergistic binding of TFIID following a putative TFIIA-induced isomerization. *Genes Dev 11*, 3007–3019.
32. Morin, S., Paradis, P., Aries, A., and Nemer, M. (2001) Serum response factor-GATA ternary complex required for nuclear signaling by a G-protein-coupled receptor. *Mol Cell Biol 21*, 1036–1044.
33. Apone, S., and Hauschka, S. D. (1995) Muscle gene E-box control elements. Evidence for quantitatively different transcriptional activities and the binding of distinct regulatory factors. *J Biol Chem 270*, 21420–21427.
34. Larkin, S. B., Farrance, I. K., and Ordahl, C. P. (1996) Flanking sequences modulate the cell specificity of M-CAT elements. *Mol Cell Biol 16*, 3742–3755.
35. Niu, Z., Li, A., Zhang, S. X., and Schwartz, R. J. (2007) Serum response factor micromanaging cardiogenesis. *Curr Opin Cell Biol 19*, 618–627.
36. Buskin, J. N., and Hauschka, S. D. (1989) Identification of a myocyte nuclear factor that binds to the muscle-specific enhancer of the mouse muscle creatine kinase gene. *Mol Cell Biol 9*, 2627–2640.
37. Blackwell, T. K., and Weintraub, H. (1990) Differences and similarities in DNA-binding preferences of MyoD and E2A protein complexes revealed by binding site selection. *Science 250*, 1104–1110.
38. Wright, W. E., Binder, M., and Funk, W. (1991) Cyclic amplification and selection of targets (CASTing) for the myogenin consensus binding site. *Mol Cell Biol 11*, 4104–4110.
39. Merika, M., and Orkin, S. H. (1993) DNA-binding specificity of GATA family transcription factors. *Mol Cell Biol 13*, 3999–4010.
40. Karasseva, N., Tsika, G., Ji, J., Zhang, A., Mao, X., and Tsika, R. (2003) Transcription enhancer factor 1 binds multiple muscle MEF2 and A/T-rich elements during fast-to-slow skeletal muscle fiber type transitions. *Mol Cell Biol 23*, 5143–5164.
41. Huang, J., Blackwell, T. K., Kedes, L., and Weintraub, H. (1996) Differences between MyoD DNA binding and activation site requirements revealed by functional random sequence selection. *Mol Cell Biol 16*, 3893–3900.
42. Weintraub, H., Davis, R., Lockshon, D., and Lassar, A. (1990) MyoD binds cooperatively to two sites in a target enhancer sequence: occupancy of two sites is required for activation. *Proc Natl Acad Sci U S A 87*, 5623–5627.
43. Rosenblatt, J. D., Lunt, A. I., Parry, D. J., and Partridge, T. A. (1995) Culturing satellite cells from living single muscle fiber explants. *In Vitro Cell Dev Biol Anim 31*, 773–779.
44. Favre, D., Cherel, Y., Provost, N., Blouin, V., Ferry, N., Moullier, P., and Salvetti, A. (2000) Hyaluronidase enhances recombinant adeno-associated virus (rAAV)-mediated gene transfer in the rat skeletal muscle. *Gene Ther 7*, 1417–1420.
45. Wozniak, A. C., and Anderson, J. E. (2005) Single-fiber isolation and maintenance of satellite cell quiescence. *Biochem Cell Biol 83*, 674–676.
46. Lee, T. C., Chow, K. L., Fang, P., and Schwartz, R. J. (1991) Activation of skeletal alpha-actin gene transcription: the cooperative formation of serum response factor-binding complexes over positive cis-acting promoter serum response elements displaces a negative-acting nuclear factor enriched in replicating myoblasts and nonmyogenic cells. *Mol Cell Biol 11*, 5090–5100.
47. Chen, C. Y., and Schwartz, R. J. (1996) Recruitment of the tinman homolog Nkx-2.5 by serum response factor activates cardiac

alpha-actin gene transcription. *Mol Cell Biol 16*, 6372–6384.
48. Groisman, R., Masutani, H., Leibovitch, M. P., Robin, P., Soudant, I., Trouche, D., and Harel-Bellan, A. (1996) Physical interaction between the mitogen-responsive serum response factor and myogenic basic-helix-loop-helix proteins. *J Biol Chem 271*, 5258–5264.
49. Gupta, M., Kogut, P., Davis, F. J., Belaguli, N. S., Schwartz, R. J., and Gupta, M. P. (2001) Physical interaction between the MADS box of serum response factor and the TEA/ATTS DNA-binding domain of transcription enhancer factor-1. *J Biol Chem 276*, 10413–10422.
50. Natesan, S., and Gilman, M. (1995) YY1 facilitates the association of serum response factor with the c-fos serum response element. *Mol Cell Biol 15*, 5975–5982.
51. Durocher, D., and Nemer, M. (1998) Combinatorial interactions regulating cardiac transcription. *Dev Genet 22*, 250–262.
52. Lee, Y., Shioi, T., Kasahara, H., Jobe, S. M., Wiese, R. J., Markham, B. E., and Izumo, S. (1998) The cardiac tissue-restricted homeobox protein Csx/Nkx2.5 physically associates with the zinc finger protein GATA4 and cooperatively activates atrial natriuretic factor gene expression. *Mol Cell Biol 18*, 3120–3129.
53. Sepulveda, J. L., Belaguli, N., Nigam, V., Chen, C. Y., Nemer, M., and Schwartz, R. J. (1998) GATA-4 and Nkx-2.5 coactivate Nkx-2 DNA binding targets: role for regulating early cardiac gene expression. *Mol Cell Biol 18*, 3405–3415.
54. Morin, S., Charron, F., Robitaille, L., and Nemer, M. (2000) GATA-dependent recruitment of MEF2 proteins to target promoters. *EMBO J 19*, 2046–2055.
55. Iwahori, A., Fraidenraich, D., and Basilico, C. (2004) A conserved enhancer element that drives FGF4 gene expression in the embryonic myotomes is synergistically activated by GATA and bHLH proteins. *Dev Biol 270*, 525–537.
56. Dai, Y. S., Cserjesi, P., Markham, B. E., and Molkentin, J. D. (2002) The transcription factors GATA4 and dHAND physically interact to synergistically activate cardiac gene expression through a p300-dependent mechanism. *J Biol Chem 277*, 24390–24398.
57. Garg, V., Kathiriya, I. S., Barnes, R., Schluterman, M. K., King, I. N., Butler, C. A., Rothrock, C. R., Eapen, R. S., Hirayama-Yamada, K., Joo, K., Matsuoka, R., Cohen, J. C., and Srivastava, D. (2003) GATA4 mutations cause human congenital heart defects and reveal an interaction with TBX5. *Nature 424*, 443–447.
58. Chen, Y., and Cao, X. (2009) NFAT directly regulates Nkx2-5 transcription during cardiac cell differentiation. *Biol Cell 101*, 335–349.
59. Bhalla, S. S., Robitaille, L., and Nemer, M. (2001) Cooperative activation by GATA-4 and YY1 of the cardiac B-type natriuretic peptide promoter. *J Biol Chem 276*, 11439–11445.
60. Stennard, F. A., Costa, M. W., Elliott, D. A., Rankin, S., Haast, S. J., Lai, D., McDonald, L. P., Niederreither, K., Dolle, P., Bruneau, B. G., Zorn, A. M., and Harvey, R. P. (2003) Cardiac T-box factor Tbx20 directly interacts with Nkx2-5, GATA4, and GATA5 in regulation of gene expression in the developing heart. *Dev Biol 262*, 206–224.
61. Molkentin, J. D., Black, B. L., Martin, J. F., and Olson, E. N. (1995) Cooperative activation of muscle gene expression by MEF2 and myogenic bHLH proteins. *Cell 83*, 1125–1136.
62. Morin, S., Pozzulo, G., Robitaille, L., Cross, J., and Nemer, M. (2005) MEF2-dependent recruitment of the HAND1 transcription factor results in synergistic activation of target promoters. *J Biol Chem 280*, 32272–32278.
63. Maeda, T., Gupta, M. P., and Stewart, A. F. (2002) TEF-1 and MEF2 transcription factors interact to regulate muscle-specific promoters. *Biochem Biophys Res Commun 294*, 791–797.
64. Armand, A. S., Bourajjaj, M., Martinez-Martinez, S., el Azzouzi, H., da Costa Martins, P. A., Hatzis, P., Seidler, T., Redondo, J. M., and De Windt, L. J. (2008) Cooperative synergy between NFAT and MyoD regulates myogenin expression and myogenesis. *J Biol Chem 283*, 29004–29010.

Chapter 2

Codon Optimization of the Microdystrophin Gene for Duchene Muscular Dystrophy Gene Therapy

Takis Athanasopoulos, Helen Foster, Keith Foster, and George Dickson

Abstract

Duchenne muscular dystrophy (DMD) is a severe muscle wasting X-linked genetic disease caused by dystrophin gene mutations. Gene replacement therapy aims to transfer a functional full-length dystrophin cDNA or a quasi micro/mini-gene into the muscle. A number of AAV vectors carrying microdystrophin genes have been tested in the mdx model of DMD. Further modification/optimization of these microgene vectors may improve the therapeutic potency. In this chapter, we describe a species-specific, codon optimization protocol to improve microdystrophin gene therapy in the mdx model.

Key words: Codon optimization, mRNA, AAV, Adeno-associated virus, Duchenne muscular dystrophy, Dystrophin, Microdystrophin, Minidystrophin, Quasidystrophin, Gene therapy, Muscle, mdx

1. Introduction

Recently, we have tested the hypothesis that the optimization of codon usage within a microdystrophin gene variant would result in increased levels of transgene expression such that the viral dose needed for effective reconstitution of dystrophin could be reduced (1). In contrast to treatment with noncodon-optimized rAAV2/8 microdystrophin, *mdx* mice treated with codon-optimized rAAV2/8 microdystrophin showed increased numbers of dystrophin – positive fibers, improved muscle function, and amelioration of dystrophic pathology. These results demonstrated for the first time that codon optimization of a microdystrophin cDNA under the control of a muscle-specific promoter can significantly improve expression levels such that reduced titers of rAAV vectors will be required for efficient systemic administration.

The genetic code uses 64 codons to encode 21 amino acids (aa); hence there are more codons per aa. The majority of

Dongsheng Duan (ed.), *Muscle Gene Therapy: Methods and Protocols*, Methods in Molecular Biology, vol. 709, DOI 10.1007/978-1-61737-982-6_2,

these 21 genetically encoded aa are coded by multiple codons (synonymous usage). Synonymous codon usage biases are associated with various biological factors, such as gene expression level, gene length, gene translation initiation signal, protein amino acid composition, protein structure, tRNA abundance, mutation frequency and patterns, and GC compositions (2). Selection may be operating on synonymous codons to maintain a more stable and ordered mRNA secondary structure, likely to be important for transcript stability and translation. Functional domains of the mRNA (5 -untranslated region (5 -UTR), CDS, and 3 -UTR) preferentially fold onto themselves, while the start/stop codon regions are characterized by relaxed secondary structures, which may facilitate initiation and termination of translation (3). The mistranslation-induced protein misfolding hypothesis predicts that selection should prefer high-fidelity codons at sites where translation errors are structurally disruptive and lead to protein misfolding and aggregation (4). De novo synthesis and codon optimization is a relatively new technology; however, it has been successfully tested previously in a variety of applications. More favored codons correspond to more abundant tRNAs; highly expressed genes are often rich in favored codons thus there is a need and practical application to codon optimize genes, according to a multiparametric series including *CAI Maximization* (measure of usage of preferred codons), *Codon sampling* (use of a set of codons with probabilities proportional to their abundance/usage in the organism), *Dicodon optimization* (adjacent codons pair nonrandomly), and *Codon frequency matching* (where native mRNA matches its target species). In addition, special expression strains may express extra copies of the rare tRNAs. Synonymous mutations do not alter the encoded protein, but they can influence gene expression.

Genes directly cloned from pathogenic organisms may not be efficiently translated in a heterologous host expression system as a consequence of codon bias. Tat genes were synthetically assembled and compared against wild-type counterparts in two different mouse strains. Codon-optimized Tat genes induced qualitatively and quantitatively superior immune responses (5). In another application, LuxA protein levels increased significantly after codon optimization whereas mRNA levels remained approximately equal. On average, bioluminescence levels were increased by more than sixfold (6). On the contrary, codon optimization had no effect on the rate of transcriptional initiation or elongation of the HIV-1 vpu mRNA. However, optimization of the vpu gene was found to stabilize its mRNA in the nucleus and enhance its export to the cytoplasm (7).

Other parameters, including conservation among species, may be of importance toward codon optimization applications.

The extent of conservation in the flanking sequence of the initiator ATG codon including Kozak's consensus sequence was an important factor in modulation of the translation efficiency as well as synonymous codon usage bias particularly in highly expressed genes (8). Bacteria and mammals prefer to use different codons so that mammalian genes frequently use codons, which are rarely employed in bacteria and vice versa. The coding sequence for the *Escherichia coli* beta-galactosidase gene was codon optimized for expression in mammalian cells resulting in the expression of beta-galactosidase at levels 15-fold higher than those resulting from an analogous construct containing the native *E. coli* gene sequence (9). The efficiency of mammalian heterologous protein production in *E. coli* can be diminished by biased codon usage. Heterologous expression of some proteins in bacteria can be improved by altering codon preference, i.e., by introducing rare codon tRNAs into the host cell (10). There is an importance of mRNA levels and RNA stability, but not necessarily tRNA abundance for efficient heterologous protein production at least in prokaryotes (11). mRNA expression is correlated to the stability of mRNA secondary structure and the codon usage bias (12). Natural selection on codon usage acts on a large variety of prokaryotic and eukaryotic genomes. The strength (S) of selected codon usage bias is low (S=0.22–0.51) in large mammalian genomes (human and mouse) for the most highly expressed genes. However, this might not be an evidence for selection in these organisms and does not properly account for nucleotide composition heterogeneity along such genomes (13). It is also suggested that mRNA folding and associated rates of translation initiation play a predominant role in shaping expression levels of individual genes, whereas codon bias influences global translation efficiency and cellular fitness (14).

In our paradigm, codon and species-specific optimization of a microdystrophin variant significantly increased levels of microdystrophin mRNA and protein after intramuscular and systemic administration of plasmid DNA or rAAV2/8; a murine based codon-optimized variant under a muscle-restricted promoter diminished immune responses following administration to skeletal muscles of the mdx mouse model of Duchenne muscular dystrophy (DMD) (1). Physiological studies further demonstrated that species-specific codon optimization of a ΔAB/Δ R3-R18/ΔCT microgene normalized specific force and protected muscle from contraction-induced injury. In the following sections, we detail the materials, methods, and notes describing the design of codon and mRNA sequence optimized microdystrophin transgenes and their application toward treatment of Duchene muscular dystrophy (DMD) gene therapy as applied in the mdx mouse model of the disease.

2. Materials

2.1. Generation of Codon-Optimized and Species-Specific Microdystrophin Sequences

1. Homo sapiens dystrophin (DMD), transcript variant Dp427m, mRNA (GeneBank ACCESSION NM_ 004006). Mus musculus dystrophin, muscular dystrophy (Dmd) mRNA (GeneBank ACCESSION NM_007868) (Figs. 1 and 2).
2. Noncodon-optimized human microdystrophin ΔAB/R3-R18/ΔCT (1).
3. Noncodon-optimized human microdystrophin sequence ΔR4-23/ΔCT (Fig. 1) (15).
4. Codon optimized sequences synthesis service (Geneart, Regensburg, Germany).

2.2. rAAV Production, Purification, and Characterization (Dot-Blot)

1. 1 M $CaCl_2$. Filter sterilized, store at –20°C.
2. HEPES buffer. Autoclave and store at RT. Stable for 3 months.
3. 0.3 M NaCl, 40 mM HEPES, pH 7.05.
4. 70 mM Na_2HPO_4. Autoclave and store at RT. Stable for 3 months.
5. Lysis buffer (500 mL): 0.15 M NaCl, 50 mM Tris–HCl, pH 8.5. Autoclave.
6. 5× PBS-MK (100 mL): Take 5 PBS Tablets (BR14a, Oxoid, UK) and add to 90 mL distilled water. Autoclave, cool, and add 0.5 mL 1 M $MgCl_2$ (5 mM) and 1.25 mL 1 M KCl (12.5 mM). It is important not to add the $MgCl_2$ and KCl before autoclaving as these salts will precipitate out of solution. Make up to 100 mL with sterile water.
7. Iodixanol (Sigma).
8. Phenol Red (0.5% in PBS) (Sigma).
9. Benzonase (Sigma).
10. DNAse I: 10 mg/mL in H_2O.
11. 2× Proteinase K buffer: 20 mM Tris–Cl (pH 8.0), 20 mM EDTA (pH 8.0), 1% (w/v) SDS. Store at room temperature.
12. Proteinase K (Sigma).
13. Glycogen (Invitrogen, UK).
14. 20× SSC: 1.5 M NaCl, 0.3 M Na_3 citrate.
15. Hybridization buffer (Amersham, UK).
16. Primary wash buffer (1 L): 360 g urea (6 M), 4 g SDS (0.4%), 25 mL 20× SSC (0.5× SSC). Store at 4°C. Stringency can be lowered by lowering SSC content (0.2× or 0.1×).

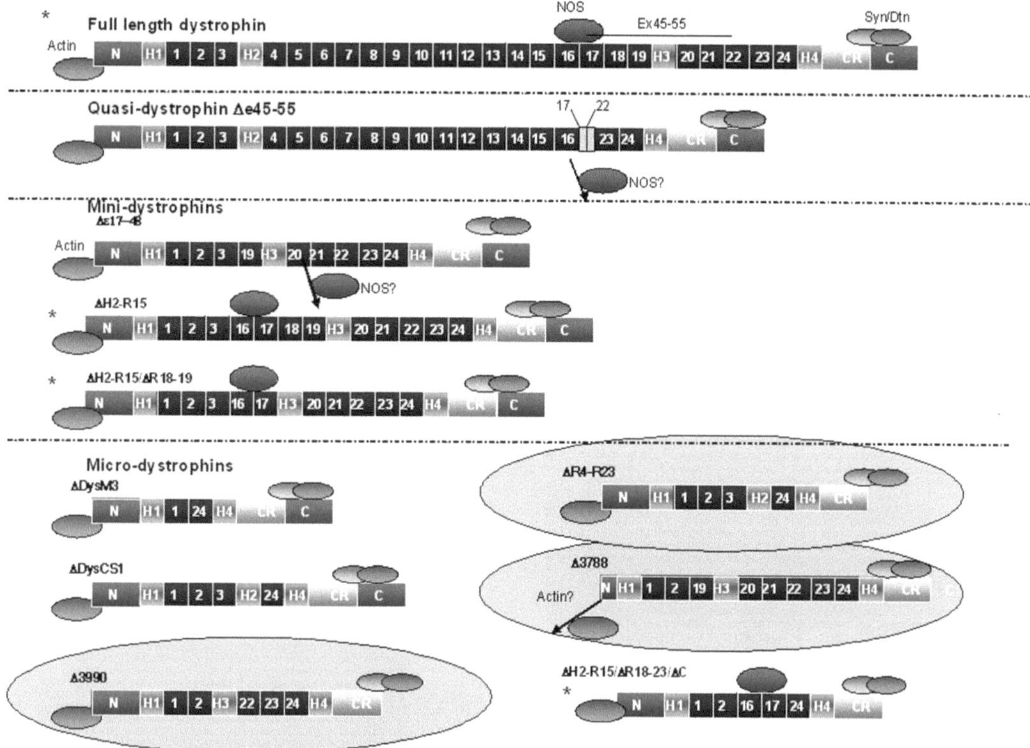

Fig. 1. De novo synthesis of functional dystrophin molecules. Dystrophin is a rod-like protein of 3,685 amino acids (aa) localized beneath the inner surface of muscle cell membrane. It functions through four major structural domains: a N-terminal domain (1-756 aa), a central rod domain (757-3122 aa), a cysteine-rich (CR) domain (3123-3409 aa), and a distal C-terminal domain (3410-3685 aa). The N-terminal domain binds to the F-actin of cytoskeletal structures, while the CR domain along with the distal C-terminal domain anchors to the cell membrane via dystrophin-associated protein (DAP) complexes, thus, dystrophin crosslinks and stabilizes the muscle cell membrane and cytoskeleton. The central rod domain contains 24 triple-helix rod repeats (R1–R24) and four hinges (H1–H4). Each repeat is approximately 109 aa long. The central rod domain presumably functions as a "shock absorber" during muscle contraction. Dystrophin crosslinks and stabilizes the muscle cell membrane and cytoskeleton. The absence of a functional dystrophin results in the loss of DAP complexes and causes instability of myofiber plasma membrane. These deficiencies in turn lead to chronic muscle damage and degenerative pathology. Examples of full-length dystrophin, quasidystrophin, minidystrophin, and microdystrophin genes that have been (or are currently in progress to be) assessed for the restoration of dystrophic muscle function via potential gene augmentation strategies are presented. Such genes have been assembled either via traditional genetic engineering approaches (PCR/RT-PCR) or via de novo oligomer synthesis incorporating codon optimization and species-specific tailor-mediated targeting. Full-length dystrophin cDNA can be delivered via plasmid, minicircles or Ad-gutted vectors, but further developments/technologies covering "full body" widespread delivery are possible to arise in the near future. Quasidystrophin, minidystrophin (Becker) genes and derivatives were developed according to various truncations and patient data deletions/mutations in dystrophin genes associated with mild dystrophinopathy, for example, BMD. These genes can be currently delivered via dual transplicing and/or overlapping AAV or lentivector approaches. Microdystrophin genes are much smaller transgenes that were generated to determine the minimum requirements for normal dystrophin function and are suitable to be packaged in rAAV or lenti-vectors. To our knowledge microdystrophin variants Δ3990, ΔR4-R23, and Δ3788, highlighted in this figure depicted under a light oval schematic, have been already codon and species-specific optimized (human, canine, murine variants) by ours and other laboratories. De novo synthesis of other variants including larger quasidystrophin forms and/or mini/microdystrophins is under way.

a Codon *Opt*imisation of microdystrophin ΔAB/R3-R18/ΔCT

'Non' is original microdystrophin ΔAB/R3-R18/ΔCT
'*Opt*' is *opt*imised microdystrophin ΔAB/R3-R18/ΔCT

```
Non   ------ATGCTTTGGTGGGAAGAAGTAGAGGACTGTTATGAAAGAGAAGATGTTCAAAAG 54
Opt   GCCACCATGCTGTGGTGGGAGGAAGTGGAGGACTGCTACGAGAGAGAGGACGTGCAGAAG 60
Non   AAAACATTCACAAAATGGGTAAATGCACAATTTTCTAAGCATTTGGAAGCTCCTGAAGAC 114
Opt   AAAACCTTCACCAAGTGGGTGAACGCCCAGTTCAGCAAGCACCTGGAGGCCCCTGAGGAC 120
Non   AAGTCATTTGGCAGTTCATTGATGGAGAGTGAAGTAAACCTGGACCGTTATCAAACAGCT 174
Opt   AAGAGCTTCGGCAGCAGCCTGATGGAGAGCGAAGTGAACCTGGACAGATACCAGACCGCC 180
Non   TTAGAAGAAGTATTATCGTGGCTTCTTTCTGCTGAGGACACATTGCAAGCACAAGGAGAG 234
Opt   CTGGAGGAAGTGCTGAGCTGGCTGCTGAGCGCCGAGGACACCCTGCAGGCCCAGGGCGAG 240
Non   ATTTCTAATGATGTGGAAGTGGTGAAAGACCAGTTTCATACTCATGAGGGGTACATGATG 294
Opt   ATCAGCAACGACGTGGAAGTGGTGAAGGACCAGTTCCACACCCACGAGGGCTACATGATG 300
Non   GATTTGACAGCCCATCAGGGCCGGGTTGGTAATATTCTACAATTGGGAAGTAAGCTGATT 354
Opt   GATCTGACCGCCCACCAGGGCAGAGTGGGCAATATCCTGCAGCTGGGCAGCAAGCTGATC 360
Non   GGAACAGGAAAATTATCAGAAGATGAAGAAACTGAAGTACAAGAGCAGATGAATCTCCTA 414
Opt   GGCACCGGCAAGCTGAGCGAGGACGAGGAGACCGAAGTGCAGGAGCAGATGAACCTGCTG 420
Non   AATTCAAGATGGGAATGCCTCAGGGTAGCTAGCATGGAAAAACAAAGCAATTTACATAGA 474
Opt   AACAGCAGATGGGAGTGCCTGAGAGTGGCCAGCATGGAGAAGCAGAGCAACCTGCACAGA 480
Non   GTTTTAATGGATCTCCAGAATCAGAAACTGAAAGAGTTGAATGACTGGCTAACAAAAACA 534
Opt   GTGCTGATGGACCTGCAGAACCAGAAGCTGAAGGAGCTGAACGACTGGCTGACCAAGACC 540
Non   GAAGAAAGAACAAGGAAAATGGAGGAAGAGCCTCTTGGACCTGATCTTGAAGACCTAAAA 594
Opt   GAGGAGCGGACCAGAAAGATGGAGGAGGAGCCCCTGGGCCCCGACCTGGAGGACCTGAAG 600
Non   CGCCAAGTACAACAACATAAGGAAACTGAAATAGCAGTTCAAGCTAAACAACCGGATGTG 654
Opt   AGACAGGTGCAGCAGCACAAGGAGACCGAGATCGCCGTGCAGGCCAAGCAGCCCGACGTG 660
Non   GAAGAGATTTTGTCTAAAGGGCAGCATTTGTACAAGGAAAAACCAGCCACTCAGCCAGTG 714
Opt   GAGGAGATCCTGAGCAAGGGCCAGCACCTGTACAAGGAGAAGCCTGCCACCCAGCCCGTG 720
Non   AAGAGGAAGTTAGAAGATCTGAGCTCTGAGTGGAAGGCGGTAAACCGTTTACTTCAAGAG 774
Opt   AAGAGAAAGCTGGAGGATCTGAGCAGCGAGTGGAAGGCCGTGAACAGACTGCTGCAGGAG 780
Non   CTGAGGGCAAAGCAGCCTGACCTAGCTCCTGGACTGACCACTATTGGAGCCTCTCCTACT 834
Opt   CTGAGAGCCAAACAGCCTGACCTGGCCCCTGGCCTGACCACCATCGGCGCCAGCCCCACC 840
Non   CAGACTGTTACTCTGGTGACACAACCTGTGGTTACTAAGGAAACTGCCATCTCCAAACTA 894
Opt   CAGACAGTGACCCTGGTGACCCAGCCTGTGGTGACCAAGGAGACAGCCATCAGCAAGCTG 900
Non   GAAATGCCATCTTCCTTGATGTTGGAGGTACCTGCTCTGGCAGATTTCAACCGGGCTTGG 954
Opt   GAGATGCCCAGCTCCCTGATGCTGGAAGTGCCCGCCCTGGCCGATTTCAATAGGGCCTGG 960
Non   ACAGAACTTACCGACTGGCTTTCTCTGCTTGATCAAGTTATAAAATCACAGAGGGTGATG 1014
Opt   ACCGAGCTGACCGATTGGCTGTCCCTGCTGGACCAGGTGATCAAGAGCCAGAGAGTGATG 1020
Non   GTGGGTGACCTTGAGGATATCAACGAGATGATCATCAAGCAGAAGGCAACAATGCAGGAT 1074
Opt   GTGGGCGATCTGGAGGACATCAACGAGATGATCATCAAGCAGAAAGCCACCATGCAGGAC 1080
Non   TTGGAACAGAGGCGTCCCCAGTTGGAAGAACTCATTACCGCTGCCCAAAATTTGAAAAAC 1134
Opt   CTGGAGCAGAGGAGACCCCAGCTGGAAGAGCTGATCACAGCCGCCCAGAACCTGAAGAAC 1140
Non   AAGACCAGCAATCAAGAGGCTAGAACAATCATTACGGATCGAATTGAAAGAATTCAGAAT 1194
Opt   AAGACCAGCAACCAGGAGGCCAGGACCATCATCACCGACCGGATCGAGAGGATCCAGAAC 1200
Non   CAGTGGGATGAAGTACAAGAACACCTTCAGAACCGGAGGCAACAGTTGAATGAAATGTTA 1254
Opt   CAGTGGGATGAAGTGCAGGAACACCTGCAGAACAGACGGCAGCAGCTGAACGAGATGCTG 1260
Non   AAGGATTCAACACAATGGCTGGAAGCTAAGGAAGAAGCTGAGCAGGTCTTAGGACAGGCC 1314
Opt   AAGGACAGCACCCAGTGGCTGGAGGCCAAGGAGGAGGCCGAGCAGGTGCTGGGCCAGGCC 1320
Non   AGAGCCAAGCTTGAGTCATGGAAGGAGGGTCCCTATACAGTAGATGCAATCCAAAAGAAA 1374
Opt   AGAGCCAAGCTGGAGTCCTGGAAGGAGGGCCCTTACACCGTGGATGCCATCCAGAAGAAG 1380
Non   ATCACAGAAACCAAGCAGTTGGCCAAAGACCTCCGCCAGTGGCAGACAAATGTAGATGTG 1434
Opt   ATCACCGAGACCAAGCAGCTGGCCAAGGACCTGAGACAGTGGCAGACCAACGTGGACGTG 1440
Non   GCAAATGACTTGGCCCTGAAACTTCTCCGGGATTATTCTGCAGATGATACCAGAAAAGTC 1494
Opt   GCCAATGATCTGGCCCTGAAGCTGCTGAGAGACTACAGCGCCGACGATACCCGGAAAGTG 1500
Non   CACATGATAACAGAGAATATCAATGCCTCTTGGAGAAGCATTCATAAAAGGGTGAGTGAG 1554
Opt   CACATGATCACAGAGAACATCAATGCTTCTTGGCGGAGCATCCACAAGAGAGTGAGCGAG 1560
Non   CGAGAGGCTGCTTTGGAAGAAACTCATAGATTACTGCAACAGTTCCCCCTGGACCTGGAA 1614
Opt   AGAGAAGCCGCCCTGGAAGAGACTCATAGGCTGCTCCAGCAGTTCCCTCTGGACCTGGAG 1620
Non   AAGTTTCTTGCCTGGCTTACAGAAGCTGAAACAACTGCCAATGTCCTACAGGATGCTACC 1674
Opt   AAGTTCCTGGCCTGGCTGACAGAGGCCGAGACCACCGCCAACGTGCTGCAGGACGCCACC 1680
Non   CGTAAGGAAAGGCTCCTAGAAGACTCCAAGGGAGTAAAAGAGCTGATGAAACAATGGCAA 1734
Opt   AGAAAGGAGAGACTGCTGGAGGATAGCAAGGGCGTGAAGGAACTGATGAAGCAGTGGCAG 1740
Non   GACCTCCAAGGTGAAATTGAAGCTCACACAGATGTTTATCACAACCTGGATGAAAACAGC 1794
Opt   GATCTGCAGGGCGAAATCGAGGCCCACACCGACGTGTACCACAACCTGGACGAGAACAGC 1800
Non   CAAAAAATCCTGAGATCCCTGGAAGGTTCCGATGATGCAGTCCTGTTACAAAGACGTTTG 1854
```

```
Opt   CAGAAGATCCTGAGAAGCCTGGAGGGCAGCGACGACGCCGTGCTGCTGCAGAGAAGGCTG 1860
Non   GATAACATGAACTTCAAGTGGAGTGAACTTCGGAAAAAGTCTCTCAACATTAGGTCCCAT 1914
Opt   GACAACATGAACTTCAAGTGGAGCGAGCTGCGGAAGAAGAGCCTGAACATCCGGAGCCAC 1920
Non   TTGGAAGCCAGTTCTGACCAGTGGAAGCGTCTGCACCTTTCTCTGCAGGAACTTCTGGTG 1974
Opt   CTGGAAGCCAGCAGCGACCAGTGGAAGAGACTGCACCTGAGCCTGCAGGAACTGCTGGTG 1980
Non   TGGCTACAGCTGAAAGATGATGAATTAAGCCGGCAGGCACCTATTGGAGGCGACTTTCCA 2034
Opt   TGGCTGCAGCTGAAGGACGACGAGCTGAGCAGACAGGCCCCCATCGGCGGCGACTTCCCC 2040
Non   GCAGTTCAGAAGCAGAACGATGTACATAGGGCCTTCAAGAGGGAATTGAAAACTAAAGAA 2094
Opt   GCCGTGCAGAAGCAGAACGACGTGCACCGGGCCTTCAAGAGGGAGCTGAAAACCAAGGAA 2100
Non   CCTGTAATCATGAGTACTCTTGAGACTGTACGAATATTTCTGACAGAGCAGCCTTTGGAA 2154
Opt   CCCGTGATCATGAGCACCCTGGAGACAGTGCGGATCTTCCTGACCGAGCAGCCCCTGGAG 2160
Non   GGACTAGAGAAACTCTACCAGGAGCCCAGAGAGCTGCCTCCTGAGGAGAGAGCCCAGAAT 2214
Opt   GGCCTGGAGAAGCTGTACCAGGAGCCCAGAGAGCTGCCCCCCGAGGAGAGAGCCCAGAAC 2220
Non   GTCACTCGGCTTCTACGAAAGCAGGCTGAGGAGGTCAATACTGAGTGGGAAAAATTGAAC 2274
Opt   GTGACCCGGCTGCTGAGAAAGCAGGCCGAGGAAGTGAATACCGAGTGGGAGAAGCTGAAT 2280
Non   CTGCACTCCGCTGACTGGCAGAGAAAAATAGATGAGACCCTTGAAAGACTCCAGGAACTT 2334
Opt   CTGCACTCCGCCGACTGGCAGAGAAAGATCGACGAGACCCTGGAACGCCTGCAGGAGCTG 2340
Non   CAAGAGGCCACGGATGAGCTGGACCTCAAGCTGCGCCAAGCTGAGGTGATCAAGGGATCC 2394
Opt   CAGGAAGCCACCGACGAGCTGGACCTGAAACTGAGGCAGGCCGAAGTGATCAAGGGCAGC 2400
Non   TGGCAGCCCGTGGGCGATCTCCTCATTGACTCTCTCCAAGATCACCTCGAGAAAGTCAAG 2454
Opt   TGGCAGCCTGTGGGCGACCTGCTGATCGATTCCCTGCAGGACCACCTGGAAAAAGTGAAG 2460
Non   GCACTTCGAGGAGAAATTGCGCCTCTGAAAGAGAACGTGAGCCACGTCAATGACCTTGCT 2514
Opt   GCCCTCAGGGGCGAGATCGCTCCTCTGAAGGAGAATGTGAGCCACGTGAACGACCTGGCC 2520
Non   CGCCAGCTTACCACTTTGGGCATTCAGCTCTCACCGTATAACCTCAGCACTCTGGAAGAC 2574
Opt   AGACAGCTGACCACCCTGGGCATCCAGCTGAGCCCCTACAACCTGAGCACACTGGAAGAT 2580
Non   CTGAACACCAGATGGAAGCTTCTGCAGGTGGCCGTCGAGGACCGAGTCAGGCAGCTGCAT 2634
Opt   CTGAACACCCGGTGGAAGCTGCTGCAGGTGGCCGTGGAGGATAGAGTGAGGCAGCTGCAC 2640
Non   GAAGCCCACAGGGACTTTGGTCCAGCATCTCAGCACTTTCTTTCCACGTCTGTCCAGGGT 2694
Opt   GAAGCCCACAGAGACTTCGGCCCTGCCAGCCAGCACTTCCTGAGCACCAGCGTGCAGGGC 2700
Non   CCCTGGGAGAGAGCCATCTCGCCAAACAAAGTGCCCTACTATATCAACCACGAGACTCAA 2754
Opt   CCCTGGGAGAGAGCCATCTCCCCCAACAAAGTGCCCTACTACATCAACCACGAGACCCAG 2760
Non   ACAACTTGCTGGGACCATCCCAAAATGACAGAGCTCTACCAGTCTTTAGCTGACCTGAAT 2814
Opt   ACCACCTGCTGGGACCACCCTAAGATGACCGAGCTGTATCAGAGCCTGGCCGACCTGAAC 2820
Non   AATGTCAGATTCTCAGCTTATAGGACTGCCATGAAACTCCGAAGACTGCAGAAGGCCCTT 2874
Opt   AATGTGCGGTTCAGCGCCTACAGAACCGCCATGAAGCTGCGGAGACTGCAGAAGGCCCTG 2880
Non   TGCTTGGATCTCTTGAGCCTGTCAGCTGCATGTGATGCCTTGGACCAGCACAACCTCAAG 2934
Opt   TGCCTGGACCTGCTGAGCCTGAGCGCCGCCTGCGACGCCCTGGACCAGCACAACCTGAAG 2940
Non   CAAAATGACCAGCCCATGGATATCCTGCAGATTATTAATTGTTTGACCACTATTTATGAC 2994
Opt   CAGAATGACCAGCCCATGGACATCCTGCAGATCATCAACTGCCTGACCACAATCTACGAT 3000
Non   CGCCTGGAGCAAGAGCACAACAATTTGGTCAACGTCCCTCTCTGCGTGGATATGTGTCTG 3054
Opt   CGGCTGGAGCAGGAGCACAACAACCTGGTGAACGTGCCCCTGTGCGTGGACATGTGCCTG 3060
Non   AACTGGCTGCTGAATGTTTATGATACGGGACGAACAGGGAGGATCCGTGTCCTGTCTTTT 3114
Opt   AATTGGCTGCTGAACGTGTACGACACCGGCAGGACCGGCAGAATCAGAGTGCTGTCCTTC 3120
Non   AAAACTGGCATCATTTCCCTGTGTAAAGCACATTTGGAAGACAAGTACAGATACCTTTTC 3174
Opt   AAGACCGGCATCATCAGCCTGTGCAAGGCCCACCTGGAGGATAAGTACCGCTACCTGTTC 3180
Non   AAGCAAGTGGCAAGTTCAACAGGATTTTGTGACCAGCGCAGGCTGGGCCTCCTTCTGCAT 3234
Opt   AAGCAGGTGGCCAGCAGCACCGGCTTCTGCGATCAGAGGAGACTGGGCCTGCTGCTGCAC 3240
Non   GATTCTATCCAAATTCCAAGACAGTTGGGTGAAGTTGCATCCTTTGGGGGCAGTAACATT 3294
Opt   GATAGCATCCAGATCCCTAGGCAGCTGGGCGAAGTGGCCAGCTTTGGCGGCAGCAACATC 3300
Non   GAGCCAAGTGTCCGGAGCTGCTTCCAATTTGCTAATAATAAGCCAGAGATCGAAGCGGCC 3354
Opt   GAGCCCTCTGTGAGGAGCTGCTTCCAGTTCGCCAACAACAAGCCCGAGATCGAGGCCGCC 3360
Non   CTCTTCCTAGACTGGATGAGACTGGAACCCCAGTCCATGGTGTGGCTGCCCGTCCTGCAC 3414
Opt   CTGTTCCTGGATTGGATGAGGCTGGAGCCCCAGAGCATGGTGTGGCTGCCTGTGCTGCAC 3420
Non   AGAGTGGCTGCTGCAGAAACTGCCAAGCATCAGGCCAAATGTAACATCTGCAAAGAGTGT 3474
Opt   AGAGTGGCCGCCGCCGAGACCGCCAAGCACCAGGCCAAGTGCAACATCTGCAAGGAGTGC 3480
Non   CCAATCATTGGATTCAGGTACAGGAGTCTAAAGCACTTTAATTATGACATCTGCCAAAGC 3534
Opt   CCCATCATCGGCTTCCGGTACAGGAGCCTGAAGCACTTCAACTACGACATCTGCCAGAGC 3540
Non   TGCTTTTTTTCTGGTCGAGTTGCAAAAGGCCATAAAATGCACTATCCCATGGTGGAATAT 3594
Opt   TGCTTTTTCAGCGGCAGAGTGGCCAAGGGCCACAAGATGCACTACCCCATGGTGGAGTAC 3600
Non   TGCACTCCGACTACATCAGGAGAAGATGTTCGAGACTTTGCCAAGGTACTAAAAAACAAA 3654
Opt   TGCACCCCCACCACCTCCGGCGAGGATGTGAGAGACTTCGCCAAAGTGCTGAAGAATAAG 3660
Non   TTTCGAACCAAAAGGTATTTTGCGAAGCATCCCCGAATGGGCTACCTGCCAGTGCAGACT 3714
Opt   TTCCGGACCAAGCGGTACTTTGCCAAGCACCCCAGGATGGGCTACCTGCCCGTGCAGACC 3720
Non   GTCTTAGAGGGGGACAACATGGAAACTGACACAATG--------- 3750
Opt   GTGCTGGAGGGCGACAACATGGAGACCGACACCATGTGATGATGA 3765
```

b Codon *Opt*imisation of microdystrophin ΔR4-R23/ΔCT

'Non' is original microdystrophin ΔR4-R23/ΔCT
'*Opt*' is *opt*imised microdystrophin ΔR4-R23/ΔCT

```
Non   ------ATGCTTTGGTGGGAAGAAGTAGAGGACTGTTATGAAAGAGAAGATGTTCAAAAG 54
Opt   GCCACCATGCTGTGGTGGGAGGAAGTGGAGGACTGCTACGAGAGAGAGGACGTGCAGAAG 60
Non   AAAACATTCACAAAATGGGTAAATGCACAATTTTCTAAGTTTGGGAAGCAGCATATTGAG 114
Opt   AAAACCTTCACCAAGTGGGTGAACGCCCAGTTCAGCAAGTTCGGCAAGCAGCACATCGAG 120
Non   AACCTCTTCAGTGACCTACAGGATGGGAGGCGCCTCCTAGACCTCCTCGAAGGCCTGACA 174
Opt   AACCTGTTCAGCGACCTGCAGGATGGCAGGAGACTGCTGGATCTGCTGGAGGGACTGACC 180
Non   GGGCAAAAACTGCCAAAAGAAAAAGGATCCACAAGAGTTCATGCCCTGAACAATGTCAAC 234
Opt   GGCCAGAAGCTGCCCAAGGAGAAGGGCAGCACCAGAGTGCACGCCCTGAACAACGTGAAC 240
Non   AAGGCACTGCGGGTTTTGCAGAACAATAATGTTGATTTAGTGAATATTGGAAGTACTGAC 294
Opt   AAGGCCCTGAGAGTGCTGCAGAACAACAACGTGGACCTGGTGAATATCGGCAGCACCGAC 300
Non   ATCGTAGATGGAAATCATAAACTGACTCTTGGTTTGATTTGGAATATAATCCTCCACTGG 354
Opt   ATCGTGGACGGCAACCACAAGCTGACCCTGGGCCTGATCTGGAACATCATCCTGCACTGG 360
Non   CAGGTCAAAAATGTAATGAAAAATATCATGGCTGGATTGCAACAAACCAACAGTGAAAAG 414
Opt   CAGGTGAAGAACGTGATGAAGAACATCATGGCCGGCCTGCAGCAGACCAACAGCGAGAAG 420
Non   ATTCTCCTGAGCTGGGTCCGACAATCAACTCGTAATTATCCACAGGTTAATGTAATCAAC 474
Opt   ATCCTGCTGAGCTGGGTGAGGCAGAGCACCAGAAACTACCCCCAGGTGAACGTGATCAAC 480
Non   TTCACCACCAGCTGGTCTGATGGCCTGGCTTTGAATGCTCTCATCCATAGTCATAGGCCA 534
Opt   TTCACCACCTCCTGGAGCGACGGCCTGGCCCTGAACGCCCTGATCCACAGCCACAGACCC 540
Non   GACCTATTTGACTGGAATAGTGTGGTTTGCCAGCAGTCAGCCACACAACGACTGGAACAT 594
Opt   GACCTGTTCGACTGGAACAGCGTGGTGTGTCAGCAGAGCGCCACCCAGAGACTGGAGCAC 600
Non   GCATTCAACATCGCCAGATATCAATTAGGCATAGAGAAACTACTCGATCCTGAAGATGTT 654
Opt   GCCTTCAACATCGCCAGATACCAGCTGGGCATCGAGAAGCTGCTGGACCCCGAGGACGTG 660
Non   GATACCACCTATCCAGATAAGAAGTCCATCTTAATGTACATCACATCACTCTTCCAAGTT 714
Opt   GACACCACCTACCCCGACAAGAAAAGCATCCTGATGTATATTACCTCTCTGTTTCAGGTG 720
Non   TTGCCTCAACAAGTGAGCATTGAAGCCATCCAGGAAGTGGAAATGTTGCCAAGGCCACCT 774
Opt   CTGCCCCAGCAGGTGTCCATCGAGGCCATCCAGGAAGTGGAAATGCTGCCCAGGCCCCCC 780
Non   AAAGTGACTAAAGAAGAACATTTTCAGTTACATCATCAAATGCACTATTCTCAACAGATC 834
Opt   AAAGTGACCAAGGAGGAGCACTTCCAGCTGCACCACCAGATGCACTATAGCCAGCAGATC 840
Non   ACGGTCAGTCTAGCACAGGGATATGAGAGAACTTCTTCCCCTAAGCCTCGATTCAAGAGC 894
Opt   ACCGTGTCCCTGGCCCAGGGCTATGAGAGAACCAGCAGCCCCAAGCCCAGATTCAAGAGC 900
Non   TATGCCTACACACAGGCTGCTTATGTCACCACCTCTGACCCTACACGGAGCCCATTTCCT 954
Opt   TACGCCTACACCCAGGCCGCCTACGTGACCACCTCCGACCCCACCAGAAGCCCCTTCCCC 960
Non   TCACAGCATTTGGAAGCTCCTGAAGACAAGTCATTTGGCAGTTCATTGATGGAGAGTGAA 1014
Opt   AGCCAGCACCTGGAGGCCCCCGAGGACAAGAGCTTCGGCAGCAGCCTGATGGAGAGCGAA 1020
Non   GTAAACCTGGACCGTTATCAAACAGCTTTAGAAGAAGTATTATCGTGGCTTCTTTCTGCT 1074
Opt   GTGAACCTGGACAGATACCAGACCGCCCTGGAGGAAGTGCTGTCTTGGCTGCTGTCCGCC 1080
Non   GAGGACACATTGCAAGCACAAGGAGAGATTTCTAATGATGTGGAAGTGGTGAAAGACCAG 1134
Opt   GAGGACACCCTGCAGGCCCAGGGCGAGATCAGCAACGACGTGGAAGTGGTGAAGGACCAG 1140
Non   TTTCATACTCATGAGGGGTACATGATGGATTTGACAGCCCATCAGGGCCGGGTTGGTAAT 1194
Opt   TTCCACACCCACGAGGGCTACATGATGGATCTGACCGCCCACCAGGGCAGAGTGGGCAAT 1200
Non   ATTCTACAATTGGGAAGTAAGCTGATTGGAACAGGAAAATTATCAGAAGATGAAGAAACT 1254
Opt   ATCCTGCAGCTGGGCAGCAAGCTGATCGGCACCGGCAAGCTGAGCGAGGACGAGGAGACC 1260
Non   GAAGTACAAGAGCAGATGAATCTCCTAAATTCAAGATGGGAATGCCTCAGGGTAGCTAGC 1314
Opt   GAAGTGCAGGAGCAGATGAACCTGCTGAACAGCAGATGGGAGTGCCTGAGAGTGGCCAGC 1320
Non   ATGGAAAAACAAAGCAATTTACATAGAGTTTTAATGGATCTCCAGAATCAGAAACTGAAA 1374
Opt   ATGGAGAAGCAGAGCAACCTGCACCGCGTGCTGATGGACCTGCAGAACCAGAAGCTGAAG 1380
Non   GAGTTGAATGACTGGCTAACAAAAACAGAAGAAAGAACAAGGAAAATGGAGGAAGAGCCT 1434
Opt   GAGCTGAACGACTGGCTGACCAAGACCGAGGAGCGGACCAGAAAGATGGAGGAGGAGCCC 1440
Non   CTTGGACCTGATCTTGAAGACCTAAAACGCCAAGTACAACAACATAAGGTGCTTCAAGAA 1494
Opt   CTGGGCCCCGACCTGGAGGACCTGAAGAGACAGGTGCAGCAGCACAAAGTGCTGCAGGAG 1500
Non   GATCTAGAACAAGAACAAGTCAGGGTCAATTCTCTCACTCACATGGTGGTGGTAGTTGAT 1554
```

```
Non   AAGACGGAAGCATGGCTGGATAACTTTGCCCGGTGTTGGGATAATTTAGTCCAAAAACTT 1974
Opt   AAAACCGAGGCCTGGCTGGACAATTTCGCCCGGTGCTGGGACAATCTGGTGCAGAAACTG 1980
Non   GAAAAGAGTACAGCACAGATTTCACAGGCTGTCACCACCACTCAGCCATCACTAACACAG 2034
Opt   GAGAAGAGCACCGCCCAGATCAGCCAGGCCGTGACCACCACCCAGCCCAGCCTGACACAG 2040
Non   ACAACTGTAATGGAAACAGTAACTACGGTGACCACAAGGGAACAGATCCTGGTAAAGCAT 2094
Opt   ACCACCGTGATGGAGACCGTGACCACAGTGACCACCAGGGAGCAGATCCTGGTGAAGCAC 2100
Non   GCTCAAGAGGAACTTCCACCACCACCTCCCCAAAAGAAGAGGCAGATTACTGTGGATACC 2154
Opt   GCCCAGGAGGAGCTGCCCCCTCCCCCCCCTCAGAAGAAGCGGCAGATCACAGTGGACACC 2160
Non   CTTGAAAGACTCCAGGAACTTCAAGAGGCCACGGATGAGCTGGACCTCAAGCTGCGCCAA 2214
Opt   CTGGAGAGACTGCAGGAGCTGCAGGAAGCCACCGACGAGCTGGACCTGAAGCTGAGACAG 2220
Non   GCTGAGGTGATCAAGGGATCCTGGCAGCCCGTGGGCGATCTCCTCATTGACTCTCTCCAA 2274
Opt   GCCGAAGTGATCAAGGGCAGCTGGCAGCCTGTGGGCGATCTGCTGATCGACAGCCTGCAG 2280
Non   GATCACCTCGAGAAAGTCAAGGCACTTCGAGGAGAAATTGCGCCTCTGAAAGAGAACGTG 2334
Opt   GACCACCTGGAGAAAGTGAAGGCCCTGCGGGGCGAGATCGCCCCCCTGAAGGAGAATGTG 2340
Non   AGCCACGTCAATGACCTTGCTCGCCAGCTTACCACTTTGGGCATTCAGCTCTCACCGTAT 2394
Opt   AGCCACGTGAACGACCTGGCCAGACAGCTGACCACCCTGGGCATCCAGCTGAGCCCCTAC 2400
Non   AACCTCAGCACTCTGGAAGACCTGAACACCAGATGGAAGCTTCTGCAGGTGGCCGTCGAG 2454
Opt   AATCTGAGCACCCTGGAAGATCTGAACACCCGGTGGAAACTGCTGCAGGTGGCCGTGGAG 2460
Non   GACCGAGTCAGGCAGCTGCATGAAGCCCACAGGGACTTTGGTCCAGCATCTCAGCACTTT 2514
Opt   GATAGAGTGAGGCAGCTGCACGAGGCCCACAGAGACTTCGGCCCTGCCTCCCAGCACTTC 2520
Non   CTTTCCACGTCTGTCCAGGGTCCCTGGGAGAGAGCCATCTCGCCAAACAAAGTGCCCTAC 2574
Opt   CTGAGCACCAGCGTGCAGGGCCCCTGGGAGAGAGCCATCTCCCCCAACAAAGTGCCCTAC 2580
Non   TATATCAACCACGAGACTCAAACAACTTGCTGGGACCATCCCAAAATGACAGAGCTCTAC 2634
Opt   TACATCAACCACGAGACCCAGACCACCTGCTGGGACCACCCTAAGATGACCGAGCTGTAC 2640
Non   CAGTCTTTAGCTGACCTGAATAATGTCAGATTCTCAGCTTATAGGACTGCCATGAAACTC 2694
Opt   CAGAGCCTGGCCGACCTGAACAATGTGCGGTTCAGCGCCTACAGAACCGCCATGAAGCTG 2700
Non   CGAAGACTGCAGAAGGCCCTTTGCTTGGATCTCTTGAGCCTGTCAGCTGCATGTGATGCC 2754
Opt   CGGAGACTGCAGAAGGCCCTGTGCCTGGACCTGCTGAGCCTGAGCGCCGCCTGCGACGCC 2760
Non   TTGGACCAGCACAACCTCAAGCAAAATGACCAGCCCATGGATATCCTGCAGATTATTAAT 2814
Opt   CTGGACCAGCACAACCTGAAGCAGAACGACCAGCCCATGGACATTCTGCAGATCATCAAC 2820
Non   TGTTTGACCACTATTTATGACCGCCTGGAGCAAGAGCACAACAATTTGGTCAACGTCCCT 2874
Opt   TGCCTGACCACCATCTACGATCGGCTGGAGCAGGAGCACAACAACCTGGTGAACGTGCCC 2880
Non   CTCTGCGTGGATATGTGTCTGAACTGGCTGCTGAATGTTTATGATACGGGACGAACAGGG 2934
Opt   CTGTGCGTGGACATGTGCCTGAATTGGCTGCTGAACGTGTACGACACCGGCAGGACCGGC 2940
Non   AGGATCCGTGTCCTGTCTTTTAAAACTGGCATCATTTCCCTGTGTAAAGCACATTTGGAA 2994
Opt   AGAATCAGAGTGCTGTCCTTCAAGACCGGCATCATCAGCCTGTGCAAGGCCCACCTGGAG 3000
Non   GACAAGTACAGATACCTTTTCAAGCAAGTGGCAAGTTCAACAGGATTTTGTGACCAGCGC 3054
Opt   GATAAGTACCGCTACCTGTTCAAGCAGGTGGCCAGCAGCACCGGCTTCTGCGATCAGAGG 3060
Non   AGGCTGGGCCTCCTTCTGCATGATTCTATCCAAATTCCAAGACAGTTGGGTGAAGTTGCA 3114
Opt   AGACTGGGCCTGCTGCTGCACGATAGCATCCAGATCCCTAGGCAGCTGGGCGAAGTGGCC 3120
Non   TCCTTTGGGGGCAGTAACATTGAGCCAAGTGTCCGGAGCTGCTTCCAATTTGCTAATAAT 3174
Opt   AGCTTTGGCGGCAGCAACATCGAGCCCTCTGTGAGGAGCTGCTTCCAGTTCGCCAACAAC 3180
Non   AAGCCAGAGATCGAAGCGGCCCTCTTCCTAGACTGGATGAGACTGGAACCCCAGTCCATG 3234
Opt   AAGCCCGAGATCGAGGCCGCCCTGTTCCTGGATTGGATGAGGCTGGAGCCCCAGAGCATG 3240
Non   GTGTGGCTGCCCGTCCTGCACAGAGTGGCTGCTGCAGAAACTGCCAAGCATCAGGCCAAA 3294
Opt   GTGTGGCTGCCTGTGCTGCACAGAGTGGCCGCCGCCGAGACCGCCAAGCACCAGGCCAAG 3300
Non   TGTAACATCTGCAAAGAGTGTCCAATCATTGGATTCAGGTACAGGAGTCTAAAGCACTTT 3354
```

Fig. 2. Codon optimization and alignment of microdystrophin variants. (**a**) Codon optimization of microdystrophin ΔAB/R3-R18/ΔCT. (**b**) Codon optimization of microdystrophin ΔR4-R23/ΔCT.

Table 1
Alphabetical list of companies providing codon optimization services

Codon optimization services/list of representative companies	Web page	Provider's service description/short summary
Biomatik	http://www.biomatik.com/service_desc.asp?id=18&classid=4	Biomatik particularly specializes in lengthy, high/low GC content, and repetitive sequences. Services include free codon optimization if required; extensive experience in high/low GC content or repetitive sequences; cloning into standard any vector of choice
Bionexus	http://www.genesynthesis.net; http://www.bionexus.net/	Bionexus has developed gene synthesis protocols that allow rapid turnaround times from optimizing the codon to confirming the sequence of the synthetic gene
BioPioneer	http://biopioneerinc.com/cs135/index.php	The research division at BioPioneer Inc. provides researchers with the following services: (1) gene synthesis (2) mouse genotyping (3) cloning/subcloning/mutagenesis (4) cell line construction (5) strain improvement (6) assay development (7) protein expression (8) site-saturation mutagenesis
Blue Heron Biotechnology	http://www.blueheronbio.com	A pioneer and leader of the gene synthesis market since its founding in 1999, clients include 19 of the top 20 pharmaceutical and biotechnology companies. Blue Heron Bio's proprietary gene design and synthesis platform synthesizes simple to highly complex sequences ranging in size from 50 to 50,000 bp in length
DNA Cloning Service	http://www.dna-cloning.com	Provides with gene cloning, gene synthesis and products for DNA handling
DNA2.0	http://www.dna20.com	Among the largest US providers of synthetic genes with a global customer base comprised of academia government and the pharmaceutical chemical agricultural and biotechnology industries
GENEART	http://www.geneart.com	First publicly listed gene synthesis company worldwide, successfully synthesizing more than 100,000 gene constructs since 1999 for customers all over the world. The company has established itself as a provider for DNA engineering and processing in the field of biotechnology and is one among the leading suppliers of synthetic genes

(continued)

Table 1 (continued)

Codon optimization services/list of representative companies	Web page	Provider's service description/short summary
GenScript Corporation	http://www.genscript.com	A biotech company and contract research organization. GenScript's custom gene synthesis service includes advanced free codon optimization tool and seamless cloning technology
Invitrogen	http://www.invitrogen.com	Invitrogen currently offers more than 25,000 products and services to support research in cellular analysis, genomics, proteomics, and drug discovery
Mr. Gene GmbH	http://www.mrgene.com	Mr. Gene is offering synthetic DNA services. Researchers can engineer and order their sequences online
NorClone	http://www.norclone.com	NorClone is a research support company that provides a broad range of molecular biology services including gene synthesis
OriGene	http://www.origene.com/	OriGene Technologies was founded in 1997 as a research tool company focused upon the creation of the largest commercial collection of full-length human cDNAs in a standard expression vector. OriGene Technologies uses high-throughput gene cloning and gene expression profiling to develop commercially available products for pharmaceutical, biotechnology, and academic research and discovery applications
Profos AG	http://www.profos.de	As an established Biotech company, Hyglos is dedicated to exploit the biochemical principles evolved by bacteriophages for healthcare applications. Company's "Gene to Protein Service" delivers the whole range from gene synthesis to protein expression and purification
RX Biosciences	http://www.rxbiosciences.com/?rxbio=GeneSynthesis.html	Rx Biosciences synthesizes gene by in vitro assembly of oligonucleotides. The genes are cloned in the vector of customer's choice and verified by double strand sequencing
ShineGene	http://www.synthesisgene.com	ShineGene Bio-Technologies, Inc., located in Shanghai, PR China, is a bio-company specialized in providing bio-technical services including gene synthesis, peptide synthesis, real time PCR, fluorogenic probe design, fluorogenic probe label, etc.

(continued)

Table 1 (continued)

Codon optimization services/list of representative companies	Web page	Provider's service description/short summary
Sigma Aldrich	http://www.sigmaaldrich.com/sigma-aldrich/home.html; http://www.sigmaaldrich.com/life-science/custom-oligos.html	A leading Life Science and High Technology company, operating in 34 countries with >6,000 employees. Sigma-Genosys offers a valuable service to the molecular biology laboratory – custom designed and assembled genes. They help with the design of synthetic genes (coding sequences) for a variety of applications, including expression, genetic manipulation for protein structure, and function studies
Viva Biotech	http://www.vivabiotech.com	Viva Biotech Limited, a Contract Research Organization (CRO), dedicates to providing preclinical drug discovery biology service and relevant products to global biotech and pharmaceutical companies

3. Methods

3.1. Generation of Codon-Optimized and Species-Specific Microdystrophin Sequences

1. Submit a sequence or multiple sequences to be codon optimized either via direct e-mail submission or by filling a Project Proposal Template and uploading to the company's website. Prior to submission, check the sequence(s) for retaining of the open reading frame (ORF), via alignment or blasting against GeneBank/PubMed (http://www.ncbi.nlm.nih.gov/pubmed/). Choose the species to have your sequence optimized to (see Notes 1–5). Provide the company a list of DNA or protein sequences as gene/protein variants or specify required mutations or directed evolution variants stating for example sequential permutation libraries, controlled randomization, combinatorial libraries, and/or SH3 phage display domain libraries. Such de novo gene synthesis enables convenient construction of rare allelic variations and combinations thereof.
2. Optimize the submitted sequence via Geneart's GeneOptimizer® software algorithms with a multiparameter optimization (see Notes 2 and 5). The software generates a total of up to 500,000 optimized variants of a target sequence in an evolutionary approach aiming for a unique sequence selection and the multiparameters incorporated include (a) database cloning,

(b) removal of introns, (c) knockout of cryptic splice sites and RNA destabilizing sequence elements, (d) increased RNA stability, (e) adaptation of codon usage, (f) extensive mutagenesis, (g) flexible combination of functional domains, (h) introduction of restriction sites, (i) epitope shuffling, (j) consideration of immune modulatory CpG motifs, and (k) different applications (Gene Transfer/Genetic Vaccination) that may require specific optimization of different critical parameters (see Note 3).

3.2. rAAV Production, Purification, and Characterization

1. Plate 293T cells at $1 \times 10^5/\text{cm}^2$ (150 cm plate = 175 cm^2 therefore plate 1.75×10^7 cells). One plate per roller bottle (RB) (Day 1).
2. Transfer cells to RB (10% FCS, DMEM), keep RB at 0.5 rpm for 24 h (Day 2).
3. After 24 h turn RB up to 1 rpm for 2 more days (Days 3–5).
4. On day 5, dilute a total of 500 mg plasmid DNA (all plasmids should be at 1:1 molar ratio for both two and three plasmid system) in 12.5 mL of 250 mM $CaCl_2$
5. Mix HEPES with 70 mM Na_2HPO_4 at 50:1 ratio in a 250 mL sterile tube.
6. Add $CaCl_2$/plasmid drop wise to the HEPES/Na_2HPO_4 buffer while constantly vortex the solution.
7. Incubate at RT for 25 min for precipitate to form.
8. Add 200 mL of 2% FCS DMEM to precipitate.
9. Decant media from RB and replace with 2% FCS DMEM and precipitate.
10. Continue culture cells in RB at 1 rpm for 3 days
11. On day 9, shake RB very vigorously to detach all cells from RB. Decant cell solution into a 250 mL centrifuge tube.
12. Centrifuge cells at $1,300 \times g$ for 10 min.
13. Decant supernatant and lyse cell pellet in 15 mL of lysis buffer.
14. Freeze/thaw lysate three times using dry ice/ethanol and 37°C water bath.
15. After three freeze–thaw cycles, add Benzonase to the final concentration of 50 U/mL. Incubate the lysate for 30 min at 37°C.
16. Calrify the lysate by centrifugation at $3,700 \times g$ at RT for 20 min and pass through a 0.45 μm filter. This is referred to as the crude lysate.
17. Prepare iodixanol gradient for ultra-centrifugation. Use a pasteur pipette, load 9 mL of 15% iodixanol/1 M NaCl in PBS-MK buffer, then 6 mL of 25% iodixanol in PBS-MK buffer containing phenol red, then 5 mL of 40% iodixanol in PBS-MK buffer, and 5 mL of 60% iodixanol containing phenol red into a 25×89 mm Beckman Quick-Seal Ultra-Clear tube.

18. Load the crude lysate over the top of the gradient.
19. Fill up the tube with lysis buffer.
20. Seal the tube and centrifuge at 255,000 × *g* 18°C for 85 min.
21. Remove 4 mL fraction from the interface between the 60 and 40% iodixanol.
22. This fraction can be stored in fridge overnight.
23. Desalting and concentration is carried out using Amicon Ultra –15 (PL100). Prerinse filter with 5 mL PBS-MK. Add 5 mL of PBS-MK to filter and centrifuge for 15 min at 4,000 × *g*.
24. Add 5 mL of PBS-MK to AAV fraction and add total volume to filter device. Centrifuge at 4,000 × *g* for 15–20 min, until volume has been reduced to ~0.5 mL.
25. Add 15 mL of PBS-MK to filter and repeat spin.
26. Repeat Step 3 two times to desalt the AAV prep.
27. During the final spin reduce volume to 300–400 mL.
28. Carefully remove the retentate from the filter. This contains your concentrated, desalted AAV.
29. Aliquot in 30–50 mL volumes and store at –80°C.
30. Take 1 and 5 μL of iodixanol fraction of rAAV mix with 5 U of DNase I in a final volume of 200 μL DMEM (without serum) and leave to digest for 1 h at 37°C.
31. An equal volume of 100 μg proteinase K in 2× proteinase K buffer (100 μL of 10 mg/mL) was added and the mixture was incubated for 1 h at 37°C.
32. Double extraction/purification of viral DNA was accomplished by using an equal volume of 25:24:1 phenol/chloroform/isoamyl alcohol, mix thoroughly, spin for 2′ at 4°C and precipitate the DNA with 1/10 volume of 3 M sodium acetate, 40 μg glycogen (2 μL), and 2.5 volumes of 100% ethanol. Incubate at –80°C for 30 min.
33. Centrifuge at 13,000 × *g* for 20 min at 4°C. Wash the pellet with 70% ethanol and dry. Resuspend in 400 μL of 0.4 M NaOH/10 mM EDTA (pH 8.0).
34. Prepare a twofold serial dilution of the rAAV vector plasmid corresponding to the rAAV virus stock to be titred in a volume of <20 μL (80–0.3125 ng). Mix the dilutions with 400 μL of 0.4 M NaOH/10 mM EDTA (pH 8.0) solution.
35. Heat viral and plasmid samples for 5 min at 100°C and cool on ice for 2 min.
36. Wet a single sheet of 3MM paper and hybridization membrane (Hybond N+) with water. Place 3MM paper and hybridization membrane onto dot-blot apparatus and secure with clips.

37. Add all denatured DNAs without vacuum and then apply vacuum.
38. Rinse wells with 400 μL of 0.4 M NaOH/10 mM EDTA (pH 8.0) and apply vacuum to dry.
39. Disassemble apparatus and rinse membrane in 2× SSC. The membrane can now be stored dry at 4°C or hybridized without further treatment.
40. Preheat the hybridization buffer to 42°C.
41. Prewet the blot in 5× SSC and place in RB or other container. Add appropriate volume of hybridization buffer (0.0625–0.125 mL/cm^2).
42. Prehybridise for a minimum of 15 min (1 h is more generally used).
43. Prepare the labeled nucleic acid probe: 100 ng of a PCR fragment. Dilute the DNA to be labeled to a concentration of 10 ng/μL in sterile water. Denature 100 ng of the DNA sample (10 μL) by heating for 5 min in a boiling water bath. Immediately cool the DNA on ice for 5 min. Spin briefly in a microcentrifuge. Add an equal volume (10 μL) of DNA labeling reagent to the cooled DNA and mix gently. Add 10 μL of glutaraldehyde solution, mix and spin in a microcentrifuge. Incubate for 10 min at 37°C. (Probe may be held on ice for 10–15 min, or for longer storage – up to 6 months – add 50% glycerol and store at –15 to –30°C.)
44. Remove 1 mL of hybridization solution from bottle and add labeled probe to it. Add to bottle and hybridize overnight at 42°C.
45. Prepare the primary wash buffer and preheat to 42°C. Transfer blot to this solution (2–5 mL/cm^2) and wash for a maximum of 20 min with gentle agitation at 42°C.
46. Repeat wash for 20 min at 42°C in primary wash buffer.
47. Remove primary wash buffer and wash 2 times for 5 min each time at room temperature in 100 mL 20× SSC. Blots can be left up to 30 min in this buffer, or stored overnight wet at 4°C. Develop blot using ECL chemiluminescence.
48. Determine AAV titer (see Note 6).

4. Notes

1. The gene optimization algorithm and thus the respective company performing the multiparameter optimization technology are very critical/important factors for the specific

species/codon optimization product requested. Certain algorithms appear to outperform other state-of-the-art optimization technologies. High G+C DNA, (inverted) terminal repeats, and secondary loop DNA or RNA structures can be difficult to assemble (or amplify) via PCR and often they do not provide with clear sequencing data. Long runs of G's can also be problematic for oligonucleotide synthesizers.

Codon optimization parameters for gene transfer products may include Codon optimization and G/C content adaptation. Inhibition of internal splicing and premature polyadenylation. Prevention of creation of stable RNA secondary structure. Introduction/avoidance of immunomodulatory CpG motifs. Avoidance of direct DNA repeats and thereby recombination events. Domain shuffling, epitope scrambling, and protein chimerization. Obstruction of formation of stable dsRNAs with host transcriptome and avoidance of possible RNAi viable genes. Optimization of tissue-specific promoter/enhancer sequences. Addition of RNA stabilizing and nuclear translocation supporting sequence elements. Validation of various gene derivatives combining the optimized gene with (a) preselected P/E sequences and (b) RNA stabilization and export signal. Elucidation of the most efficient cell line/vector combination. Quantification of gene expression by state-of-the-art immunological analysis such as FACS or WB analysis.

Codon optimization parameters for genetic vaccination products may include:

Increased genetic stability of vector constructs. RNA and codon optimization. Facilitated nuclear mRNA export and increased translational efficiency. Domain shuffling or scrambling of epitope fragments for safe and efficient expression of selected immunogens. Precise immune modulation through introduction of immune stimulating sequences in DNA vaccine constructs. Immune silencing through introduction of immune silencing sequences. Significantly reduced homology to wild-type sequences counteracting the risk of homologous recombination events. Enhanced efficacy by combining transgenes with a variety of on stock cytokine/chemokine DNA expression constructs (e.g., GMCSF, IL-15, MIP1a, etc.), available as wild-type c-DNA or expression optimized synthetic variants.

In our experimental application setting, human or murine microdystrophins were optimized for *Homo Sapiens* (human) or *Mus-musculus* (mouse) mammalian species based on available GenBank human and murine dystrophin sequences (Fig. 2c) and according to existing proprietary to GENEART respective algorithms/software databases, respectively. cDNA sequences were modified to include an optimum consensus

Kozak sequence for translation initiation, codon usage was optimized based on transfer RNA frequencies in human/murine species, and GC content was increased to promote RNA stability.

Optimization of microdystrophin cDNA sequences by Geneart resulted in GC content being increased from 46.4 to 60.6% in microdystrophin ΔAB/R3-R18/ΔCT and from 45.2 to 60.7% in microdystrophin ΔR4-R23/ΔCT whereas 59% of codons being modified in microdystrophin ΔAB/R3-R18/ΔCT and 63% in microdystrophin ΔR4-R23/ΔCT, respectively (Fig. 2).

Stretches of oligomer fragments are de novo synthesized and assembled together constituting the final codon-optimized sequence, which is subsequently ligated into a final plasmid vector(s) of choice by restriction endonuclease (RE) mediated sub-cloning. The final sequence can be verified by precise RE mapping and normally is fully sequenced as part of the service provided. The progress of the de novo synthesis for individual oligo-fragments and the final sequence assembly (which can take from days for a fast track uncomplicated project to months depending on the complexity of the requested synthesis – for e.g., depending on the presence of internal loops, secondary structures, size, etc.) can be observed on line under a specific customer's website supplied by the company. Full sequence data and corresponding files are normally provided in CD-ROM or e-mail for the customer following the final assembly of the de novo synthesized oligonucleotide sequences after verification of the final construct by sequencing. De novo synthesized species/codon-optimized cDNAs were supplied by GENEART as lyophilized plasmid material, reconstituted in water upon arrival and further downstream processed as required.

2. Gene optimization and gene synthesis services are normally based on a pure fee-for-service model. The company should disclaim any rights whatsoever on any intellectual property and know-how including any delivered gene sequences and material supplied by the customer. Upon request, the company responsible to perform the optimization will prepare or negotiate nondisclosure agreements.

3. Innovative medical approaches such as DNA vaccination or gene therapy rely on both the careful design of transgenes and the choice of the appropriate vector. Plasmid, viral, and bacterial vectors must promote an efficient gene transfer and at the same time ensure subsequent high-level expression of the transgene in vitro or in vivo. However, the expression level is not exclusively regulated by vector properties but is also tremendously influenced by the composition and qualities of the transgene. In particular, many in vitro/in vivo applications

in the fields of target validation, drug discovery, and evaluation, DNA vaccine development or protein biochemistry rely on efficient, durable expression in mammalian cells or different tissues. However, expression in mammalian cells is very often hampered by transcriptional silencing, low mRNA half-life, alternative splicing events, premature polyadenylation, inefficient nuclear translocation, and availability of rare tRNAs pools. Transgene's mRNA and its regulation on many different levels, including posttranscriptional steps such as RNA stability, nuclear RNA translocation, translational efficiency, and protein half-life are important factors.

4. Human dmd sequence is derived from GeneBank ACCESSION NM_ 004006 submitted sequence or patented (for e.g., the sequence for ΔAB/R3-R18/ΔCT mdys is described under http://www.patentstorm.us/patents/6869777/fulltext.html) (known) or other externally/internally filled for patent sequences and they generally consist of known exonic locus stretches or combinations thereof. Normally, a DNA or protein sequence or a PubMed accession number can be submitted. The sequence can be either submitted without flanking restriction sites and stop codon(s) or including a 5 and/or 3 restriction sites/UTRs and/or stop codon(s). Certain genes and sequences cannot be assembled directly using a standard, highly automated gene synthesis process. In cases where there are very high GC contents, direct/indirect repeats or long polypurine/pyrimidine runs, companies requested to codon-optimize a selected sequence have to use specially modified and purified oligonucleotides and special gene synthesis methods. Moreover, complex operon structures and very long natural sequences require highly sophisticated assembly strategies. Such nonstandard synthesis projects may be priced and dealt individually and may succumb to long synthesis delays.

5. Certain algorithms may differ (slightly) in their choice of multiparametric models toward codon optimization of a gene; thus certain companies may be more "successful" than others in optimizing the same transgene cassette. For e.g., OptimumGene™ algorithm optimizes a variety of parameters that are critical to *transcription* (GC content, CpG dinucleotides content, Cryptic splicing sites, Negative CpG islands, SD sequence, TATA boxes, Terminal signal), *translation* efficacy (Codon usage bias, GC content, mRNA secondary structure, Premature PolyA sites, Internal chi sites and ribosomal binding sites, RNA instability motif (ARE), inhibition sites (INS), Stable free energy of mRNA), and *protein folding* (Codon usage bias, Interaction of codon and anticodon, Codon-context).

6. We first need to calculate the MW in daltons of the control double stranded DNA molecule. This is done by multiplying the size of molecule in bp by 650 (average MW in daltons of 1 bp), e.g., for pX (5,410 bp) MW = $5,410 \times 650 = 3.52 \times 10^6$ Da. We then use Avogadro's number (6.023×10^{23}) to calculate the number of particles present. For pX, there are 6.023×10^{23} molecules in 3.52×10^6 g of this DNA, which translates as 1.71×10^{17} pX molecules per gram DNA ($6.023 \times 10^{23}/3.52 \times 10^6$). To calculate the number of molecules in an ng of this DNA we divide 1.71×10^{17} by 10^9, which gives 1.71×10^8 molecules per ng of pX DNA. We then compare the intensity of the control and sample dots, and based on this make an estimation of the amount of DNA in ng in the sample dots. This value is then multiplied by the calculated number of molecules/ng control DNA and then by 2 (as rAAV is single stranded and the control plasmid is double stranded) to give an estimated number of particles in the sample rAAV. If several different rAAV molecules are blotted at the same time, then differences in size can be adjusted by dividing the control rAAV MW by the test rAAV MW and then multiplying the calculated amount of particles by this amount.

References

1. Foster, H., Sharp, P.S., Athanasopoulos, T., Trollet, C., Graham, I.R., Foster, K., Wells, D.J., and Dickson, G. (2008) Codon and mRNA sequence optimization of microdystrophin transgenes improves expression and physiological outcome in dystrophic mdx mice following AAV2/8 gene transfer. *Mol Ther* 16, 1825–1832.
2. Angellotti, M.C., Bhuiyan, S.B., Chen, G., and Wan, X.F. (2007) Codon O: codon usage bias analysis within and across genomes. *Nucleic Acids Res* 35, W132–W136.
3. Shabalina, S.A., Ogurtsov, A.Y., and Spiridonov, N.A. (2006) A periodic pattern of mRNA secondary structure created by the genetic code. *Nucleic Acids Res* 34, 2428–2437.
4. Zhou, T., Weems, M., and Wilke, C.O. (2009) Translationally optimal codons associate with structurally sensitive sites in proteins. *Mol Biol Evol* 26, 1571–1580.
5. Ramakrishna, L., Anand, K.K., Mohankumar, K.M., and Ranga, U. (2004) Codon optimization of the tat antigen of human immunodeficiency virus type 1 generates strong immune responses in mice following genetic immunization. *J Virol* 78, 9174–9189.
6. Patterson, S.S., Dionisi, H.M., Gupta, R.K., and Sayler, G.S. (2005) Codon optimization of bacterial luciferase (lux) for expression in mammalian cells. *J Ind Microbiol Biotechnol* 32, 115–123.
7. Nguyen, K.L., Ilano, M., Akari, H., Miyagi, E., Poeschla, E.M., Strebel, K., and Bour, S. (2004) Codon optimization of the HIV-1 vpu and vif genes stabilizes their mRNA and allows for highly efficient Rev-independent expression. *Virology* 319, 163–175.
8. Sakai, H., Washio, T., Saito, R., Shinagawa, A., Itoh, M., Shibata, K., Carninci, P., Konno, H., Kawai, J., Hayashizaki, Y., and Tomita, M. (2001) Correlation between sequence conservation of the 5′ untranslated region and codon usage bias in Mus musculus genes. *Gene* 276, 101–105.
9. Anson, D.S., and Limberis, M. (2004) An improved beta-galactosidase reporter gene. *J Biotechnol* 108, 17–30.
10. Burgess-Brown, N.A., Sharma, S., Sobott, F., Loenarz, C., Oppermann, U., and Gileadi, O.

(2008) Codon optimization can improve expression of human genes in Escherichia coli: a multigene study. *Protein Expr Purif* 59, 94–102.

11. Wu, X., Jornvall, H., Berndt, K.D., and Oppermann, U. (2004) Codon optimization reveals critical factors for high level expression of two rare codon genes in Escherichia coli: RNA stability and secondary structure but not tRNA abundance. *Biochem Biophys Res Commun* 313, 89–96.
12. Jia, M., and Li, Y. (2005) The relationship among gene expression, folding free energy and codon usage bias in Escherichia coli. *FEBS Lett* 579, 5333–5337.
13. Dos, R.M., and Wernisch, L. (2009) Estimating translational selection in eukaryotic genomes. *Mol Biol Evol* 26, 451–461.
14. Kudla, G., Murray, A.W., Tollervey, D., and Plotkin, J.B. (2009) Coding-sequence determinants of gene expression in Escherichia coli. *Science* 324, 255–258.
15. Harper, S.Q., Hauser, M.A., DelloRusso, C., Duan, D., Crawford, R.W., Phelps, S.F., Harper, H.A., Robinson, A.S., Engelhardt, J.F., Brooks, S.V., and Chamberlain, J.S. (2002) Modular flexibility of dystrophin: implications for gene therapy of Duchenne muscular dystrophy. *Nat Med* 8, 253–261.

Chapter 3

Monitoring Duchenne Muscular Dystrophy Gene Therapy with Epitope-Specific Monoclonal Antibodies

Glenn Morris, Nguyen thi Man, and Caroline A. Sewry

Abstract

Several molecular approaches to Duchenne muscular dystrophy (DMD) therapy are at or near the point of clinical trial and usually involve attempts to replace the missing dystrophin protein. Although improved muscle function is the ultimate measure of success, assessment of dystrophin levels after therapy is essential to determine whether any improved function is a direct consequence of the treatment or, in the absence of improved function, to determine whether new dystrophin is present, though ineffective. The choice of a monoclonal antibody (mAb) to distinguish successful therapy from naturally occurring "revertant" fibres depends on which dystrophin exons are deleted in the DMD patient. Over the past 20 years, we have produced over 150 "exon-specific" mAbs, mapped them to different regions of dystrophin and made them available through the MDA Monoclonal Antibody Resource for research and for clinical trials tailored to individual patients. In this protocol, we describe the use of these mAb to monitor DMD gene therapy.

Key words: Duchenne muscular dystrophy, Dystrophin, Utrophin, Revertant fibres, Muscle biopsy, Immunofluorescence microscopy, Immunoperoxidase, Monoclonal antibody, Epitope mapping

1. Introduction

Most patients with Duchenne muscular dystrophy (DMD) have mutations in the dystrophin gene that result in the absence of dystrophin protein from muscle (1, 2). The majority of mutations are genetic deletions that remove one or more dystrophin-encoding exons from the dystrophin gene. Such deletions usually cause a frameshift in the dystrophin mRNA which inevitably results in early termination of translation. The dystrophin fragment produced by this early termination is usually unstable and fails to accumulate. In the rare cases where an amino-terminal dystrophin fragment does accumulate, it may not ameliorate the clinical features and, indeed, they may be even more severe than classic

Dongsheng Duan (ed.), *Muscle Gene Therapy: Methods and Protocols*, Methods in Molecular Biology, vol. 709, DOI 10.1007/978-1-61737-982-6_3, © Springer Science+Business Media, LLC 2011

DMD (3). Sometimes the genetic deletion removes exons that leave the reading frame unchanged and in this case a slightly smaller, internally deleted dystrophin protein is produced, usually in reduced amounts (1, 2). The resulting disorder, Becker Muscular Dystrophy, is more variable and less severe than DMD, though cardiac problems become more evident as the patients live longer (2).

Point mutations that result in early stop codons ("nonsense" mutations) have much the same effect, as do splice-site mutations that also result in frameshifts in the dystrophin mRNA (4).

Several different approaches have been used in the attempt to replace the missing dystrophin protein, and all of the following examples have used the anti-dystrophin monoclonal antibodies (mAbs) described in this chapter. Attempts have been made to introduce normal muscle-cell precursors into DMD muscle (myoblast cell therapy) (5). Bone-marrow-derived stem cell transplantation can also give rise to dystrophin-positive fibres in DMD muscle (6). Viral vectors or DNA alone have been used to introduce normal dystrophin-encoding DNA permanently into muscle cells (gene therapy) (7). Drugs that cause "read-through" of nonsense mutations may be effective in those few cases with this type of mutation (8). Oligonucleotide-based drugs have been used to cause targeted exon-skipping that restores the reading frame when the dystrophin mRNA is produced from the primary RNA transcript (9). Many of these approaches have potential problems with effective delivery and with the immune response (against cells, DNA and even dystrophin itself) (10). In some, the aim is to "convert" the DMD phenotype into a BMD phenotype, rather than to effect a complete cure. Finally, drugs that further upregulate utrophin, the non-muscle "isoform" of dystrophin, in DMD muscle may have a beneficial effect (11, 12), although it is unclear whether utrophin can effectively replace all dystrophin functions. The anti-dystrophin mAbs have also been widely used in animal model studies of therapies for DMD (see Note 1).

To assess whether any of these treatments has been effective in clinical trials, it is necessary to use antibodies against dystrophin to determine whether dystrophin has been produced during the therapy. Of course, improved muscle function is the ultimate aim, but it is important to confirm that such improvement is due to new dystrophin production or, if no improvement was observed, to determine whether new dystrophin is present, though ineffective. Where treatments are performed by local injection, it is normal to take a muscle biopsy near the site of injection and to determine dystrophin expression by immunohistochemistry. This technique can detect even a small number of dystrophin-positive fibres in a biopsy, whereas western blotting may only work well when the proportion of positive fibres is quite high.

Although the great majority of muscle fibres in biopsies from DMD patients appear to lack dystrophin completely, there is usually a small proportion (1–5%) that has near-normal levels of dystrophin (13). Studies with exon-specific anti-dystrophin mAbs have shown that these "revertant" fibres may arise from exon-skipping events in dystrophin transcripts, so that they produce an internally deleted dystrophin (14, 15). In DMD patients with a genetic deletion, the extra skipping is usually only of one or two exons, the minimum required to restore the reading frame (14). These revertant fibres may appear in groups with the same skipping pattern, suggesting a clonal origin. In the mdx mouse, there is a nonsense point mutation in exon 23. However, very large skippings of many exons encompassing the mutation have been observed (15), though it is not clear whether this is a property of mice or of the specific mutation. In addition to this small number of high expressing revertant fibres, there may also be a very low level of dystrophin expression in all fibres, detectable only with the strongest dystrophin antibodies (Fig.1a). The molecular nature of this low level dystrophin immunostaining is unknown.

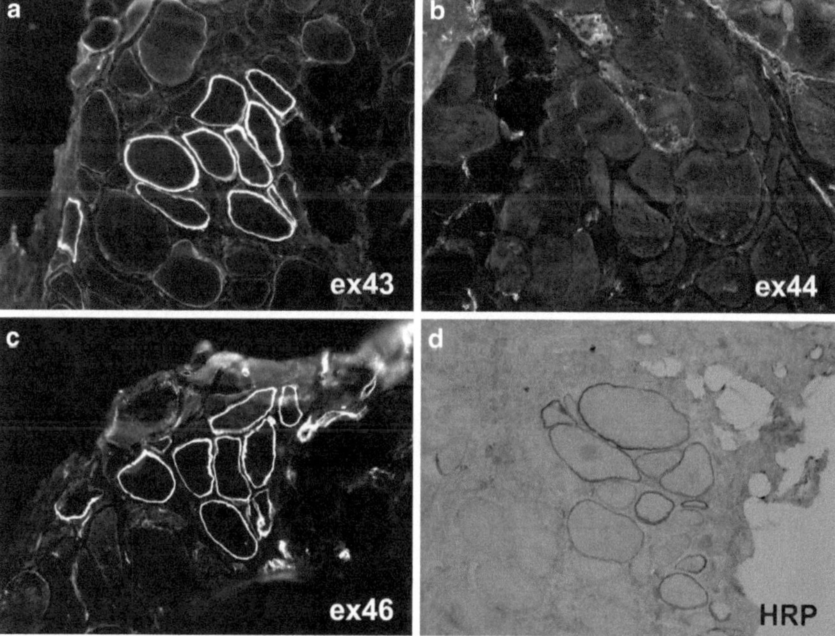

Fig. 1. Revertant fibres in a muscle biopsy from a DMD patient with a deletion of exon 45 in the dystrophin gene. This group of nine fibres is dystrophin-positive at the fibre membrane when mAbs against exon 43 (**a**) or exon 46 (**c**) are used, but is dystrophin-negative when mAbs against exon 44 (**b**) are used. This shows that the dystrophin in this group of fibres has been produced by skipping exon 44 to restore the reading frame disrupted by the absence of exon 45. (We do not have a good mAb for exon 45, but all revertant dystrophins arising from the exon 45 deletion will also lack exon 44 and/or exon 46.) Note also that MANDYS106 mAb against exon 43 shows lower level staining of most fibres. Although utrophin is present on DMD muscle membranes, this particular mAb does not appear to cross-react with utrophin. A different biopsy stained by the immuno-peroxidase method (Section 3.2) is illustrated in (**d**).

The appearance of less than 10% dystrophin-positive fibres after treatment could either be a result of the treatment or due to revertant fibres present in the muscle before treatment. This is not a mere theoretical possibility, since there have been occasions in the past when revertant fibres appear to have been wrongly interpreted as successful therapy (16). Fortunately, it is possible to distinguish these two possibilities with dystrophin antibodies. It is quite impossible for revertant fibres to contain dystrophin encoded by the deleted exon(s). In a patient with a deletion of exon 45, for example, any dystrophin contained within revertant fibres must lack the part of the protein encoded by exon 45. Consequently, an antibody directed against exon 45 of dystrophin protein would detect only "new" fibres resulting from cell therapy using normal myoblast precursors or using gene therapy with full-length dystrophin sequence. Clearly, antibodies against each of the 79 dystrophin exons, or at least against those exons that are commonly deleted in DMD patients, are required.

Over the years, we have produced over 150 mAbs against different regions of dystrophin. These antibodies have been mapped to specific exons using a variety of epitope mapping techniques. These are listed in Table 1 and can also be found on the website http://www.glennmorris.org.uk/mabs.htm. In some cases, exons in mutation hotspots are targeted, though other mAbs are intended for different purposes, such as characterization of the shorter isoforms of dystrophin. Figure 1 illustrates how these mAbs can be used to identify and characterise revertant fibres in DMD patient muscle biopsies. This patient had a deletion of exon 45. A group of dystrophin-positive fibres was found in his muscle biopsy with mAbs against exons 43 and 46 (MANEX43 and MANEX46). Interestingly, these fibres were negative for exon 44 specific mAb. Since deletion of both exons 44 and 45 restores the open reading frame. Our results confirmed that these fibres were revertant fibres.

Therapies involving exon-skipping of the dystrophin RNA transcript result in an internally deleted dystrophin protein which may be indistinguishable from the dystrophin protein in revertant fibres. Clearly, mAbs against the deleted exon cannot be used to demonstrate full-length dystrophin in such cases. Only if the number of dystrophin-positive fibres after treatment is much higher (e.g. >15%) than typical revertant fibre levels can one reliably conclude that the treatment has had any effect. Dystrophin expression in revertant fibres is usually quite high (Fig. 1) but, if exon-skipping therapies result in dystrophin levels that are variable or lower than revertant levels, it may be more difficult to identify therapy-derived fibres. In this case, it may be necessary to use mAbs that can detect very low levels of dystrophin. Here we present detailed immunoperoxidase labelling and immunofluorescence protocols for detecting dystrophin expression with these mAbs (see Note 2).

Table 1
Mouse monoclonal antibodies against human dystrophin and utrophin

Antibody name	Clone number	Type	Epitope	Use for	Species	Cross-reactions	References
MANEX1A*	4C7	G2a	ex1, 1–10	wb, if	hu, mo, dog	300K (brain)	(17, 18)
MANEX1B*	1D6	G2b	ex1, 1–10	wb, if	hu, mo, dog	300K (brain)	(17, 18)
MANEX6	4H4	G1	ex6	wb, if(w)	hu		(19)
MANEX7B*	8E11	G1	ex7/8	wb, if	hu, dog (xxe)		(19)
MANEX7C	6F7	G1	ex7/8	wb, if(w)	hu, dog (xxe)	Fibre type internal staining	(19)
MANEX8A	5D12	G1	ex8, 219–223	wb(w)	hu		(19)
MANHINGE1A	6F11	G1	ex8, 254–267	wb, if	hu, dog		(20)
MANHINGE1B	10F9	G1	ex8, 254–267	wb, if	hu, dog		(20)
MANHINGE1C	5D12	G1	ex8, 281–296	wb	hu, (xdog)	Internal fibre staining only	(20)
MANEX1011A	8A12	G1	ex10/11	wb, if	hu, mo, dog, xe	Some internal fibre staining	(19)
MANEX1011B*	1C7	G2a	ex10/11	wb, if	hu, mo, dog, xe		(19)
MANEX1011C*	4F9	G1	ex10/11	wb, if	hu, mo, dog, xe		(19)
MANEX1011D	7G5	G1	ex10/11	wb, if	hu, mo, dog, xe		(19)
MANEX1011E	8H7	G2a	ex10/11	wb, if(w)	hu, mo, dog, xe	Utrophin	(19)
MANEX1216B	6B11	G2a	ex12–16	wb (xif)	hu, mo (xxe)		(19)
MANEX1216D	8D11	G1	ex12–16	wb (xif)	hu (xmo, xxe)		(19)
MANEX1216E*	2G10	G1	ex12–16	wb, if	hu (xmo, xxe)		(19)
MANEX1216A*	5A4	G2a	ex14, 557–564	wb if(w)	hu (xmo, xxe)		(19)

(continued)

Table 1 (continued)

Antibody name	Clone number	Type	Epitope	Use for	Species	Cross-reactions	References
MANEX1216C	8C8	G1	ex14, 557–564	wb (xif)	hu, mo, xe	Utrophin?	(19)
MANHINGE2A	8B11	G1	ex17, 683–698	wb, if	hu, (xdog)		(20)
MANHINGE2B*	9C11	G1	ex17, 683–698	wb, if	hu, (xdog)		(20)
MANHINGE2C	8D11	G1	ex17, 699–714	wb if(w)	hu, (xdog)		(20)
MANDYS19*	8F6	G1	ex20/21, 856–915	wb, if	hu, mo (xch)		(21, 22)
MANDYS18*	5H9	G2a	ex26, 1181–1187	wb, if	hu, mo		(21, 22)
MANDYS17	3B12	G1	ex26/27, 1187–1205	wb, if	hu, mo	Some internal fibre staining	(21, 22)
MANDYS16*	1B12	G2b	ex27/28, 1226–1245	wb, if	hu, mo		(21, 22)
MANDYS1**	3B7	G2a	ex31/32, 1431–1505	wb, if, ip	hu, mo, dog, rb, pig (xxe)		(21, 22)
MANDYS2*	5G4	G1	ex31/32, 1431–1505	wb, if	hu, mo, xe		(21, 22)
MANDYS3	6A9	G1	ex31/32, 1431–1505	wb, if	hu, mo, xe		(21, 22)
MANDYS4	8H4	G1	ex31/32, 1431–1505	wb(w) if	hu, mo (xxe)		(21, 22)
MANDYS5	7F10	G2a	ex31/32, 1431–1505	wb, if	hu, mo (xxe)		(21, 22)
MANDYS6	7H7	G2b	ex31/32, 1431–1505	wb(w) if	hu, mo (xxe)		(21, 22)
MANDYS7	8D2	G2b	ex31/32, 1431–1505	wb, if	hu, mo (xxe)		(21, 22)
MANDYS8*	8H11	G2b	ex31/32, 1431–1505	wb, if	hu, mo (xxe)		(21, 22)
MANDYS9	4G10	G2a	ex31/32, 1431–1505	wb, if	hu, mo, dog (xxe)		(21, 22)
MANDYS10	2E10	G2a	ex31/32, 1431–1505	wb, if	hu, mo, dog (xxe)		(21, 22)

(continued)

Table 1 (continued)

Antibody name	Clone number	Type	Epitope	Use for	Species	Cross-reactions	References
MANDYS11*	1D6	G2b	ex31/32, 1431–1505	wb, if	hu, mo (xxe)		(21, 22)
MANDYS12	3E1	G1	ex31/32, 1431–1505	wb, if	hu, mo, dog (xxe)	Internal fibre staining	(21, 22)
MANDYS13	6G1	G2a	ex31/32, 1431–1505	wb, if	hu, mo (xch)	Some internal fibre staining	(21, 22)
MANDYS14	5E10	G1	ex31/32, 1431–1505	wb(w) if	hu, mo (xch)		(21, 22)
MANDYS15*	6B11	G1	ex31/32, 1431–1505	wb, if	hu, mo, dog (xch, xxe)		(21, 22)
MANDYS141	4A12	G2b	ex38, 1800–1815	wb (xif)	hu (xxe)	Actinin (hu, mo, dog, fi)	(23, 24)
MANDYS142	6D4	G2a	ex38, 1800–1815	wb (xif)	hu	Actinin (hu, mo, fi)	(23, 24)
MANDYS110*	3H10	G1	ex38/39, 1774–1856	wb(w) if	hu, mo (xch)		(23)
MANDYS121	8C9	G1	ex38/39, 1774–1856	wb(w) if	hu, mo, ch, fi, xe	Internal fibre staining	(23)
MANDYS123	4C3	G1	ex38/39, 1774–1856	wb if(w)	hu, mo, xe (xdog, xch)		(23)
MANDYS125	4F6	G2b	ex38/39, 1774–1856	wb if	hu, mo, ch, xe		(23)
MANDYS126	3E5C9	G1	ex38/39, 1774–1856	wb if(w)	hu, mo, dog, xe (xch)		(23)
MANDYS127	2G2	G1	ex38/39, 1774–1856	wb if(w)	hu, mo, xe (xch)		(23)
MANDYS129*	3H4	G1	ex38/39, 1774–1856	wb if	hu, mo, xe (xch)		(23)
MANDYS130	2H2	G1	ex38/39, 1774–1856	wb if(w)	hu, mo, xe (xch)		(23)
MANDYS104*	7F7	G2a	ex39–46, 1841–2254	wb, if	hu, dog (xmo, xxe)		(23)

(continued)

Table 1 (continued)

Antibody name	Clone number	Type	Epitope	Use for	Species	Cross-reactions	References
MANDYS105	8A4	G1	ex39–46, 1841–2254	if (xwb)	hu, mo, ch, dog, xe		(23)
MANDYS108	4D8	G2a	ex39–46, 1841–2254	wb, if	hu, dog (xmo, xxe, xch)		(23)
MANDYS109	2H5	G1	ex39–46, 1841–2254	if (xwb)	hu (xmo, xxe, xch)		(23)
MANDYS143	1A11D2	G2b	ex39–46, 1841–2254	if (xwb)	hu (xmo, xxe, xch)	Fast fibre cyto-plasm	(23)
MANDYS101*	7D12	G2b	ex40/41, 1874–1978	wb(w) if	hu (xmo, xch)		(23)
MANDYS107*	4H8	G2b	ex40/41, 1874–1978	wb if	hu, dog (xmo, xxe, xch)		(23)
MANDYS111	4F12	G2a	ex40/41, 1874–1978	wb if	hu (xmo, xxe, xch)	All fibre cyto-plasm	(23)
MANDYS124	3H7	G1	ex40/41, 1900–1960	wb(w) (xif)	hu, mo (xxe, xch)		(23)
MANDYS128	3D5	G1	ex40/41, 1900–1960	wb(w) (xif)	hu, mo (xdog, xxe, xch)		(23)
MANDYS131	3D6	G1	ex40/41, 1900–1960	wb if	hu, mo (xch)	Internal fibre staining	(23)
MANDYS122	8D11	G1	ex41, 1938–1960	wb if	hu, mo (xch)		(23)
MANDYS102*	7D2	G2a	ex43, 2047–2105	wb, if	hu (xmo, xxe, xch)	Utrophin in if (weak)	(23)
MANDYS103*	8E11	G2a	ex43, 2047–2105	wb, if	hu (xmo, xxe, xch)		(23)
MANDYS106*	2C6	G2a	ex43, 2063–2078	wb, if	hu (xmo, xxe, xch)		(23)
MANEX44A*	5B2		ex44	wb, if	hu (xxe)		(14)
MANEX44B*	10G5		ex44	wb, if	hu (xxe)	Utrophin	(14)

(continued)

Table 1 (continued)

Antibody name	Clone number	Type	Epitope	Use for	Species	Cross-reactions	References
MANEX45A	8F10	G1	ex45 (ex46?)	wb, if	hu, mo (xxe)		(25)
MANEX45B	7F8	G1	ex45	wb(w) (xif)	hu (xxe)	Utrophin	(25)
MANEX46A	6A6	G1	ex46, 2206–2255	wb, if	hu, mo (xxe)		(25)
MANEX46B*	7G1	G1	ex46	wb, if	hu, mo (xxe)		(25)
MANEX46C	3H12	G1	ex46	wb, if	hu, mo (xxe)		(25)
MANEX46D	8F2	G1	ex46	wb, if	hu, mo (xxe)	Some internal fibre staining	(25)
MANEX46E*	8G10	G1	ex46	wb, if	hu, mo (xxe)		(25)
MANEX46F	6G1	G1	ex46	wb, (xif)	hu, mo (xxe)		(25)
MANEX47*	8B12	G1	ex47, 2277	wb, if	hu (xmo, xxe)		(25)
MANEX4748A	8C1	G2b	ex47/48	wb, if	hu, mo (xxe)		(25)
MANEX4748B	8E6	G1	ex47/48	wb, if	hu, mo (xxe)		(25)
MANEX4748C	5G11	G2a	ex47/48	wb, if	hu (xmo, xxe)		(25)
MANEX4748D	7A5	G2b	ex47/48	wb, if	hu (xxe)		(25)
MANEX4850A	5E3	G2b	ex48–50	wb, if	hu (xmo, xxe)		(25)
MANEX4850B	6A7	G1	ex48–50	wb, if	hu, mo (xxe)		(25)
MANEX4850C	7B8	G1	ex48–50	wb, if	hu, mo (xxe)		(25)
MANEX4850D	7A10	G2b	ex48–50	wb, if	hu (xmo, xxe)	wb bgd	(25)
MANEX4850E	8C5	G2a	ex48–50	wb,if	hu, mo (xxe)		(25)
MANEX4850F	5G2	G2a	ex48–50	wb, if	hu (xmo, xxe)		(25)
MANEX50*	6A9	G1	ex50	wb,if	hu (xxe)		(25)
MANHINGE3A	6D6	G1	ex51, 2454–2468	wb (xif)	hu		(20)
MANHINGE3B	12D2	G2a	ex51, 2454–2468	wb (xif)	hu		(20)

(continued)

Table 1 (continued)

Antibody name	Clone number	Type	Epitope	Use for	Species	Cross-reactions	References
MANHINGE3C	1E13	M	ex51, 2454–2468	wb(w) if(w)	hu	Internal fibre staining	(20)
MANHINGE4A*	5C11	G1	ex62, 3058–3064	wb, if	hu	Some internal fibre staining	(20)
MANHINGE4B	9B1	G2b	ex62, 3058–3064	wb, if	hu		(20)
MANHINGE4C	10F2	M	ex63, 3070–3086	wb(w), if(w)	hu	Strong internal fibre staining	(20)
MANCHO19	5D12	G1	ex70, 3389–3392	wb, if	hu	Utrophin	(11, 26)
MANCHO11	9F2	G1	ex72/73, 3436–3450	wb, if	hu	Utrophin	(11, 26)
MANCHO12	1H1	G1	ex72/73, 3436–3450	wb, if	hu	Utrophin	(11, 26)
MANCHO13	2G12	G1	ex72/73, 3436–3450	wb, if	hu	Utrophin	(11, 26)
MANCHO14	4E1	G1	ex72/73, 3436–3450	wb, if	hu	Utrophin	(11, 26)
MANEX7374A*	10A11	G1	ex73/74, 3445–3517	wb, if	hu, mo, xe		(27)
MANEX7374B*	8G2	G1	ex73/74, 3445–3517	wb, if	hu, mo (xxe)		(27)
MANEX7374C	7F2	G1	ex73/74, 3445–3517	wb, if(w)	hu, mo (xxe)		(27)
MANEX7374D	4G10	G2b	ex73/74, 3445–3517	wb, if	hu, mo (xxe)	Several extra bands on blot	(27)
MANEX7374E*	3D8	G1	ex73/74, 3445–3517	wb, if	hu, mo (xxe)		(27)
MANEX7374F	8E10	G1	ex73/74, 3445–3517	wb(w), (xif)	hu, mo (xxe)	Internal fibre staining only	(27)

(continued)

Table 1 (continued)

Antibody name	Clone number	Type	Epitope	Use for	Species	Cross-reactions	References
MANEX7374G*	8B4	G1	ex73/74, 3445–3517	wb, if	hu, mo, xe		(27)
MANEX7374H	8G1	G1	ex73/74, 3445–3517	wb, if(w)	hu, mo (xxe)	Internal fibre staining	(27)
MANEX7374I*	6D5	G1	ex73/74, 3445–3517	wb, if	hu, mo (xxe)		(27)
MANEX7374J**	8F10	G1	ex73/74, 3445–3517	wb, if	hu, mo, xe		(27)
MANEX7374K	9A1	G1	ex73/74, 3445–3517	wb, if	hu, mo, xe		(27)
MANEX7374L	7H9	G1	ex73/74, 3445–3517	wb, if	hu, mo, xe		(27)
MANEX7374M	7E10	G1	ex73/74, 3445–3517	wb, if	hu, mo, xe		(27)
MANEX7374N	10C4	G1	ex73/74, 3445–3517	wb, if	hu, mo (xxe)		(27)
MANEX7374O	9H8	G1	ex73/74, 3445–3517	wb, if(w)	hu, mo, xe		(27)
MANEX7374P	7E11	G1	ex73/74, 3445–3517	wb, if	hu, mo (xxe)		(27)
MANEX7374Q	10B12	G1	ex73/74, 3445–3517	wb, if(w)	hu, mo (xxe)		(27)
MANEX7374R*	8A12	G1	ex73/74, 3445–3517	wb, if	hu, mo (xxe)		(27)
MANEX7374S	10A6	G2b	ex73/74, 3445–3517	wb(w), if(w)	hu, mo (xxe)		(27)
MANEX7374T	9A7	G1	ex73/74, 3445–3517	wb, if	hu, mo		(27)
MANEX7374U*	9F5	G1	ex73/74, 3445–3517	wb, if	hu, mo, xe		(27)
MANDRA9*	4A2	G1	n.d. (ex70–79)	wb, if	hu, mo, dog, to (xfi, xxe)		(23, 26)
MANDRA2	7G12	G1	n.d. (ex70–79)	wb, if	hu, mo, dog, fi (xto)		(23, 26)
MANDRA17	5C3	G1	ex74/75, 3513–3523	wb, if(w)	hu, mo, dog, fi, to	Utrophin	(23, 26)

(continued)

Table 1 (continued)

Antibody name	Clone number	Type	Epitope	Use for	Species	Cross-reactions	References
MANDRA14	2F2	G1	ex75, 3520–3527	wb, if(w)	hu, mo, dog (xfi, xto, xxe)		(23, 26)
MANDRA15*	3G11	G2b	ex75, 3520–3527	wb, if	hu, mo, dog, fi, xe (xto)		(23, 26)
MANDRA6*	8D11	G1	ex75, 3520–3590	wb, if	hu, mo, dog, fi, to, dm (xxe)		(23, 26)
MANDRA7*	5A8	G1	ex75, 3520–3590	wb, if	hu, mo, dog, fi, to (xxe)		(23, 26)
MANDRA8	6H6	G1	ex75, 3520–3590	wb, if	hu, mo, dog (xfi, xto, xxe)		(23, 26)
MANDRA10	8E9	G1	ex75, 3520–3590	wb, if	hu, mo, dog (xfi, xto, xxe)		(23, 26)
MANDRA11*	8B11	G1	ex75, 3520–3590	wb, if	hu, mo, dog, fi, to (xxe)		(23, 26)
MANDRA12	5C5	M	ex75, 3553–3558	wb, if	hu, mo, dog, fi, to (xxe)		(23, 26)
MANDRA13	7E5	M	ex75, 3520–3558	wb, if(w)	hu, mo, dog, fi, to (xxe)		(23, 26)
MANDRA16	4B11	G1	ex75, 3584–3595	wb, if	hu, mo, fi, to, dm (xxe)	Utrophin	(23, 26)
MANCHO15	5C3	G1	ex75, 3584–3595	wb, if	hu, mo, dog, fi, to (xxe)	Utrophin	(23, 26)
MANCHO16	9C2	G3	ex75, 3584–3595	wb, if	hu, mo, fi (xto)	Utrophin	(23, 26)
MANCHO17	4C8	G2a	ex75, 3584–3595	wb, if	hu, mo, fi, to	Utrophin	(23, 26)
MANDRA3*	3F4	G1	ex75, 3590–3600	wb, if	hu, mo, fi, to, dm	Utrophin	(23, 26)
MANCHO18	3C1	G1	ex75, 3602–3614	wb, if	hu, to, xe	Utrophin	(23, 26)
MANDRA5*	2E2	G2a	ex76/77, 3634–3644	wb, if	hu, mo, dog, fi, to (xxe)		(23, 26)

(continued)

Table 1
(continued)

Antibody name	Clone number	Type	Epitope	Use for	Species	Cross-reactions	References
MANDRA4	5H7	G1	ex77, 3644–3649	wb, if	hu, mo, fi (xto, xxe)		(23, 26)
MANDRA18*	7H7	G1	ex77, 3644–3649	wb, if	hu, mo, dog, fi, to (xxe)	Utrophin?	(23, 26)
MANDRA19	5G6	G1	ex77, 3644–3649	wb, if	hu, mo, dog, fi, to (xxe)	Utrophin?	(23, 26)
MANDRA1**	7A10	G1	ex77, 3667–3671	wb, if	hu, mo, fi, xe (xto)		(23, 26)
Utrophin							
MANNUT16	4H12		13–22		hu (xxe)		(28)
MANNUT17	6H7		1–113		hu (xxe)		(28)
MANNUT18	7C7		1–113		hu (xxe)		(28)
MANNUT19	8B5		1–113		hu (xxe)		(28)
MANNUT3*	4E4	G2b	194–200	wb	hu, dog (xxe)		(28)
MANNUT4	6H4	G2a	194–200	wb	hu, dog (xxe)		(28)
MANNUT14	8A3	G2b	113–261	wb	hu (xxe, xdog)		(28)
MANNUT1*	3B6	G1	261-371	wb, if	hu, rat (xxe)		(28)
MANNUT2	5C9	G2b	261–371	wb	hu, rat (xxe)		(28)
MANNUT5	8C11	G1	261–371	wb	hu, dog (xxe)		(28)
MANNUT6	10D4	G2a	261–371	wb	hu (xxe, xdog)		(28)
MANNUT7	6A10	G1	261–371	(xwb)	hu (xxe)		(28)
MANNUT8	10A7	G2a	261–371	wb	hu, rat (xxe)		(28)
MANNUT9	7C11	G1	261–371	(xwb)	hu (xxe)		(28)
MANNUT10	3C10	n.d.	261–371	wb	hu (xxe, xdog)		(28)
MANNUT11	3B3	G1	261–371	wb	hu (xxe, xdog)		(28)
MANNUT12*	4B12	G2a	265–280	wb	hu (xxe, xdog)		(28)

(continued)

Table 1 (continued)

Antibody name	Clone number	Type	Epitope	Use for	Species	Cross-reactions	References
MANNUT13	2A5	G2a	265–280	wb	hu, dog (xxe)		(28)
MANNUT15	5E10	G2b	261–371	wb	hu, dog (xxe)		(28)
MANCHO19	5D12	G1	3146–3149	wb, if	hu, mo (xfi, xxe)	Dystrophin	(23, 26)
MANCHO10	9E10	G2a	3164–3184	wb, if	hu, mo (xfi, xxe)		(23, 26)
MANCHO11	9F2	G1	3193–3207	wb, if	hu, mo, fi (xxe)	Dystrophin	(23, 26)
MANCHO12	1H1	G1	3193–3207	wb, if	hu, mo, fi (xxe)	Dystrophin	(23, 26)
MANCHO13	2G12	G1	3193–3207	wb, if	hu, mo, fi (xxe)	Dystrophin	(23, 26)
MANCHO14	4E1	G1	3193–3207	wb, if	hu, mo, fi, dm (xxe)	Dystrophin	(23, 26)
MANCHO1	2C7	G1	3232–3273	wb, if	hu, mo, dog, xe (xfi)		(23, 26)
MANCHO2	9G3	G1	3232–3273	wb, if	hu, mo, dog, xe (xfi)		(23, 26)
MANCHO3**	8A4	G1	3232–3273	wb, if	hu, mo, dog, xe (xfi)		(23, 26)
MANCHO4	2G5	G1	3232–3273	wb, if	hu, mo, dog, xe (xfi)		(23, 26)
MANCHO5	4F11	G1	3232–3273	wb, if	hu, mo, dog, xe (xfi)		(23, 26)
MANCHO6	5A6	G1	3232–3273	wb, if	hu, mo, dog, xe (xfi)		(23, 26)
MANCHO7*	7F3	G2a	3273–3314	wb, if	hu, mo (xdog, xfi, xxe)		(23, 26)
MANCHO8	5C5	G2a	3273–3314	wb, if	hu, mo (xdog, xfi, xxe)		(23, 26)
MANCHO9	8C10	G2a	3273–3314	wb, if	hu, mo (xdog, xfi, xxe)		(23, 26)

(continued)

Table 1 (continued)

Antibody name	Clone number	Type	Epitope	Use for	Species	Cross-reactions	References
MANCHO15	5C3	G1	3340–3351	wb, if	hu, mo, fi, xe, dm	Dystrophin	(23, 26)
MANCHO16	9C2	G3	3340–3351	wb, if	hu, mo (xfi, xxe)	Dystrophin	(23, 26)
MANCHO17	4C8	G2a	3340–3351	wb, if	hu, mo, fi, dm (xxe)	Dystrophin	(23, 26)
MANCHO18	3C1	G1		wb, if	hu, mo, fi, xe	Dystrophin	(23, 26)

Each antibody name is followed by the clone number and Ig subtype (note that clone numbers should be avoided, since several mAbs have the same clone number). Best mAbs for routine use are indicated by ** and mAbs that are the better within a group of similar mAbs are indicated by *. Relevant applications tested include western blotting (wb) and immunolocalization using frozen sections (if). Some mAbs may be weak (w) or negative (xif) in one application. All mAbs recognise the human antigens (hu) and may recognise other species (*mo* mouse; *dog, pig, xe* Xenopus laevis; *ch* chicken; *rb* rabbit; *fi* zebrafish; *to* torpedo; *dm* drosophila). Species that have been tested but not recognised are indicated, e.g. by (xdog), (xmo), etc., but have otherwise not been tested (except drosophila where only these mAbs indicated are OK). Known cross-reactions with other (non-dystrophin) proteins are indicated, but this is not exhaustive and may differ with non-human species. The references are to papers in which the epitope mapping of each mAb was done, where possible, since this is the most important factor for exon specificity. To understand epitope mapping methods, see refs (34, 35)

2. Materials

2.1. Preparations of Muscle Sections

1. Isopentane.
2. Liquid nitrogen (Caution: hazardous).
3. Prepared slides (see Note 3).
4. Cryostat (Leica Microsystems, Bannockburn, IL, USA).

2.2. Immunoperoxidase Labelling

1. Phosphate-buffered saline (PBS): pH 7.2. Any formulation is suitable. One example is 8 g/l NaCl, 2 g/l KCl, 1.44 g/l $Na_2HPO_4 \cdot 2H_2O$, 0.2 g/l KH_2PO_4.
2. Hydrophobic PAP pen (Invitrogen, Paisley, UK).
3. One percent hydrogen peroxide in PBS to block endogenous peroxidase (optional).
4. Biotinylated secondary antibody, e.g. anti-mouse Ig for mouse mAbs and anti-rabbit Ig for rabbit anti-sera, can be purchased from many sources (Vector Laboratories, Burlingame, CA, USA) (see Note 4).
5. Normal rabbit, goat, horse, sheep or donkey sera, depending on the animal in which the biotinylated secondary antibody was raised. (This is optional for labelling of human muscle.)

6. Streptavidin conjugated to peroxidase (DAKO, Carpinteria, CA, USA).
7. Diaminobenzidine (DAB, Sigma, St Louis, MO, USA). Stock solution 32 mg/ml in water or PBS (can be stored frozen). Commercial DAB tablets are also available from several companies. Caution: DAB may be carcinogenic and the powder should be exposed only in a fume hood. Add powder to a weighed universal tube in the fume hood, close the cap and remove it from the fume hood for weighing again. Return it to the fume hood and add the correct amount of water or PBS to make the stock solution at 32 mg/ml. This stock should be diluted to 1 mg/ml working solution when used in staining.
8. DAB-peroxidase solution: 0.1% H_2O_2 added to DAB working solution just before use.
9. Hematoxylin.
10. Graded alcohols: 70, 90 and 100% ethanol (in water) and xylene.
11. DPX mounting medium (Sigma).
12. A good quality photomicroscope with 20× and 40× objectives (e.g. Leica, Olympus, Zeiss, Nikon, etc.). A digital image capture system is essential and is usually supplied with the microscope, although third-party equipment to convert older film-based microscopes is also available. Images are usually captured at 72 dpi, so it is best to set the capture size at 512×512 or 1,024×1,024 pixels. These can be converted to 300 dpi for publication using Adobe Photoshop or similar software. Backups can be made to CD, DVD or an external hard-drive, although automatic backups may be made if using networked university or hospital facilities.

2.3. Immunofluorescence Labelling

1. PBS.
2. Hydrophobic PAP pen.(Invitrogen).
3. Biotinylated secondary antibodies (Vector Laboratories).
4. Normal serum as in Subheading 2.2 (optional).
5. Streptavidin conjugated to Texas red, fluorescein (FITC), rhodamine (TRITC), Cy or ALEXA (Invitrogen).
6. Diamidinophenylindole (DAPI): 10 mg/ml stock, 1 pg/ml final concentration (Sigma).
7. Hydromount (Merck, Darmstadt, Germany).
8. Photomicroscope as in Subheading 2.2. A fluorescence system and filter sets appropriate for the fluorophore to be used (plus UV filter for DAPI staining) are also required. Although biopsy sections are relatively thick, high magnifications are not usually required, so a confocal microscopy system is not essential.

3. Methods

3.1. Preparation of Muscle Sections

1. Mount the muscle sample (≤5 mm diameter) on a cork under a dissecting microscope to insure that the microtome blade will cut transversely across the fibres. Rapidly freeze the sample in isopentane cooled in liquid nitrogen. Once frozen, tissue blocks can be stored indefinitely at −80°C or in liquid nitrogen (see Note 5).
2. Cut 5–7 μm muscle cryosections at about −23°C. Collect sections onto prepared slides (see Note 3). Air-dry sections for about 20 min before use or before storage at −20 or −80°C for long-term storage. For storage, wrap slides back-to-back in Clingfilm and seal in foil. If only a few sections from a package are required, the remainder should not be allowed to thaw. Before immunolabelling, allow sections to warm to room temperature (RT) for about 15–20 min.

3.2. Immuno-peroxidase Labelling (see Note 6)

1. Air-dry sections on the bench for 15 min. Place slides or cover slips in a moist atmosphere (e.g. Petri dish with damp filter paper; slides or cover slips are placed flat, and drops of each reagent placed on top).
2. Optional (see Note 7): Cover each section with methanol or acetone for 10 min (RT, unless specified by supplier). Then drain and air-dry.
3. Circle each section with hydrophobic PAP pen.
4. Optional: Soak in PBS for 10 min.
5. Optional: Incubate with 1% hydrogen peroxide in PBS at RT for 20 min. This is to destroy endogenous peroxidase activity.
6. Rinse with PBS.
7. Optional: Incubate with 5% normal serum of the same species in which the secondary Ab is raised, for 20 min (approximately 20–25 μl for each section). Drain, but do not wash (see Note 8).
8. Incubate with diluted primary Ab (approximately 20–25 μl) at RT for 1 h (see Note 9).
9. Wash with PBS for 5 min, 3×.
10. Incubate with diluted biotinylated secondary Ab (1:100–1:200 in PBS) at RT for 1 h.
11. Wash with PBS for 5 min, 3×.
12. Incubate with diluted streptavidin-peroxidase (1:200) at RT for 1 h (see Note 10).
13. Wash slides with PBS for 5 min, 3×.

14. Prepare DAB-peroxidase solution.
15. Incubate in DAB-peroxide solution (enough to cover whole section) at RT for 4 min.
16. Rinse with distilled water.
17. Counterstain with hematoxylin for 30–60 s.
18. Rinse thoroughly with tap water and soak for 5–10 min.
19. Dehydrate through graded alcohols, 70, 90 and 100%. Clear in two changes of xylene. Leave in xylene until ready to mount.
20. Cover each section with a drop of DPX mounting medium and a cover-slip.

3.3. Immunofluorescence Labelling (see Note 6)

1. Air-dry unfixed cryostat sections and place slides or cover slips in a moist atmosphere (e.g. Petri dish with damp filter paper; slides or cover slips are placed flat, and drops of each reagent placed on top). Optional methanol or acetone treatment can be used as described in Subheading 3.2.
2. Circle each section with a hydrophobic PAP pen.
3. Dilute the primary Ab with PBS. Incubate muscle section with diluted primary Ab (approximately 20–25 μl) at RT for 30 min (see Note 8).
4. Rinse in PBS: Dip-wash each slide 10 s, 3×; or place in racks, and wash in dishes for 5 min, 3×.
5. Incubate with secondary biotinylated Ab from appropriate species (1:200) at RT for 30 min.
6. Rinse in PBS as in Step 4.
7. Incubate with streptavidin-Texas Red diluted in PBS (1:200) at RT, 15 min (see Note 11).
8. Rinse in PBS as in Step 4.
9. Mount in Hydromount. Carefully blot the slides with filter paper and pressure slightly to ensure even distribution of the mounting media. Store mounted slides in the dark.

4. Notes

1. Although raised in mice against human dystophin, most of the mAbs also react with mouse and dog dystrophin and have been widely used in animal model studies of DMD therapy (29–32). The method requires modification for mouse muscle sections because of the high background staining when anti-(mouse Ig) secondary antibodies are used on mouse tissues. This can be prevented by a pre-treatment method (33)

or by direct labelling of the primary mAb with Zenon Antibody Labelling reagents (Invitrogen).

2. Besides immunostaining, one also needs to confirm the results by western blotting, even though levels of dystrophin may be quite low. Any proteolysis occurring in the biopsy will further reduce the amount of detectable dystrophin, so protease inhibitors must be used. Ideally, part of the biopsy should be kept unmounted for western blotting purposes. If it is necessary to use a biopsy mounted for sectioning, it is essential to remove all traces of mountant before extracting the muscle. As before, it is best to choose a mAb that does not recognise revertant dystrophins (i.e. it recognises an epitope within the deleted dystrophin region for that particular patient). If that mAb does not give a dystrophin band on western blots, it may be necessary to do a parallel blot with a mAb, such as MANDYS1, that can detect very small amounts of dystrophin. Ideally, one should aim to use SDS-PAGE conditions that give good separation of dystrophin (427 kD) and utrophin (395 kD). Utrophin is over-expressed in Duchenne muscle and any slight cross-reaction of the anti-dystrophin mAb with utrophin could be misleading. Another parallel blot with an utrophin-specific mAb, such as MANCHO3, can be used to locate the utrophin band.
3. Prepared slides are made by (a) spreading 10 μl of 1 mg/ml poly-L-lysine over the slide, using a second slide, or (b) dipping slides into 2% 3-aminopropyl-triethoxysilane in acetone for 10 s, rinsing in acetone and then in water and allowing to dry. Superfrost or Superfrost Plus slides (Sigma) are designed for good adhesion and do not require coating.
4. Make sure the anti-mouse Ig antibody recognises all Ig, including IgM, and not IgG alone. Alternatively, you can buy a separate anti-mouse IgM and IgG antibodies.
5. Not only is tissue architecture better in frozen material than in fixed material, but many anti-dystrophin mAbs only work on unfixed tissue. The most common cause of poor results is poor tissue or section quality. This may be caused by poor initial freezing or subsequent handling, resulting in bad freezing artefact, or repeated thawing and freezing of stored sections. If only a few sections are required from a package stored in the freezer, the remainder must not be allowed to thaw. Samples containing much fat and connective tissue, or very necrotic tissue, may also cause adhesion problems.
6. Controls for immunocytochemistry are essential and should include both sample and Ab controls. To ensure that working procedures and reagents are correct, at least one sample known to be positive should be included. Abs to β-spectrin

can be used to give an indication of the preservation of the plasma membrane. Necrotic fibres usually lose their plasma membrane and appear negative with all Abs to the sarcolemma proteins, but may retain laminins.

7. Although acetone or methanol fixation may be used for studies of muscle sections, in the authors' experience, this is of no advantage for human muscle.
8. Some polyclonal primary Abs may produce high background, which may be reduced by blocking the sections with a high percentage (up to 50%) of normal serum of the species in which the secondary Ab was raised. Sections should not be allowed to dry out during the immunostaining procedure, because this also causes non-specific binding of Abs, resulting in high background. Short incubation times and a maximum dilution of the primary Ab may also reduce background. There is usually no advantage in using normal serum when using mAbs for human muscle
9. The dystrophin and utrophin monoclonal antibodies described in this chapter are available from the MDA Monoclonal Antibody Resource (http://www.glennmorris.org.uk/mabs.htm), supported by the Muscular Dystrophy Association (USA). The appropriate dilution must be established by titration studies
10. Peroxidase detected with DAB has traditionally been the most-used enzyme label. This produces a brown end product, which is stable for long periods at RT. Additional contrast and enhancement of the DAB end product can be achieved using nickel or silver to produce a black product. Other colours, such as red, with aminoethyl carbazole, or blue, with 4-chloro-l-napthol (Sigma), can also be obtained with different substrates. A common alternative to peroxidase is alkaline phosphatase, which, with the appropriate substrate, can be visualised as a red end product. β-galactosidase is useful for double-labelling and provides good contrast against the brown from peroxidase or red from alkaline phosphatase. There are a number of commercial kits available for amplification with peroxidase labelling (e.g. the MenaPath system from Menarini Diagnostics)
11. The most commonly used fluorochromes are fluorescein, rhodamine, Texas Red, AMCA, Alexa, Cy3 and Cy5. The choice is often of personal preference but factors such as cost, intensity, filters fitted to the microscope and fading also need to be considered. Fluorescein fades particularly fast on excitation. Labels such as the Cy range and Alexa give a particularly high signal and their popularity is increasing. Development and improvements in a technique have obvious advantages, but the pathological assessment of muscle often requires

comparisons of intensity, and new baselines may have to be established if methods are changed.

Acknowledgments

We thank the Muscular Dystrophy Campaign (UK) for support during the production of dystrophin and utrophin mAbs and Muscular Dystrophy Association (USA) for supporting the Monoclonal Antibody Resource with a Translational Research Infrastructure Grant (TRIG).

References

1. Koenig, M., Beggs, A.H., Moyer, M., Scherpf, S., Heindrich, K., Bettecken, T., et al. (1989) The molecular basis for Duchenne versus Becker muscular dystrophy: correlation of severity with type of deletion. *Am J Hum Genet* 45, 498–506.
2. Emery, A.E. and Muntoni, F. (2003) Duchenne Muscular Dystrophy, 3rd Edn, O.U.P., Oxford.
3. Helliwell, T.R., Ellis, J.M., Mountford, R.C., Appleton, R.E. and Morris, G.E. (1992) A truncated dystrophin lacking the C-terminal domains is localized at the muscle membrane. *Am J Hum Genet* 50, 508–514.
4. Roberts, R.G., Gardner, R.J. and Bobrow M. (1994) Searching for the 1 in 2,400,000: a review of dystrophin gene point mutations. *Hum Mutat* 4, 1–11.
5. Mendell, J.R., Kissel, J.T., Amato, A.A., King, W., Signore, L., Prior, T.W., et al. (1993) Myoblast transfer in the treatment of Duchenne's Muscular Dystrophy. *New Engl J Med* 333, 832–838.
6. Gussoni, E., Bennett, R.R., Gilgoff, I., Stein, J., Chan, Y., Muskiewicz, K.R., et al. (2002) Detection of donor nuclei in the muscles of a Duchenne muscular dystrophy patient 13 years following bone marrow transplantation. *J Clin Invest* 110, 807–814.
7. Romero, N.B., Braun, S., Benveniste, O., Leturcq, F., Hogrel, J.Y., Morris, G.E., et al (2004) Phase I study of dystrophin plasmid-based gene therapy in Duchenne/Becker muscular dystrophy. *Hum Gene Ther* 15, 1065–1076.
8. Welch, E.M., Barton, E.R., Zhuo, J., Tomizawa, Y., Friesen, W.J., Trifillis, P., et al. (2007) PTC124 targets genetic disorders caused by nonsense mutations. *Nature* 447, 87–91.
9. Kinali, M., Arechavala-Gomeza, V., Feng, L., Cirak, S., Hunt, D., Adkin, C., Guglieri, M., Ashton, E., Abbs, S., Nihoyannopoulos, P., Garralda, M.E., Rutherford, M., McCulley, C., Popplewell, L., Graham, I.R., Dickson, G., Wood, M.J., Wells, D.J., Wilton, S.D., Kole, R., Straub, V., Bushby, K., Sewry, C., Morgan, J.E. and Muntoni, F. (2009) Local restoration of dystrophin expression with the morpholino oligomer AVI-4658 in Duchenne muscular dystrophy: a single-blind, placebo-controlled, dose-escalation, proof-of-concept study. *Lancet Neurol* 8, 918–928.
10. Braun, S., Thioudellet, C., Rodriguez, P., Ali-Hadji, D., Perraud, F., Accart, N., et al. (2000) Immune rejection of human dystrophin following intramuscular injections of naked DNA in mdx mice. *Gene Ther* 7, 1447–1457.
11. Nguyen, T.M., Ellis, J.M., Love, D.R., Davies, K.E., Gatter, K.C., Dickson, G. and Morris, G.E. (1991) Localization of the DMDL-gene-encoded dystrophin-related protein using a panel of 19 monoclonal antibodies. Presence at neuromuscular junctions, in the sarcolemma of dystrophic skeletal muscle, in vascular and other smooth muscles and in proliferating brain cell lines. *J Cell Biol* 115, 1695–1700.
12. Burton, E.A., Tinsley, J.M., Holzfeind, P.J., Rodrigues, N.R. and Davies, K.E. (1999) A second promoter provides an alternative target for therapeutic up-regulation of utrophin

in Duchenne muscular dystrophy. *Proc Natl Acad Sci USA* 96, 14025–14030.

13. Klein, C.J., Coovert, D.D., Bulman, D.E., Ray, P.N., Mendell, J.R. and Burghes, A.H. (1992) Somatic reversion/suppression in Duchenne muscular dystrophy (DMD): evidence supporting a frame-restoring mechanism in rare dystrophin-positive fibers. *Am J Hum Genet* 50, 950–959.
14. Le, T.T., Nguyen, T.M., Helliwell, T.R. and Morris, G.E.(1995) Characterization of revertant muscle fibres in Duchenne Muscular Dystrophy using exon-specific monoclonal antibodies against dystrophin. *Am J Hum Genet* 56, 725–731.
15. Lu, Q.L., Morris, G.E., Wilton, S.D., Ly, T., Artem'yeva, O.V., Strong, P. and Partridge, T.A. (2000) Massive idiosyncratic exon skipping corrects the nonsense mutation in dystrophic mouse muscle and produces functional revertant fibres by clonal expansion. *J Cell Biol* 148, 985–996.
16. Partridge, T.A., Lu, Q., Morris, G. E. and Hoffmann, E. (1998) Is myoblast transplantation effective? *Nat Med* 4, 1208–1209.
17. Le, T.T., Nguyen, T.M., Love, D.R., Helliwell, T.R., Davies, K.E. and Morris, G.E (1993) Monoclonal antibodies against the muscle specific N-terminus of dystrophin: characterizaton of dystrophin in a muscular dystrophy patient with a frameshift deletion of exons 3-7. *Am J Hum Genet* 53, 131–139.
18. Morris, G. E., Nguyen, C. and Nguyen, T.M. (1995) Specificity and VH sequence of two monoclonal antibodies against the N-terminus of dystrophin. *Biochem J* 309, 355–359.
19. Bartlett, R.J., Stockinger, S., Denis, M.M., Bartlett, W.T., Inverardi, L., Le, T.T., et al. (2000) In vivo targeted repair of a point mutation in the canine dystrophin gene by a chimeric RNA/DNA oligonucleotide. *Nat Biotechnol* 18, 615–622.
20. Ahmed, N., Nguyen, T.M. and Morris, G. E. (1998) Flexible hinges in dystrophin. *Biochem Soc Trans* 26, S310.
21. Nguyen, T.M., Cartwright, A. J., Morris, G. E., Love, D.R., Bloomfield, J. F. and Davies, K. E. (1990) Monoclonal antibodies against defined regions of the muscular dystrophy protein, dystrophin. *FEBS Lett* 262, 237–240.
22. Nguyen, T.M. and Morris, G.E. (1993) Use of epitope libraries to identify exon-specific monoclonal antibodies for characterization of altered dystrophins in muscular dystrophy. *Am J Hum Genet* 52, 1057–1066.
23. Nguyen, T.M., Ginjaar, H.B., van Ommen, G.J.B. and Morris, G.E. (1992) Monoclonal antibodies for dystrophin analysis: epitope mapping and improved binding to SDS-treated muscle sections. *Biochem J* 288, 663–668.
24. James, M., Nguyen, T.M., Edwards, Y.H. and Morris, G.E. (1997) The molecular basis for cross-reaction of an anti-dystrophin antibody with alpha-actinin. *Biochim Biophys Acta* 1360, 169–176.
25. Le, T.T., Nguyen, T.M., Hori, S., Sewry, C.A., Dubowitz, V. and Morris, G.E. (1995) Characterization of genetic deletions in Becker Muscular Dystrophy using monoclonal antibodies against a deletion-prone region of dystrophin. *Am J Med Genet* 58, 177–186.
26. Morris, G.E., Sedgwick, S. G., Ellis, J. M., Pereboev, A., Chamberlain, J. S. and Nguyen, T.M. (1998) An epitope structure for the C-terminal domain of dystrophin and utrophin. *Biochemistry* 37, 11117–11127.
27. Simmons, C. (1997) Monoclonal antibody studies of an alternatively-spliced region of dystrophin. *M. Phil. thesis* (University of Salford).
28. Morris, G. E., Nguyen, T.M., Nguyen, T.N., Pereboev, A., Kendrick-Jones, J. and Winder S. J. (1999) Disruption of the utrophin-actin interaction by monoclonal antibodies and prediction of an actin-binding surface of utrophin. *Biochem J* 337, 119–123.
29. Dunckley, M.G., Love, D.R., Davies, K.E., Walsh, F.S., Morris, G.E. and Dickson, G. (1992) Retroviral-mediated transfer of a dystrophin minigene into mdx mouse myoblasts in vitro. *FEBS Lett* 296, 128–134.
30. Bartlett, R.J., Stockinger, S., Denis, M.M., Bartlett, W.T., Inverardi, L., Le, T.T., Nguyen, T.M., Morris, G.E., Bogan, D.J., Metcalf-Brogan, J. and Kornegay, J.N. (2000) In vivo targeted repair of a point mutation in the canine dystrophin gene by a chimeric RNA/DNA oligonucleotide. *Nat Biotechnol* 18, 615–622.
31. Lu, Q.L., Mann, C.J., Lou, F., Bou-Gharios, G., Morris, G.E., Xue, S.A., Fletcher, S., Partridge, T.A. and Wilton, S.D. (2003) Functional amounts of dystrophin produced by skipping the mutated exon in the mdx dystrophic mouse. *Nat Med* 9, 1009–1014.
32. Bostick, B., Yue, Y., Long, C., Marschalk, N., Fine, D.M., Chen, J. and Duan, D. (2009) Cardiac expression of a mini-dystrophin that normalizes skeletal muscle force only partially

restores heart function in aged mdx mice. *Mol Ther* 17, 253–261.

33. Lu, Q.L. and Partridge, T.A. (1998) A new blocking method for application of murine monoclonal antibody to mouse tissue sections. *J Histochem Cytochem* 46, 977–984.
34. Morris, G.E. (ed.) (1996) "*Epitope Mapping Protocols*" (Methods in Mol. Biol. vol. 66) Humana Press, Totowa.
35. Morris, G.E. (2008) "Epitope Mapping" in "*Molecular Biomethods Handbook*" (Rapley R. and Walker, J.M. eds.) Humana Press, Totowa, pp. 683–696.

Chapter 4

Methods for Noninvasive Monitoring of Muscle Fiber Survival with an AAV Vector Encoding the mSEAP Reporter Gene

Jérôme Poupiot, Jérôme Ausseil, and Isabelle Richard

Abstract

Muscular dystrophies (MD) are a group of genetically and phenotypically heterogeneous inherited disorders characterized by the progressive degeneration of the skeletal muscle tissue. In the last decade, a tremendous amount of studies were performed to test therapeutic strategies in animal models. Evaluation of such strategies requires the use of criteria predictive of their therapeutic relevance. Here we describe a simple, noninvasive assay to monitor muscle degenerative process. An adeno-associated vector encoding a secreted form of murine embryonic alkaline phosphatase (mSEAP) reporter gene is administrated at the time of treatment. The amount of circulating mSEAP will reflect the level of myofiber survival. We tested this assay with therapeutic gene transfer. We found a strong correlation between therapeutic gene expression/muscle disease amelioration and the circulating levels of mSEAP. The assay will be very useful for monitoring muscle cell survival after therapeutic intervention.

Key words: AAV, mSEAP, Muscular dystrophy, Therapy, Noninvasive monitoring

1. Introduction

Muscular dystrophies (MD) are a group of inherited disorders characterized by a progressive primary degeneration of the skeletal muscle tissue, leading to motor deficiency and muscle atrophy. The clinical features such as the time of onset, severity, progression, and pattern of involvement fluctuate tremendously among different MDs. The current classification is principally based on clinical and genetic features. The commonly seen MD includes Duchenne/Becker muscular dystrophies (DMD and BMD), limb-girdle muscular dystrophy (LGMD), congenital muscular dystrophy (CMD), facioscapulohumeral (FSH) muscular dystrophy, myotonic

Dongsheng Duan (ed.), *Muscle Gene Therapy: Methods and Protocols*, Methods in Molecular Biology, vol. 709, DOI 10.1007/978-1-61737-982-6_4, © Springer Science+Business Media, LLC 2011

muscular dystrophy (DM), oculopharyngeal muscular dystrophy (OPMD), and Emery-Dreifuss muscular dystrophy (EDMD). Except for some dominant forms of LGMD and one congenital MD, the genetic defect is known for most MD. It should also be noted that some patients are still excluded from all these diagnosis, indicating that other genes remain to be identified.

In MD, the genetic defect leads to the necrosis of structurally mature muscle fibers. This degeneration is associated with the liberation of soluble enzymes such as creatine kinase in blood. Necrosis is immediately followed by macrophage infiltration and, secondarily, by fiber regeneration mediated through satellite cell proliferation. Through these successive degeneration/regeneration cycles, the muscle tissue is progressively replaced by fibrotic and adipose tissues. At the histological level, muscles present dystrophic signs such as variation in fiber size, hyper-contracted fibers, necrotic fibers, fibers with centrally located nuclei, and inflammatory and fatty infiltrations. The current treatment is mostly palliative. However, new therapeutic perspectives are rapidly emerging in the fields of cell, gene, and drug therapies. Indeed, two of these strategies, including exon skipping and stop-codon read-through, are currently being evaluated in clinical trials in DMD patients (1–4).

Determining the efficacy of a therapeutic strategy is the foremost step in preclinical studies. In particular, the improvement of the dystrophic histopathology, cellular functions, and muscle force is often used to validate the efficiency of the treatment in animal models. With the exception of the in vivo force measurement, all these methods require the sacrifice of the animals, hence impeding longitudinal follow-up. Noninvasive methods taking advantage of imaging methods such as computed tomography (CT), magnetic resonance imaging (MRI), or single-photon–emission CT (SPECT) are additional useful tools. However, a major drawback is that these powerful techniques require expensive equipments and specialized expertise. As for circulating marker, though creatine kinase is released subsequent to fiber necrosis, this protein is nonspecific to muscle damage. Further, the sensitivity is too low to reflect the efficiency of local treatment, and the results are hardly reproducible in mice.

Considering that the main pathophysiological mechanism in MD is the degeneration/regeneration process, we reasoned that the expression of a reporter gene injected in the muscle could reflect the extent of cell necrosis over time. Consequently, if this reporter was administered at the time of treatment, it would reflect the cell survival-related recovery of the muscle. To test this hypothesis, we chose to use the murine secreted embryonic alkaline phosphatase (mSEAP) because this very sensitive and nonimmunogenic reporter protein is easily measured in blood samples and detected in muscle section (5). As a demonstration,

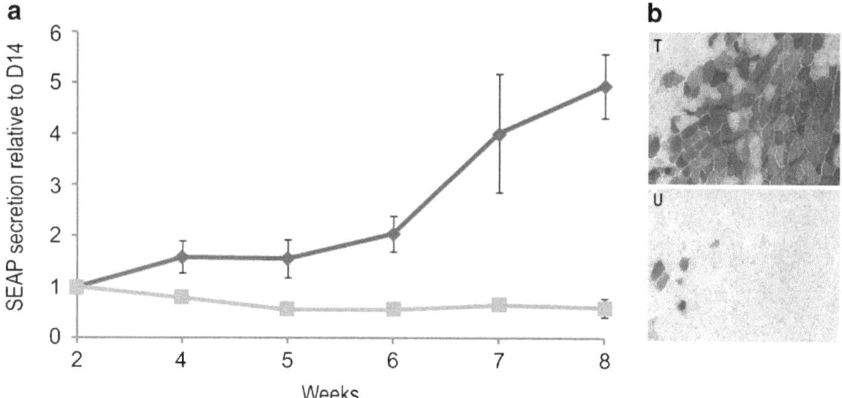

Fig. 1. mSEAP detection in blood and muscle tissue. Detection of the alkaline phosphate activity is carried out with the Phospha-Light Kit following the manufacturer's instructions, except for modifications of dilution buffer, optimized assay buffer and time of inactivation of endogenous phosphatase (13). (**a**) Levels of mSEAP were measured in blood at weeks (W) 2, 4, 5, 6, 7, and 8. Values of control (*gray line*) and treated (*black line*) mice are means of four different mice and are expressed as a ratio of the reference mean value at W2. (**b**) Representative pictures of TA muscle sections of untreated (U) and treated (T) model of muscular dystrophy (AAV-mediated transfer of α-sarcoglycan in *Sgca*-null mice). On these sections, mSEAP was observed by a histo-enzymatic staining.

we applied this method in two gene transfer experiments in MD animal models including adeno-associated virus (AAV)-mediated exon skipping for dystrophin deficient *mdx* mice and AAV-mediated transfer of α-sarcoglycan in *Sgca*-null mice (6). We demonstrated that mSEAP is a reliable surrogate marker for the correction of the underlying necrotic process (Fig. 1). Beside these two examples, the technique could be applied for evaluation of disease evolution or therapeutic strategies in other MD or even other disease types involving death of postmitotic tissues such as brain. Adaptation of this technique with integrated vectors is also possible for monitoring the efficiency of cell therapy or for the measurement of tumor growth.

2. Materials

2.1. Generating the Cis-Plasmid for Recombinant AAV Packaging

1. The coding sequence of mSEAP. The coding sequence is derived from the *mus musculus* alkaline phosphatase (mEAP) cDNA (GenBank AY054302.1). The secreted form of EAP (mSEAP) is deleted of the last 22 codons and obtained from the plasmid pXL3937 (5).
2. A muscle-expressing promoter (see Note 1). We chose to place the mSEAP cDNA under the control of the synthetic promoter SPc5-12, as it is a strong promoter with a relatively good muscle specificity (7).

3. An AAV backbone vector. In our experiments, we used pAAV-SPc5-12-MCS. It was constructed by replacing the CMV promoter of pAAV-CMV-MCS, an AAV-based pSMD2-derived vector carrying type 2 AAV ITRs (8), with the SPc5-12 muscle-specific promoter.

2.2. Production of the rAAV Reporter Vector

1. The AAV cis-plasmid carrying the reporter mSEAP cassette.
2. An adenoviral helper plasmid: pXX6 (9).
3. A packaging plasmid expressing the Rep and Cap genes required for Recombinant AAV (rAAV) replication and packaging (see Note 2). For the generation of type 1 AAV, the plasmid used is pAAV1pLT_RC02 (10).
4. HEK293 cells (ATCC, Teddington, UK). This is an adenovirus El gene-transformed human kidney cell line. It is used as a packaging cell line for rAAV production.
5. Seeding Medium: Dubelcco's Modified Eagle's Medium (DMEM), 4.5 g/L glucose with Glutamax (Invitrogen, Carlsbad, CA, USA) supplemented with 10% Fetal Bovine Serum (FBS; Hyclone, Thermo Fisher Scientific, Waltham, MA, USA). Store at 4°C.
6. Transfection Medium: DMEM, 4.5 g/L glucose with Glutamax (Invitrogen) supplemented with 1% FBS. Store at 4°C.
7. Production Medium: DMEM, 1 g/L glucose with Glutamax (Invitrogen) supplemented with 10% FBS. Store at 4°C.
8. Polyethylenimine (PEI) 25 kDa, 10 mM, pH7 (SIGMA-Aldrich, St.Louis, MO, USA).
9. 150 mM sodium chloride (NaCl). Store at room temperature.
10. Lysis Buffer: 50 mM HEPES pH7.6, 150 mM NaCl, 1 mM magnesium dichloride ($MgCl_2$), and 1 mM calcium dichloride ($CaCl_2$).
11. Benzonase 1×10^4 U/mL (Merck, Whitehouse station, NJ, USA).
12. Saturated ammonium sulfate (SIGMA-Aldrich). Store at 4°C.
13. PBS with $CaCl_2$ and $MgCl_2$ (Invitrogen).
14. Cesium chloride (SIGMA-Aldrich) dissolved in PBS with calcium magnesium to obtain solutions at 1.35, 1.41, and 1.5 g/mL, respectively.
15. OptiSeal™ tube, Polyallomer, 8.9 mL, 16 × 60 mm (Beckman Coulter, Fullerton, CA, USA).
16. Slide-A-Lyzer 10K MWCO Dialysis Cassettes (Interchim Inc., San Pedro, CA, USA).

2.3. In Vivo Vector Delivery into Muscle Tissue

1. Anesthesia solution: 2 mg/mL of Xylazine (Centravet, Dinan, France) and 20 mg/mL of ketamine 500FL (Centravet) in 150 mM NaCl.
2. TERUMO 1 mL sterile syringe (VWR–international, West Chester, PA, USA) with 27G*3/4 needle (20 mm length) for intraperitoneal injection of anesthesia solution.
3. Hamilton 50 μL syringe (NH Bio, Massy, France) with 30 G needle (27 mm length) for intramuscular injection of rAAV preparations.

2.4. mSEAP Secretion Measurement

1. Phospha-Light™ Reaction Buffer Diluent. (Phospha-Light™ System, Applied Biosystems). Store at +4°C.
2. CSPD® chemiluminescent substrate and Emerald™ luminescence enhancer. (Phospha-Light™ System, Applied Biosystems, Carlsbad, CA, USA). Store at +4°C.
3. Optimized Assay Buffer: 2 M diethanolamine (pH 10.5–11), 2 mM $MgCl_2$, 20 mM l-homoarginine. Store at +4°C.
4. Dilution buffer: 0.05 M Tris–HCl pH7.4, 150 mM NaCl. Store at room temperature.
5. Positive control: purified human placental alkaline phosphatase, 0.3 ng/μL in 150 mM Tris-HCl pH7.8, 50 mM NaCl, 80% glycerol (Phospha-Light™ System, Applied Biosystems). Store at +4°C.
6. Blood collection capillary tubes with heparin (VWR International).
7. Polycarbonate Amplitrack PCR plate (Dominique Dutscher, Brumath, France).
8. Reacti-Bind™ White opaque 96-well EIA Plates (Thermo Fisher Scientific).
9. Luminometer such as Victor luminometer with Wallac 1,420 Workstation software (PerkinElmer Life and Analytical Sciences, Inc, Wellesley, MA, USA).

2.5. Histological and mSEAP Histo-Enzymatic Revelation

1. Liquid nitrogen-cooled isopentane.
2. Glutaraldehyde 0.5% in PBS.
3. PBS without calcium and magnesium.
4. Solution A: of 5-bromo-4-chloro-3-indoyl phosphate (BCIP) 165 μg/mL and of nitroblue tetrazolium (NBT) 330 μg/mL in 100 mM Tris–HCl pH9.5, 100 mM NaCl, 50 mM $MgCl_2$ (NBT/BCIP, Promega, Madison, WI, USA). Store at 4°C several months. (See Note 3 for alternative methods).

5. Nuclear Fast Red Solution: Kernechtrot 1 mg/mL, aluminum sulfate 50 mg/mL in distillated water. Filtrate and store at room temperature.
6. Absolute Ethanol.
7. Xylene.
8. Eukitt (Labonord, Templemars, France).
9. Cork specimen mounts with 6% gum tragacanth (SIGMA).
10. Cryostat Leica CM 3050. (Leica, Solms, Germany).
11. Slide superfrost® Plus (LaboNord).
12. A CCD camera with the Cartograph (Microvision, Evry, France) software for capturing and assembling of digital images into a full muscle section image.

3. Methods

3.1. Generating the Cis Plasmid for Recombinant AAV Packaging

1. Release a *HindIII-XhoI* fragment containing the mSEAP coding sequence isolated from pXL3937 (5).
2. Using standard molecular cloning methods, Subclone the *HindIII-XhoI* fragment into pAAV-SPc5-12-MCS to make pAAV-SPc5-12-mSEAP.

3.2. Production of the rAAV Reporter Vector

The reporter AAV vector is produced by a method requiring the transfection of the three plasmids (pAAV-SPc5-12-mSEAP, pXX6, and pAAV1pLT_RC02) into HEK293 cells, followed by a cesium chloride gradient purification and dialysis. Triple-transfection of thirty 150 mm plates gives about 3 mL of a viral preparation with an expected titer of 1×10^{11} viral genomes (vg) to 2×10^{11} vg/mL (see Note 4). Efficient muscle transduction is usually obtained with intramuscular injection of 5×10^9 vg of the tibialis anterior muscle (TA). The quantity indicated in this method is for thirty 150 mm plates.

1. Two days before viral production, seed 5×10^6 HEK293 cells in 20 mL of seeding medium per 150 mm culture plates.
2. Prepare transfection solution extemporaneously. For thirty 150 mm plate, 3.37 mL of PEI is diluted in 30 mL of 150 mM NaCl.
3. Mix the plasmids: pXX6 (750 μg), pAAV1pLT_RC02 (375 μg), and pAAV-SPc5-12-mSEAP (375 μg) in 30 mL of 150 mM NaCl and dispense the mix by 5 mL in 50 mL Falcon tube (see Note 5). The reduction of the volume by this splitting step increases the efficiency of the complex formation.
4. For each tube, add drops by drops 5 mL of Transfection Solution to the 5 mL of DNA/NaCl mixture. Incubate at

room temperature for 20 min. Pool the DNA/PEI complexes (60 mL) and add 360 mL of transfecting medium. Discard the seeding medium of each culture plate and replace by 14 mL of transfection medium with the DNA/PEI mix. Incubate 6 h at 37°C, 5% CO_2, then finally add 12 mL of production medium and incubate 3 days at 37°C, 5% CO_2.

5. Harvest cells by scrapping and centrifuge 10 min at 500 g. Discard supernatants and suspend cell pellets with 30 mL of lysis buffer. Freeze/thaw cell lysate 4 times using dry ice/ethanol (10 min) and a 37°C water bath (15 min). Centrifuge 15 min at 1,300 g at 4°C. Digest the cell lysate supernatant with Benzonase 60 U/mL at 37°C for 30 min. Centrifuge 20 min at 7,300 × *g* at 4°C.
6. Keep supernatant and add cold saturated ammonium sulfate. Incubate in ice for 1 h to precipitate virus. Centrifuge 30 min at 10,000 × *g* at 4°C, then suspend the pellet in 2 mL PBS with $MgCl_2$ and $CaCl_2$.
7. Process for purification (see Note 6). Load crude viral particles onto a two-tier CsCl gradient (1.35 g/mL and 1.50 g/mL prepared in PBS) in an Optiseal tube. Overlay with PBS and spin the gradient by centrifugation at 67,000 × *g* for 13 h at 8°C in the 90TI Beckman fixed angle rotor. Discard the first mL of the lower part of the tube, then collect the following 4–5 mL which correspond to the viral particles and load it on a second CsCl gradient (1.41 g/mL prepared in PBS). Spin the gradient at 278,282 × *g* for 13 h at 8°C. Collect 12 fractions of 0.5 mL each and select the virus-rich fractions by Flash Dot (see Note 7). Pool selected fractions and fill a dialysis cassette to desalt the virus-rich fractions by five cycles of dialysis with 1.5 L of PBS during 2 h at 4°C.
8. After titration (see Note 8), aliquot the viral preparation adequately (see Note 9). Store at –80°C (up to several months).

3.3. In Vivo Vector Delivery into Muscle Tissue

In this method, the viral preparation of the mSEAP rAAV vector injected in the muscles as a reporter of muscle fiber survival is administered by IntraMuscular (IM) injection (see Notes 10 and 11) (Fig. 2).

Handle mice according to legal guidelines for the humane care and use of experimental animals.

1. Anesthetize mice by intra peritoneal injection of anesthesia solution at the dose of 100 μL per 20 g of body weight in order to reach a final concentration of 10 mg/kg xylazine and 100 mg/kg ketamine.
2. Dilute viral preparations in 150 mM NaCl at 5×10^9 vg/30 μL for each TA.

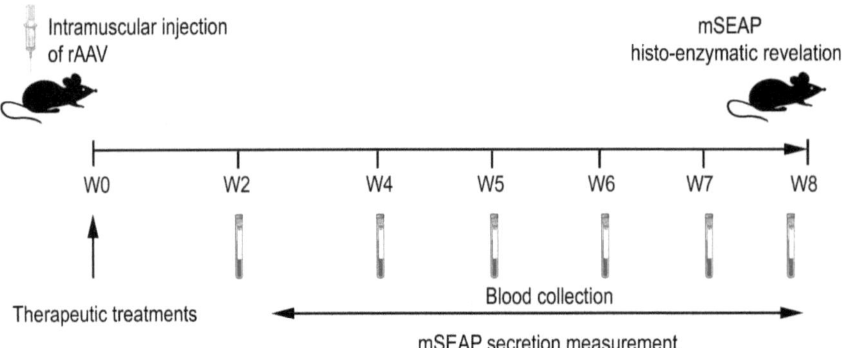

Fig. 2. Schematic representation of the protocol. At week 0 (W0), an intramuscular coadministration of the mSEAP vector together with the therapeutic vector is performed. Plasma concentration of mSEAP is monitored every week from week 1 (W1) to week 8 (W8). At W8, mice are sacrificed and alkaline phosphatase activity is measured on serial muscle cryosections.

3. Inject the muscle of interest with the viral preparation (in our case, 30 μL in the TA) with a 50-μL Hamilton syringe equipped with a 30-G needle.

3.4. mSEAP Secretion Measurement (see Note 12)

1. Collect blood samples by retro-orbital puncture of anesthetized animals. For a mouse of about 20 g, the volume usually obtained is ~200 μL.
2. Centrifuge blood samples at 2,655 × *g* for 5 min at 4°C.
3. Collect supernatant (plasma) in a fresh tube.
4. Store plasma at −20°C until analysis.
5. Dilute sufficient CSPD® substrate with Reaction Buffer Diluent (1 vol:20 vol) to obtain the reaction buffer (50 μL/well). The total volume of reaction buffer necessary for an assay is: 50 μL × N ($N = n$ samples + 5 dilutions of positive control + 1 negative control + 5).
6. Equilibrate the Optimized Assay Buffer (50 μL/well) and reaction buffer prepared above to room temperature.
7. Prepare samples by diluting 15 μL of plasma with 45 μL of Dilution Buffer in a PCR plate. Successive half dilutions of the positive control (from 300 pg/μL to 18.75 pg/μL) are also performed in Dilution Buffer.
8. Using a thermocycler, heat the samples plate at 65°C for 5 min, then cool to room temperature.
9. Transfer 50 μL of the diluted samples to Reacti-Bind™ White opaque 96-well EIA Plates.
10. Add 50 μL of Optimized Assay Buffer per well and incubate for 5 min.
11. Add 50 μL of reaction buffer per well and incubate for 20 min.
12. Place the microplate in the luminometer and measure for 10 s/well. The results are then expressed in Relative Light Units per

second. The serial dilution of the positive control is used to verify that the samples fall within the linear range of the assay.

3.5. mSEAP Histo-Enzymatic Revelation

In addition to the measurement of mSEAP in blood, it is useful to perform in situ histoenzymatic revelation to correlate the level of mSEAP secretion and the number of transduced fibers (Fig. 1b).

1. Anesthetize mice by intra peritoneal injection of xylazine (10 mg/kg) and ketamine (100 mg/kg). Inject 100 μL of anesthesia solution for 20 g of body weight.
2. Remove muscles and transversally mount them on cork specimen mounts with 6% gum tragacanth. Quickly froze the muscles in liquid nitrogen-cooled isopentane.
3. Prepare cryosections (6–8 μm thickness) from frozen muscles. Stock at −80°C until use.
4. Dry and then rehydrate in PBS frozen muscle cryosections.
5. Fix cryosections in 0.5% glutaraldehyde for 10 min.
6. Wash 3 times in PBS.
7. Incubate the sections at 65°C for 30 min to inactivate endogenous alkaline phosphatase activity.
8. Wash twice in PBS.
9. Overlay with 0.5 mL of Solution A at 37°C for 30 min to 2 h. Stop reaction when the purple blue color has developed.
10. Wash twice in PBS.
11. Incubate 1 min with Nuclear Fast Red solution.
12. Wash briefly with cold water.
13. Dehydrate cryosections by successive washes in ethanol (70, 95, and 100%).
14. Incubate 2 times for 2 min in xylene to change solvent for Eukitt mounting.
15. Mount with Eukitt.
16. Visualize SEAP-expressing fibers as purple staining using white light microscopy.

4. Notes

1. In our studies, we utilized the SPc5-12 promoter. It is possible to work with other promoters as long as its specificity corresponds to the same tissue than the therapeutic vector or the tissue targeted by the therapeutic molecule.
2. Type 1 AAV serotype has been proven a good serotype for IM muscle transduction (10). However, other AAV serotypes such

as AAV6, AAV8, and AAV9 have also been shown to transduce efficiently the muscle tissue after IM injection and are even more efficient for systemic delivery (11). Though the same protocol of production can be used, the packaging plasmid expressing the Cap gene of the corresponding serotype should be used.

3. Fast red chromogen (Roche, Neuilly-sur-Seine, France) can be used as an alternative staining of transfected fiber instead of NBT/BCIP.
4. Usually, for thirty 150 mm transfected culture plates, 3 mL of rAAV preparation can be expected using the conditions of production described in the method. This represents a number of particles varying between 3×10^{11} and 3×10^{12} vg. Yield can vary according to different parameters such as transfection efficiency, quality of the transfected plasmids, and transgene toxicity.
5. It is recommended to use the optimal ratio of 2:1:1 for the adenoviral helper, rep/cap packaging, and AAV cis-plasmids to maximize the level of full AAV particles.
6. In this protocol, we describe a CsCl ultracentrifugation-based protocol for rAAV purification. However, purification of rAAV can also be carried out by iodixanol gradient ultracentrifugation, ion exchange, or immunoaffinity columns. The last two methods are preferred for large scale production. These methods could be preceded by a diafiltration to concentrate the preparation.
7. Flash Dot is a simple technique to quantify the viral particles. Bulk DNA samples are immobilized on a nylon membrane, usually using a manifold attached to a suction device. Hybridization analysis is then carried out with an ITR probe to determine the relative abundance of target sequences in the blotted DNA preparations.
8. After successive treatments with DNAse I and proteinase K, viral genomes (vg) are quantified by a TaqMan real-time PCR assay using primers and probes corresponding to the ITR region (12). The primer pairs and TaqMan MGB probes used for ITR amplification are: AAV22mers.F: 5′ CTCCATCACTA GGGGTTCCTTG 3′; AAV18mers.R 5′ GTAGATAAGTAG CATGGC 3′; and AAV_MGB.P: 5′ TAGTTAATGATTAAC CC 3′. A series of successive dilutions of a control plasmid is used to calculate the number of vg.
9. Aliquoting the viral production is recommended in order to avoid repeated freeze-thawing cycles that are deleterious for the virus stability.
10. Using IM injection, it is possible to target the expression of the reporter to one muscle, preferentially the most seriously affected muscle of the model considered. Systemic injection of

the reporter is also possible and is especially relevant if many muscles are affected by the pathology. However, the risk to transduce other organs such as the liver and to dilute the modification of mSEAP secretion increases consequently. In that case, the use of a very strict muscular promoter is mandatory.

11. We previously demonstrated the efficacy of this method in two gene therapy strategies: (1) coinjection of the mSEAP reporter together with a rAAV2/1 vector expressing, through an U7 small nuclear modified RNA, an antisense sequence targeting exon 23 for skipping in the dystrophin deficient mice and (2) injection of a rAAV2/1 vector encoding for the α-sarcoglycan in the *Sgca*-null mice. In case of an AAV-based therapy, the injection of the reporter at the time of the injection of the therapeutic AAV is essential in order to prevent the potential immune response against the capsids prompted by two successive AAV transductions. It is also possible to evaluate the efficacy of cell therapy or drug treatment with this method, though in these cases, it is recommended to start the treatment few days before the injection of the mSEAP reporter.
12. In order to determine the evolution of secretion of mSEAP in the plasma, blood samples are collected by retro-orbital puncture on mice once every week until the end of the experiments (Fig. 2). The difference of kinetics of the reporter expression between treated and control animals will reflect the potential therapeutic effect. Because of the interindividual heterogeneity of muscle necrosis/regeneration, the level of initial transduction, and therefore the secretion of SEAP, can vary dramatically. In order to overcome this problem, the mSEAP secretion is expressed as a ratio of the values measured at the week considered (referred to as Wx) and the maximum value of the control (or, if appropriate, with the value obtained when mSEAP secretion has reached a level of expression reflecting the transduction efficiency) (i.e., about 2 weeks for *Sgca*-null and mdx models; W_X/W_2) (Fig. 1a).

Acknowledgments

We thank Dr. Marc Bartoli for helpful discussion and Dr. Nathalie Daniele for critical reading of the manuscript. We are grateful to Laetitia Van Wittenberghe, Isabelle Brelière, Simon Jimenez, Isabelle Adamski, Frederic Barnay Toutain, and Cyril Dupuis for excellent technical help. This work was supported by the French Association against the Myopathies (AFM).

References

1. van Deutekom, J. C., Janson, A. A., Ginjaar, I. B., Frankhuizen, W. S., Aartsma-Rus, A., Bremmer-Bout, M., den Dunnen, J. T., Koop, K., van der Kooi, A. J., Goemans, N. M., de Kimpe, S. J., Ekhart, P. F., Venneker, E. H., Platenburg, G. J., Verschuuren, J. J., and van Ommen, G. J. (2007) Local dystrophin restoration with antisense oligonucleotide PRO051. *N Engl J Med* 357, 2677–2686
2. Welch, E. M., Barton, E. R., Zhuo, J., Tomizawa, Y., Friesen, W. J., Trifillis, P., Paushkin, S., Patel, M., Trotta, C. R., Hwang, S., Wilde, R. G., Karp, G., Takasugi, J., Chen, G., Jones, S., Ren, H., Moon, Y. C., Corson, D., Turpoff, A. A., Campbell, J. A., Conn, M. M., Khan, A., Almstead, N. G., Hedrick, J., Mollin, A., Risher, N., Weetall, M., Yeh, S., Branstrom, A. A., Colacino, J. M., Babiak, J., Ju, W. D., Hirawat, S., Northcutt, V. J., Miller, L. L., Spatrick, P., He, F., Kawana, M., Feng, H., Jacobson, A., Peltz, S. W., and Sweeney, H. L. (2007) PTC124 targets genetic disorders caused by nonsense mutations. *Nature* 447, 87–91
3. Le Roy, F., Charton, K., Lorson, C. L., and Richard, I. (2009) RNA-targeting approaches for neuromuscular diseases. *Trends Mol Med* 15, 580–591
4. Kinali, M., Arechavala-Gomeza, V., Feng, L., Cirak, S., Hunt, D., Adkin, C., Guglieri, M., Ashton, E., Abbs, S., Nihoyannopoulos, P., Garralda, M. E., Rutherford, M., McCulley, C., Popplewell, L., Graham, I. R., Dickson, G., Wood, M. J., Wells, D. J., Wilton, S. D., Kole, R., Straub, V., Bushby, K., Sewry, C., Morgan, J. E., and Muntoni, F. (2009) Local restoration of dystrophin expression with the morpholino oligomer AVI-4658 in Duchenne muscular dystrophy: a single-blind, placebo-controlled, dose-escalation, proof-of-concept study. *Lancet Neurol* 8, 918–928
5. Wang, M., Orsini, C., Casanova, D., Millan, J. L., Mahfoudi, A., and Thuillier, V. (2001) MUSEAP, a novel reporter gene for the study of long-term gene expression in immunocompetent mice. *Gene* 279, 99–108
6. Bartoli, M., Poupiot, J., Goyenvalle, A., Perez, N., Garcia, L., Danos, O., and Richard, I. (2006) Noninvasive monitoring of therapeutic gene transfer in animal models of muscular dystrophies. *Gene Ther* 13, 20–28
7. Li, X., Eastman, E. M., Schwartz, R. J., and Draghia-Akli, R. (1999) Synthetic muscle promoters: activities exceeding naturally occurring regulatory sequences. *Nat Biotechnol* 17, 241–245
8. Snyder, R. O., Miao, C. H., Patijn, G. A., Spratt, S. K., Danos, O., Nagy, D., Gown, A. M., Winther, B., Meuse, L., Cohen, L. K., Thompson, A. R., and Kay, M. A. (1997) Persistent and therapeutic concentrations of human factor IX in mice after hepatic gene transfer of recombinant AAV vectors. *Nat Genet* 16, 270–276
9. Xiao, X., Li, J., and Samulski, R. J. (1998) Production of high-titer recombinant adeno-associated virus vectors in the absence of helper adenovirus. *J Virol* 72, 2224–2232
10. Riviere, C., Danos, O., and Douar, A. M. (2006) Long-term expression and repeated administration of AAV type 1, 2 and 5 vectors in skeletal muscle of immunocompetent adult mice. *Gene Ther* 13, 1300–1308
11. Inagaki, K., Fuess, S., Storm, T. A., Gibson, G. A., McTiernan, C. F., Kay, M. A., and Nakai, H. (2006) Robust systemic transduction with AAV9 vectors in mice: efficient global cardiac gene transfer superior to that of AAV8. *Mol Ther* 14, 45–53
12. Fougerousse, F., Bartoli, M., Poupiot, J., Arandel, L., Durand, M., Guerchet, N., Gicquel, E., Danos, O., and Richard, I. (2007) Phenotypic correction of alpha-sarcoglycan deficiency by intra-arterial injection of a muscle-specific serotype 1 rAAV vector. *Mol Ther* 15, 53–61
13. Cullen, B. R., and Malim, M. H. (1992) Secreted placental alkaline phosphatase as a eukaryotic reporter gene. *Methods Enzymol* 216, 362–368

Chapter 5

Monitoring Murine Skeletal Muscle Function for Muscle Gene Therapy

Chady H. Hakim, Dejia Li, and Dongsheng Duan

Abstract

The primary function of skeletal muscle is to generate force. Muscle force production is compromised in various forms of acquired and/or inherited muscle diseases. An important goal of muscle gene therapy is to recover muscle strength. Genetically engineered mice and spontaneous mouse mutants are readily available for preclinical muscle gene therapy studies. In this chapter, we outlined the methods commonly used for measuring murine skeletal muscle function. These include ex vivo and in situ analysis of the contractile profile of a single intact limb muscle (the extensor digitorium longus for ex vivo assay and the tibialis anterior muscle for in situ assay), grip force analysis, and downhill treadmill exercise. Force measurement in a single muscle is extremely useful for pilot testing of new gene therapy protocols by local gene transfer. Grip force and treadmill assessments offer body-wide evaluation following systemic muscle gene therapy.

Key words: Skeletal muscle, Twitch force, Tetanic force, Eccentric contraction, Grip strength, Treadmill, Gene therapy

1. Introduction

Skeletal muscles comprise about half of the body weight. The force generated by skeletal muscle is essential to voluntary actions such as talking, lifting, and running. Skeletal muscles distribute the loads along the skeleton. A skeletal muscle consists of myofibers that are bundled together by the endomysium. Each myofiber contains numerous myofibrils. Myofibrils are the force generating apparatus and they comprise about 80% of the muscle fiber volume and extend along the length of the myofiber. The basic unit of a myofibril is the sarcomere, a structure of highly organized contractile proteins (1). Muscle force is reduced in diseases that affect sarcomere configurations and/or myofibril integrity.

Dongsheng Duan (ed.), *Muscle Gene Therapy: Methods and Protocols*, Methods in Molecular Biology, vol. 709, DOI 10.1007/978-1-61737-982-6_5,

The force generated in the myofibrils is transmitted either longitudinally to the bone through the muscle tendon junction (MTJ) or laterally to adjacent myofibers (2, 3). In either case, the force is transferred to the surrounding tissues. The sarcolemma plays an important role in force transfer. The integrity of the sarcolemma is maintained by various transmembrane protein complexes, such as the integrin complex and the dystrophin-associated glycoprotein complex (4, 5). Defects in any one of these transmembrane linkages may destabilize the sarcolemma, compromise force transmission, and eventually decrease muscle strength (6). In summary, force reduction is a major physiological consequence of muscle diseases.

Many pharmacological treatments have been developed to modulate signaling and other cellular events (such as inflammatory responses) in various muscle diseases. Although limited clinical benefit has been documented, it is understood that such treatments cannot fully compensate for the function of the missing structural proteins. Gene therapy offers the hope of correcting the genetic defect at the molecular level. Murine models of various muscle diseases have been developed. These models open the door for preclinical testing of different muscle gene therapy strategies. A critical aspect in evaluating muscle gene therapy efficacy is to determine muscle force changes. This can be achieved by measuring the force of a single intact muscle, force generated from the entire limb, or the exercise capacity of the animal. In this chapter we present protocols for (a) ex vivo analysis of muscle forces generated from an intact extensor digitorum longus (EDL) muscle, (b) in situ analysis of muscle forces of the tibialis anterior (TA) muscle, (c) measurement of whole limb force by a grip strength meter, and (d) treadmill performance in conscious mice (7–12).

2. Materials

Get the approval from relevant authorities (such as the institutional animal care and use committee) prior to in vivo animal studies.

2.1. Ex Vivo Analysis of Muscle Forces of an Intact EDL Muscle

1. Anesthetic cocktail: 25 mg/mL ketamine, 2.5 mg/mL xylazine, and 0.5 mg/mL acepromazine dissolved in 0.9% NaCl. Keep the cocktail away from light and store at 4°C.
2. A custom-fabricated plexiglass dissection board (Fig. 1a). The dissection board is used for positioning the animal during muscle dissection (see Note 1).

Fig. 1. (continued) *asterisk*, the electrode connecting cable. (*b*) a closer view of the dotted boxed region of the *panel a*. *Top inserts*, electrodes with (*right panel*) and without (*left panel*) the attached sciatic nerve (*arrowhead*); *bottom insert*, a closer view showing the knee secured to the metal pin. (*c*) The suture loop on the patella tendon. (*d*) The suture loop on the distal TA muscle tendon. (*e*) The hind limb with skin removed. *Arrow*, the location of the sciatic never. (*f*) The sciatic nerve (*asterisk*). (*g*) Isolation of the sciatic nerve.

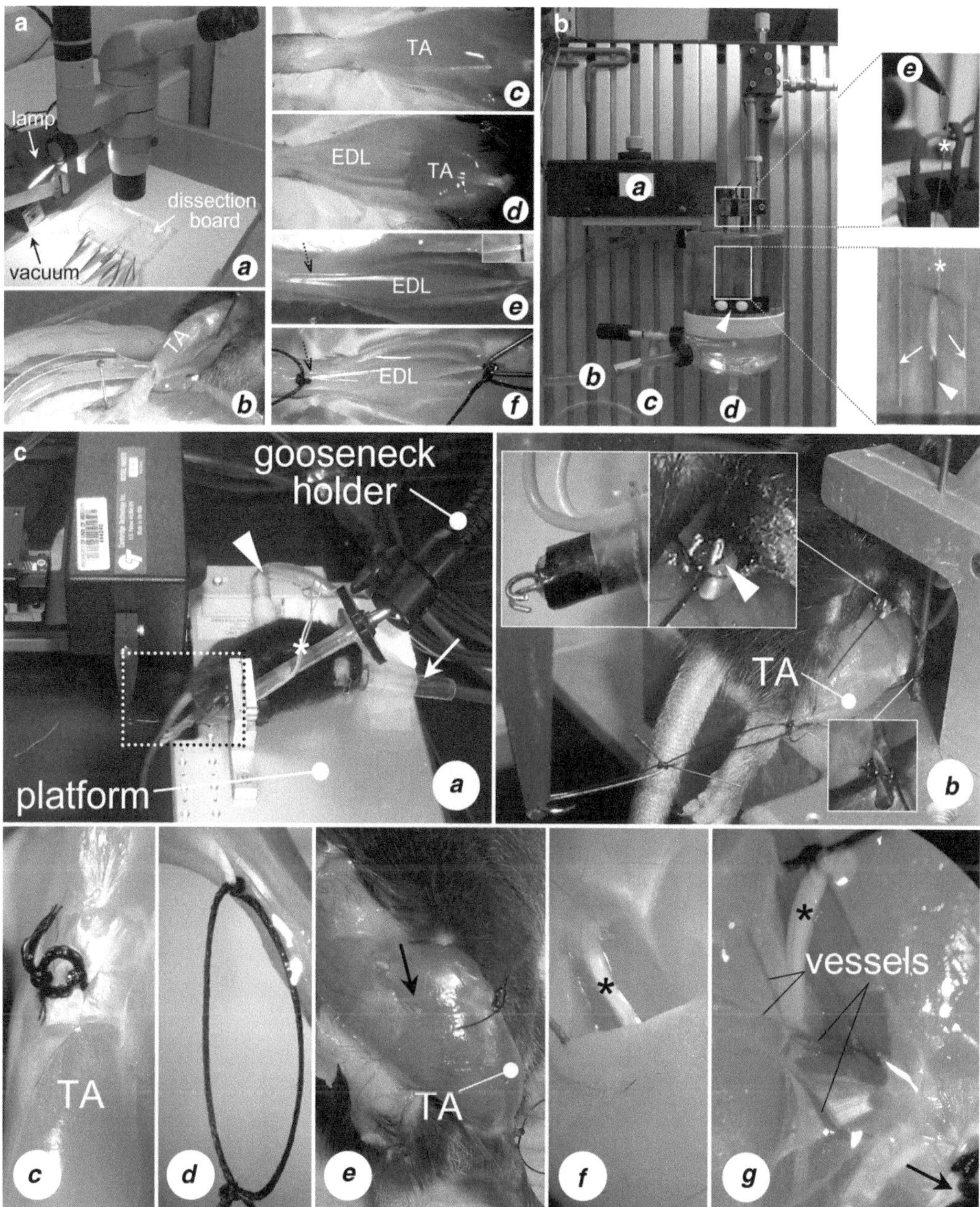

Fig. 1. Evaluating the EDL and TA muscle function. (**a**) The EDL muscle dissection system. (*a*) The dissection setup (heating lamp, dissection board, and vacuum line are marked). (*b*) The hind limb secured on the sylgard ring. (*c*) The TA muscle. (*d*) The TA muscle is partially lifted showing the underlying EDL muscle. (*e*) The EDL muscle. Note the distal EDL tendon is composed of four independent tendons while the proximal EDL tendon is not (see *insert*). (*f*) The EDL tendons are tied to the sutures and the muscle is ready to be taken out. The *dotted arrows* in (*e*, *f*) mark the peroneus brevis tendon. (**b**) The EDL muscle force measurement system. The EDL muscle is mounted on the system. The *left panel* shows the entire system. (*a*) The transducer. (*b*) The tubing connecting the circulating water bath and the tissue bath. (*c*) The 95% oxygen supply tubing. (*d*) The draining tubing. (*e*) The level arm. *Arrowhead*, the clamp. The *top right panel* shows a closer view of the level arm and the stainless steel hook (*asterisk*). The *bottom right panel* shows a closer view of the EDL muscle inside the organ bath. *Asterisk*, the stainless steel hook; *arrow*, the platinum electrode; *arrowhead*, the suture connecting the distal EDL MTJ and the clamp. (**c**) The TA muscle force measurement system. (*a*) An image of an experimental mouse mounted on the footplate apparatus. The gooseneck holder and the thermo-controlled platform are marked. *Arrow*, the oxygen cone; *arrowhead*, the circulating water bath inlet;

3. Heating lamp (Tensor Lighting Company, Boston, MA, USA).
4. Regular dissecting forceps and scissors (World Precision Instruments, Sarasota, FL, USA).
5. Microdissecting scissors and forceps (straight and 45°angled) (Fine Science Tools, Foster City, CA, USA).
6. Bread silk suture #4-0 (SofSilk USSC Sutures, Norwalk, CT, USA).
7. Stereo dissecting microscope (Nikon, Melville, NY, USA).
8. Ringer's Buffer: 1.2 mM NaH_2PO_4, 1 mM $MgSO_4$, 4.83 mM KCl, 137 mM NaCl, 24 mM $NaHCO_3$, 2 mM $CaCl_2$, and 10 mM Glucose, pH 7.4. Store at 4°C (see Note 2).
9. Circulating water bath (Fisher Scientific, Waltham, MA, USA).
10. Oxygen tank containing 95% O_2 and 5% CO_2 (Airgas National, Charlotte, NC, USA) (see Note 3).
11. Digital caliper (Fisher Scientific).
12. Microbalance (Fisher Scientific).
13. In vitro muscle function assay system (Aurora Scientific, Aurora, ON, Canada). The system consists of a 300B or 305B dual mode transducer that measures/controls the muscle force and length, a 701A stimulator, a 604B signal interface, and an 805A in vitro test apparatus equipped with a vertically mounted tissue organ bath (Fig. 1b) (see Note 4).
14. A custom-made 2-in.-long stainless steel hook and a custom-made 2.5-in.-long stainless steel hook (see Note 5).
15. Dynamic muscle control (DMC) software (Aurora Scientific). The software controls the force-length transducer and the stimulator. It allows real-time acquisition of the force and length data.
16. Dynamic muscle control data analysis (DMA) software (Aurora Scientific). The software analyzes the contractile data acquired by the DMC software.

2.2. In Situ TA Muscle Force Analysis

1. Physiological saline (0.9% NaCl).
2. In situ muscle function assay system (Aurora Scientific). The system consists of a 305C-LR dual mode transducer, a 701B stimulator, a 604A signal interface, and an 809A footplate apparatus (Fig. 1c).
3. A custom-made 0.5-in.-long stainless steel hook (see Note 5).
4. Custom-made 25G platinum electrodes (see Note 6) (Fig. 1c).
5. Circulating water bath, heating lamp, sutures, and dissection instruments (same as in Subheading 2.1).
6. Gooseneck holder (SK Science Kit, Tonawanda, NY, USA).
7. Oxygen tank containing 100% oxygen (Airgas National).

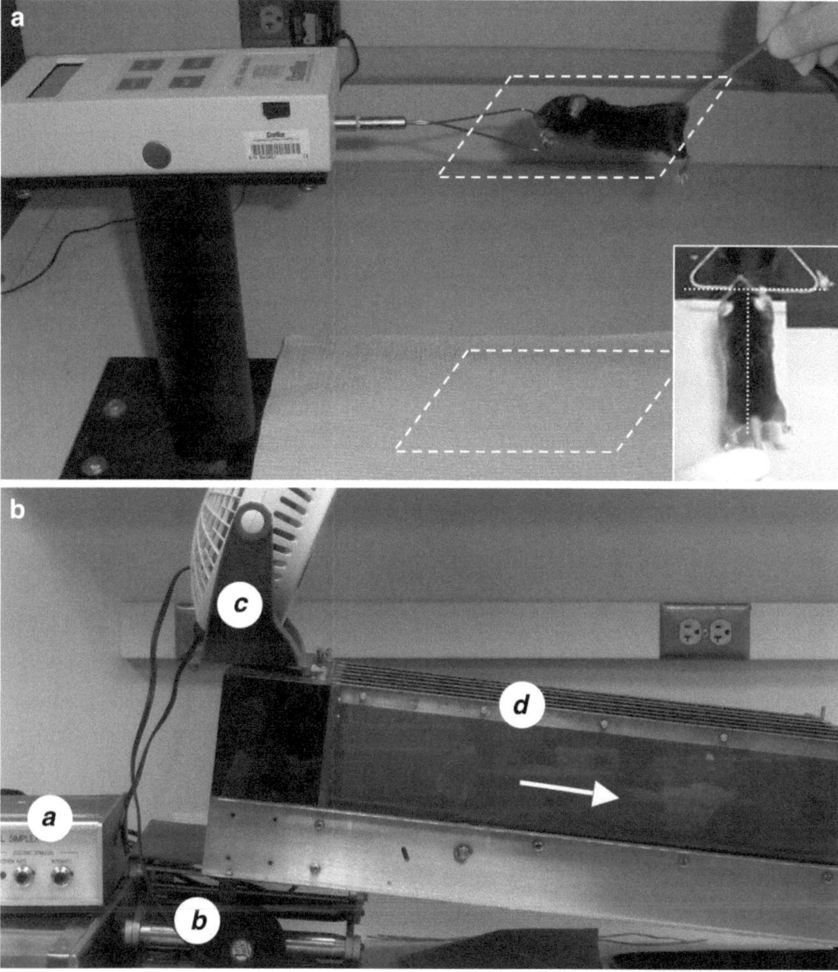

Fig. 2. Grip force study and treadmill exercise. (**a**) A side view image of a mouse undergoing grip force assay. *Dashed squares mark* the horizontal planes. The *insert* shows the position of the mouse perpendicular to the grip bar as marked by the *dotted lines*. (**b**) The downhill treadmill setup. *Arrow* marks the mouse running direction. (*a*) Control unit; (*b*) jack; (*c*) cooling fan; (*d*) running track.

2.3. Grip Force Measurement

1. Grip strength meter (Columbus Instruments, Columbus, OH, USA). The apparatus includes a pulling bar attached to a force transducer and a digital display (Fig. 2a).

2.4. Downhill Treadmill Assay

1. Exer 3/6 treadmill system (Columbus Instruments). The system includes a treadmill with an electrical stimulus and a control unit. The control unit regulates the treadmill speed and the intensity of the electrical shock. The system allows simultaneous analysis up to six animals in separate compartments (Fig. 2b).
2. 9-in. Cooling fan (Lasko, West Chester, PA, USA).
3. Jiffy jack (Cole Parmer, Vernon Hills, IL, USA).

3. Methods

3.1. Ex Vivo Analysis of Muscle Forces of an Intact EDL Muscle

1. Fill the organ bath with the Ringer's buffer. Turn on the circulating water bath to 30°C and start equilibrating the Ringer's buffer with 95% O_2 and 5% CO_2 (see Notes 3 and 7).
2. Load the DMC program and set the parameters at the dynamic muscle control and data acquisition panel as following: test duration, 3 s; update frequency, 500 Hz; sampling frequency, 1,000 Hz; input style, rise/fall; first stimulation, 1 s; last stimulation, 2 s; stimulation delay, 6 ms; pulse width, 200 µs. Turn on the dual mode level system. Turn on the simulator and set the stimulator parameters as following: trigger mode, follow; pulse phase, biphase; current multiplier, ×100; current adjust, 8.5 (see Note 8).
3. Record mouse and experimental information (body weight, date of birth, gender, strain, ear tag number, project title).
4. Anesthetize the mouse with intraperitoneal injection of 2.5 µL/g body weight of the anesthetic cocktail.
5. Set the heating lamp at the low intensity and adjust the lamp position to ~6″ above the mouse body. Position the mouse face up on the dissection board and carefully peel off the skin to expose the hind limb muscles. Secure the hind limb on the sylagard ring using regular dressmaker pins. Place the first pin between the metatarsal bones and the second pin in the distal end of the gracilis muscle. Constantly superfuse the exposed muscles with the Ringer's buffer (see Note 9).
6. Dissect the remaining skin towards the metatarsal bones to expose the distal tendon of the TA muscle. Gently peel off the fascia on the surface of the TA muscle. Cut off the extensor ligament that covers the distal TA tendon. Lift up the distal TA tendon with a pair of 45° microdissection forces and separate it from the distal EDL tendon. Cut off the TA tendon and slowly peel off the TA muscle towards the knee to expose the EDL muscle. Cut out the TA muscle from its proximal attachment near the knee (see Note 10) (Fig. 1).
7. Identify the EDL tendons and clear them from the surrounding connective tissue and fat. Tie a double square knot with a bread silk suture around the distal EDL tendon. Tie a double square knot around the proximal EDL tendon and tie another double square knot using the same suture to secure the proximal EDL tendon to the distal end of the steel hook. Cut the proximal EDL tendon superior to the suture knot. Gently pull off the EDL muscle from the proximal end. Use a pair of microdissection scissors to separate the EDL muscle from connective tissue and vessels beneath the muscle. Cut the distal EDL tendon

inferior to the suture knot and remove the intact EDL muscle. Attach the proximal end of the steel hook to the level arm of the transducer. Vertically position the EDL muscle between two platinum electrodes before securing the distal EDL tendon to a fixed clamp (see Note 11) (Fig. 1b).

8. Submerge the EDL muscle in the Ringer's buffer and adjust the muscle length to generate a 1 g resting tension. Allow the muscle to equilibrate for 10 min.
9. Adjust the muscle length to generate a resting tension of 0.5 g. Stimulate the muscle at 1 Hz (twitch stimulation). Record muscle force. Rest for ~15 s. Repeat twitch stimulation at muscle lengths that generate 0.8, 1.0, 1.2, 1.5, 1.8, and 2.0 g of resting tension. The length that generates the maximal twitch force is defined as the optimal muscle length (Lo). Record Lo with a digital caliper (see Note 12).
10. Rest muscle for 3 min. Set the muscle length to Lo and stimulate muscle with a single pulse. Record the force as the absolute twitch force (see Note 13).
11. Rest muscle for 3 min. Set the muscle length to Lo and stimulate muscle at 50, 80, 120, and 150 Hz. Allow 1 min rest between each contraction. Record the forces as the absolute tetanic forces under each stimulation frequency.
12. Rest muscle for 10 min. Set the muscle length to Lo and apply 10 cycles of eccentric contraction. For each cycle, stimulate muscle at 150 Hz for 700 ms. During the last 200 ms of stimulation, stretch the muscle for 10% Lo (0.5 Lo/s). After each stimulus, restore the muscle length to Lo at a speed of 0.5 Lo/s. Allow a 2 min rest between each cycle of eccentric contraction. Record the tetanic force of the first 500 ms stimulation for each cycle. The tetanic force of the first cycle is arbitrarily designated as 100%. This force equals the tetanic force obtained at the 150 Hz stimulation frequency in step 11.
13. Detach the EDL muscle and the steel hook from the force transducer. Cut the proximal and distal EDL tendons at the MTJ. Blot the muscle with Kimwipes twice and record the muscle weight (see Note 14).
14. Calculate the EDL muscle cross-sectional area using the equation of cross-sectional area = (muscle mass, in gram)/ [1.06 g/cm^3 × (optimal fiber length, in cm)]. 1.06 g/cm^3 is the muscle density. The optimal fiber length is calculated as 0.44 × Lo. 0.44 represents the ratio of the fiber length to the Lo of the EDL muscle (13).
15. Load the DMA program and open the file containing the force data. Analyze the data using the software. Export the data to an Excel file. The force value obtained from the DMA program represents the absolute force. Calculate the specific muscle

force by dividing the absolute force with the cross-sectional area. Calculate the force drop after eccentric contraction with the formula of Force drop % = $(F_1 - Fn)/F_1$. F_1 is the tetanic force obtained during the first cycle. *Fn* is the tetanic force obtained during the *n*th cycle.

3.2. In Situ TA Muscle Force Analysis

1. Load the DMC program and set the parameters at the dynamic muscle control and data acquisition panel as following: test duration, 3 s; update frequency, 1,000 Hz; sampling frequency, 10,000 Hz; input style, rise/fall; first stimulation, 0.5 s; last stimulation, 0.8 s; stimulation delay, 4 ms; pulse width, 200 μs. Turn on the dual mode level system and the simulator. Set the circulating water bath to 37°C.
2. Record mouse/experimental information and anesthetize the mouse as described in Subheading 3.1.
3. Position the mouse face up under the heating lamp. Carefully remove the hind limb skin to expose the TA muscle. Identify the patella tendon at the knee. Tie two double square knots around the patella tendon leaving a ~2 mm loop between two knots (Fig. 1c). Tie two double square knots around the distal TA tendon leaving a ~10 mm loop between two knots (Fig. 1c). Make an incision on the lateral side of the thigh. Expose the biceps femoris muscle. Cut open the muscle to reveal the sciatic nerve. Tie a double square knot around the nerve and cut the nerve superior to the knot (Fig. 1c). Constantly superfuse exposed muscles with saline.
4. Gently transfer the mouse to the thermo-controlled platform of the footplate apparatus. Supply oxygen to the mouse by placing the oxygen cone to the mouse nose. Secure the knee position by running the metal pin through the patella tendon suture loop. Attach the distal TA tendon suture loop to the level arm of the force transducer. Position the gooseneck holder to secure the sciatic nerve to the electrodes (Fig. 1c). Set the muscle length at a resting tension of 5 g for 5 min. Periodically apply prewarmed (37°C) physiological saline to the exposed tendon and muscle to keep moist. Stimulate the TA muscle by applying electric pulses to the sciatic nerve. Determine Lo as the muscle length at which the maximal twitch force is elicited. Record Lo with a digital caliper (see Note 15).
5. Rest muscle for 2 min. Set the muscle length to Lo and stimulate muscle with a single pulse. Record the force as the absolute twitch force.
6. Rest muscle for 3 min. Set the muscle length to Lo and stimulate muscle at 50, 100, 150, and 200 Hz. Allow 1 min rest between each contraction. Record the forces as the absolute tetanic forces under each stimulation frequency.

7. Rest muscle for 10 min. Set the muscle length to Lo and apply 10 cycles of eccentric contraction. At each cycle of eccentric contraction, stimulate the muscle for a total of 350 ms at 200 Hz. During the last 200 ms, stretch the muscle by 10% of Lo at the speed of 0.5 Lo/s. At the end of each cycle, readjust the muscle length to Lo. The maximal isometric tetanic force developed during the first 150 ms of stimulation of the first cycle was designated as 100%.
8. Carefully dissect out the TA muscle. Blot the muscle with Kimwipes twice and record the muscle weight (see Note 14).
9. Calculate the TA muscle cross-sectional area as described in step 14 of Subheading 3.1 except using a TA muscle-specific fiber length to the Lo ratio of 0.6 (14).
10. Load the DMA program and analyze the data. Calculate specific muscle force and force drop after eccentric contraction as described in step 15 of Subheading 3.1.

3.3. Grip Force Measurement

1. Record mouse/experimental information as described in Subheading 3.1. Check to make sure there is no visible sore in the forelimbs or toes. Turn on the grip meter and set it to measure the peak tension.
2. Grab the mouse by tip of the tail and gently lower the mouse to allow the forepaws to grab the metal grip bar (see Note 16) (Fig. 2).
3. Swiftly pull the mouse away from the grip bar by the tail. Record the reading when the mouse can no longer grasp the bar (see Note 17).
4. Rest the mouse for 30 s. Repeat the procedure five times and record each reading. Take the mean of the three highest readings as the absolute grip force. Divide the absolute grip force with the body weight to get the normalized grip force.

3.4. Downhill Treadmill Assay

1. Record mouse/experimental information as described in Subheading 3.1. Check to make sure there is no visible injury in the limbs and toes. Record room temperature. Set up the electric shocker at the intensity of 7 and the repetition rate of 9. Adjust the inclination of the treadmill platform with the Jiffy jack. Turn on the fan to the low setting and have the air blow in the same direction as the mouse is running (see Note 18) (Fig. 2).
2. Acclimate the mouse to the treadmill for 5 days. On day 1, place the mouse on an unmoving flat treadmill for 2 min, an unmoving 15° downhill treadmill for 5 min, a 15° downhill treadmill at the speed of 5 m/min for 15 min, and a 15° downhill treadmill at the speed of 10 m/min for 5 min. On day 2, place the mouse on an unmoving 15° downhill

treadmill for 2 min, a 15° downhill treadmill at the speed of 5 m/min for 5 min, a 15° downhill treadmill at the speed of 10 m/min for 15 min, and a 15° downhill treadmill at the speed of 12 m/min for 5 min. On day 3, perform the same type of training as on day 2 except extend the last step from 5 to 10 min. On day 4, perform the same type of training as on day 3 except add an additional session on a 15° downhill treadmill at the speed of 15 m/min for 5 min. On day 5, perform the same type of training as on day 4 (see Note 19).

3. On day 6 (the test day), subject the mouse to a single bout of 15° downhill running starting at the speed of 10 m/min. Twenty min later, increase the treadmill speed 1 m/min every 2 min until the mouse is exhausted. Continuously nudge the mouse to keep it stay on the track. Exhaustion is defined as the point at which the mouse spends more than 10 s on the shocker without attempting to resume running when nudged (see Note 20).
4. Record the running distance.

4. Notes

1. The dissection board (9 in. wide and 7 in. long) is made from a ½ inch thick plexiglass. At the lower portion of the board, a 5 in. (width) × 3 in. (length) × 0.25 in. (depth) reservoir is grooved to collect excess of buffer during muscle superfusion. The excess buffer is removed from the reservoir through a vacuum line. In the middle of the reservoir, a sylgard ring of a ~1.5 in. diameter and 0.5 in. height is glued for securing the hind limb (Fig. 1a).
2. We strongly suggest using the freshly prepared Ringer's buffer. Discard the buffer if it is more than 2 weeks old. Some investigators have included 6–25 μM tubocurarine chloride in their Ringer's buffer (15, 16). However, we found this is not necessary.
3. Visit the company website www.airgas.com to find a local supplier. Oxygen tension profoundly influences force generation (17). We recommend using Blood Gas Mix 2Pt 210 CF (catalog number 730237). It is important to not use 100% pure oxygen. 5% CO_2 is essential to maintain the pH of the Ringer's buffer at 7.4 during the assay.
4. The 300B and the 305B transducers yield the same results (10). However, the larger force range of the 305B transducer allows a full on-screen graphic view of the force profile during eccentric contraction. The tissue organ bath should be cleaned

periodically. For cleaning, disassemble the glassware (including the oxygen disperser tube, Teflon valve, and Radnoti organ bath) and hand wash with the Alconox detergent. Rinse with tap water and soak in 1 N HCl for 1 min. Rinse again with excess of tap water and then rinse with deionized water. Air dry.

5. The 2-in-long hook is made of S/S 304V (0.18 in. diameter) stainless steel wire (Small Parts, Inc. http://www.smallparts.com). It connects the proximal end of the EDL muscle to the level arm of the 305B transducer. The 2.5-in-long hook connects the proximal end of the EDL muscle to the level arm of the 300B transducer. It is made of S/S 304V (0.006 in. diameter) stainless steel wire (Small Parts, Inc). The 0.5-in-long hook is made of the same material as that of the 2-in-long hook. It connects the distal tendon of the TA muscle to the level arm of the 305C-LR transducer.
6. To make the electrode, solder two 0.016-in.-thick platinum wires to two 24G electric wires (Chalgren Enterprises, Gilroy, CA), respectively. Bend the tip of the wire into a hook shape. Carefully pack the wire with the electric heat shrink wrap into the empty barrel of a 1 mL syringe. Attach the assembled electrode to the gooseneck holder. Connect the other ends of the 24G electric wire to the stimulator via a pair of binding posts and a BNC connector.
7. The contractility and fatigability of isolated mouse muscles are affected by temperature (18, 19). We found that 30°C yielded consistent force output without inducing apparent fatigue. The Ringer's buffer should be equilibrated with 95% O_2 and 5% CO_2 for at least 20 min before each use. During force measurement, one should adjust oxygen valve to allow for a steady flow of gas without generating large bubbles.
8. A single DMC panel can be used to simultaneously control more than one muscle assay systems. Define each system with a different bath number. Make sure to select the correct bath number. If you are using an earlier version of the DMC software (such as version 3.1.2), you will also need to set the parameter for "delay after the protocol" as 10 s. However, this is not necessary for the newer version of the software. There are several other parameters on the stimulator (including pulse width, PPS rate, and PPS multiplier). Under the "follow" trigger mode, the setting of these parameters on the stimulator is not functioning. Instead, the DMC software will control these parameters as defined by the investigator.
9. To preserve the contractility of the contralateral EDL muscle, it is critical to keep the core body temperature of an anesthetized mouse at 37°C. The investigator may need to empirically test the position of the heating lamp while monitoring the core temperature with a rectal thermometer (Thermocouple

thermometer, Fisher Scientific). We usually obtain comparable muscle forces from the left and right EDL muscles. During the dissection procedure, there will be minor bleedings from small surface vessels, constantly superfuse with the Ringer's buffer to keep the dissection area clean.

10. Take extreme care to avoid damaging the EDL muscle during dissection. The rupture of the vessels at the proximal end of the TA muscle may cause bleeding. This can be stopped with a thin piece of Ringer's buffer-soaked cotton.
11. To clearly expose the proximal EDL tendon, one may need to cut open the distal biceps femoris muscle. While taking the EDL muscle out, there will be bleeding from the vessels beneath the muscle. This can be stopped with a thin piece of Ringer's buffer-soaked cotton. Tie the knot as close to the MTJ as possible. When mounting the EDL muscle, make sure it is positioned right in the middle of two electrodes.
12. We usually obtain the optimal muscle length at a resting tension of 1 g.
13. Prior to the twitch and tetanic force measurement, we usually apply three rounds of 0.5 s tetanic stimulation at 150 Hz to warm up the muscle. Rest the muscle for 1 min between each warm-up stimulus. This treatment stabilizes muscle and allows consistent muscle force output during subsequence measurements (20). Make sure to adjust the resting tension to 1 g after each contraction.
14. When measuring the muscle weight, make sure to remove the tendons and sutures. We recommend cut right at the position of the MTJ.
15. The TA muscle is innervated by the peroneal nerve, one of the two branches of the sciatic nerve. Alternatively, the TA muscle can be stimulated by applying electric pulse directly to the peroneal nerve. We typically get the optimal TA muscle length at a resting tension of 5 g. After measuring the Lo, we usually apply three preliminary tetanic stimuli (150 Hz, 300 ms duration, 1-min rest between contractions) to warm up the muscle. Adjust the resting muscle tension to 5 g after each round of stimulus.
16. We usually let the mouse acclimate to the test apparatus for ~5 min. When positioning the mouse, make sure it has a firm hold on the grip bar. Both forepaws should face forward and space evenly apart on the grip bar. Mice that have undergone toe clipping have difficulties to grasp. We recommend not using these mice.
17. The mouse must be in the horizontal plane and perpendicular to the grip bar during the pulling procedure (Fig. 2). It is imperative that the mouse has a good grip on the bar and is

willing to pull. Sometimes the mouse will merely let go of the bar without pulling. If this happens, let the mouse adjust its grip on the bar and then pull. Occasionally, the mouse may squirm, bump, and even flip over the grip bar. This may yield an abnormally high reading. In this case, let the mouse rest for 5–10 min and then repeat the assay. In general, the difference between three highest readings should be less than 20 g.

18. To test skeletal muscle function, we recommend using 15° downhill treadmill running. The strain, age, and sex influence treadmill performance (21). In the C57BL/6 strain, the body weight normalized running distance is shorter in male than that in female (12). The ideal room temperature for treadmill running is ~70°F. If higher than 80°F, mouse performance will be affected. A mild electric shock is used as a negative stimulus to discourage the mouse from standing/sitting at the end of the belt rather than running on the track. If the electric shock is too strong, it may damage the mouse. Since toe clipping may affect the gait, we recommend not using mice that have undergone toe clipping.

19. Each training session takes about 30–40 min. We recommend performing training once per day. It is important to always perform the training at the same time of the day (e.g., always at 9 am in the morning). This should also be the time of the test. The mouse should be continuously monitored throughout the entire session. When speed reaches 10 m/min, some mice may stop running. This usually does not mean these mice are exhausted. Coax the mice to run by gently brushing the tail. Occasionally, a mouse may run towards the direction of the electric shocker. If this happens, stop the treadmill, manually place the mouse to the right direction, and restart. It may take several rounds for some mice to learn. At the end of each session, check the mouse for feet or toe damage. Exclude the mice that are either injured or reluctant to run. Up to 20% mice may fail to acclimate.

20. Caution should be taken when using the time spent on the electric shocker as an indicator for exhaustion (22, 23). Individual animal may display different running styles. Mice that do not run willingly or do not show a consistent running style should be excluded from the study. For aged mice (≥15-m-old), instead of starting with a 20 min session of 10 m/min downhill running, we recommend starting the test with the speed of 5 m/min for 5 min, then 10 m/min for 15 min, then increase the speed 1 m/min every 5 min. It is important to point out that different laboratories may use different settings (e.g., 10 or 20° downhill instead of 15° downhill). To demonstrate therapeutic efficacy, one should always include age- and gender-matched untreated controls and normal

mouse controls. We recommend starting with at least 7 mice for each group.

Acknowledgments

The studies are supported by grants from the National Institutes of Health AR-49419 and the Muscular Dystrophy Association. We thank Drs. Rob Grange, Frank Booth, Steve Yang, and Ron Terjung for helpful discussion during the protocol development.

References

1. Clark, K. A., McElhinny, A. S., Beckerle, M. C., and Gregorio, C. C. (2002) Striated muscle cytoarchitecture: an intricate web of form and function. *Annu Rev Cell Dev Biol* 18, 637–706.
2. Huijing, P. A. (1999) Muscle as a collagen fiber reinforced composite: a review of force transmission in muscle and whole limb. *J Biomech* 32, 329–345.
3. Grounds, M. D., Sorokin, L., and White, J. (2005) Strength at the extracellular matrix-muscle interface. *Scand J Med Sci Sports* 15, 381–391.
4. Burkin, D. J., and Kaufman, S. J. (1999) The alpha7beta1 integrin in muscle development and disease. *Cell Tissue Res* 296, 183–190.
5. Ervasti, J. M., and Sonnemann, K. J. (2008) Biology of the striated muscle dystrophin-glycoprotein complex. *Int Rev Cytol* 265, 191–225.
6. Campbell, K. P. (1995) Three muscular dystrophies: loss of cytoskeleton-extracellular matrix linkage. *Cell* 80, 675–679.
7. Bostick, B., Yue, Y., Long, C., Marschalk, N., Fine, D. M., Chen, J., and Duan, D. (2009) Cardiac expression of a mini-dystrophin that normalizes skeletal muscle force only partially restores heart function in aged Mdx mice. *Mol Ther* 17, 253–261.
8. Dellorusso, C., Crawford, R. W., Chamberlain, J. S., and Brooks, S. V. (2001) Tibialis anterior muscles in mdx mice are highly susceptible to contraction-induced injury. *J Muscle Res Cell Motil* 22, 467–475.
9. Harper, S. Q., Hauser, M. A., DelloRusso, C., Duan, D., Crawford, R. W., Phelps, S. F., Harper, H. A., Robinson, A. S., Engelhardt, J. F., Brooks, S. V., and Chamberlain, J. S. (2002) Modular flexibility of dystrophin: implications for gene therapy of Duchenne muscular. *Nat Med* 8, 253–261.
10. Liu, M., Yue, Y., Harper, S. Q., Grange, R. W., Chamberlain, J. S., and Duan, D. (2005) Adeno-associated virus-mediated microdystrophin expression protects young mdx muscle from contraction-induced injury. *Mol Ther* 11, 245–256.
11. Li, D., Yue, Y., and Duan, D. (2008) Preservation of muscle force in mdx3cv mice correlates with low-level expression of a near full-length dystrophin protein. *Am J Pathol* 172, 1332–1341.
12. Li, D., Lai, Y., Yue, Y., Rabinovitch, P. S., Hakim, C., and Duan, D. (2009) Ectopic catalase expression in mitochondria by adeno-associated virus enhances exercise performance in mice. *PLoS ONE* 4, e6673.
13. Brooks, S. V., and Faulkner, J. A. (1988) Contractile properties of skeletal muscles from young, adult and aged mice. *J Physiol* 404, 71–82.
14. Burkholder, T. J., Fingado, B., Baron, S., and Lieber, R. L. (1994) Relationship between muscle fiber types and sizes and muscle architectural properties in the mouse hindlimb. *J Morphol* 221, 177–190.
15. Ebihara, S., Guibinga, G. H., Gilbert, R., Nalbantoglu, J., Massie, B., Karpati, G., and Petrof, B. J. (2000) Differential effects of dystrophin and utrophin gene transfer in immunocompetent muscular dystrophy (mdx) mice. *Physiol Genomics* 3, 133–144.
16. Lynch, G. S., Hinkle, R. T., Chamberlain, J. S., Brooks, S. V., and Faulkner, J. A. (2001) Force and power output of fast and slow skeletal muscles from mdx mice 6-28 months old. *J Physiol* 535, 591–600.

17. Eu, J. P., Hare, J. M., Hess, D. T., Skaf, M., Sun, J., Cardenas-Navina, I., Sun, Q. A., Dewhirst, M., Meissner, G., and Stamler, J. S. (2003) Concerted regulation of skeletal muscle contractility by oxygen tension and endogenous nitric oxide. *Proc Natl Acad Sci U S A* 100, 15229–15234.
18. Faulkner, J. A., Zerba, E., and Brooks, S. V. (1990) Muscle temperature of mammals: cooling impairs most functional properties. *Am J Physiol* 259, R259–R265.
19. Segal, S. S., Faulkner, J. A., and White, T. P. (1986) Skeletal muscle fatigue in vitro is temperature dependent. *J Appl Physiol* 61, 660–665.
20. Grange, R. W., Gainer, T. G., Marschner, K. M., Talmadge, R. J., and Stull, J. T. (2002) Fast-twitch skeletal muscles of dystrophic mouse pups are resistant to injury from acute mechanical stress. *Am J Physiol Cell Physiol* 283, C1090–C1101.
21. Lightfoot, J. T., Turner, M. J., Debate, K. A., and Kleeberger, S. R. (2001) Interstrain variation in murine aerobic capacity. *Med Sci Sports Exerc* 33, 2053–2057.
22. Handschin, C., Summermatter, S., LeBrasseur, N. K., Lin, J., and Spiegelman, B. M. (2010) For a pragmatic approach to exercise studies. *J Appl Physiol* 108, 223–223; author reply 26.
23. Booth, F. W., Laye, M. J., and Spangenburg, E. E. (2010) Gold standards for scientists who are conducting animal-based exercise studies. *J Appl Physiol* 108, 219–221.

Chapter 6

Phenotyping Cardiac Gene Therapy in Mice

Brian Bostick, Yongping Yue, and Dongsheng Duan

Abstract

Heart disease is the leading health problem of industrialized countries. The development of gene therapies tailored towards the heart has grown exponentially over the past decade. Murine models of heart diseases have played a pivotal role in testing novel cardiac gene therapy approaches. Unfortunately, the small body size and rapid heart rate of mice present a great challenge to heart function evaluation. Here we outline the commonly used cardiac phenotyping methods of treadmill exercise regimen, full 12-lead electrocardiographic assay and left ventricular catheterization hemodynamic assay. Application of these protocols will allow critical testing of gene therapy efficacy in mouse models of heart diseases.

Key words: Heart, Electrocardiography, Treadmill, Left ventricular catheterization, Heart disease, Cardiomyopathy, PV loop, Cardiac hemodynamics

1. Introduction

Heart disease is the leading health problem of industrialized countries. Cardiomyopathy is also a common complication of muscular dystrophies (1). Increased attentiveness toward cardiac dysfunction has accelerated the design of gene therapies specifically tailored for the heart (2, 3). Mouse models play a critical role in the development of cardiac gene therapy (4, 5). However, the small size and rapid heart rate of the mouse have created a daunting challenge for evaluation of cardiac function (6).

The development of appropriate anesthetic regimens has played a pivotal role in murine heart function examination (7). Numerous anesthetics have been tested in mice with significant variations in heart rate and blood pressure (Table 1). When compared to conscious telemetry, isoflurane and a triple anesthetic regimen of morphine, etomidate, and urethane have produced the least alterations in hemodynamics (10, 16). Isoflurane is

Dongsheng Duan (ed.), *Muscle Gene Therapy: Methods and Protocols*, Methods in Molecular Biology, vol. 709,
DOI 10.1007/978-1-61737-982-6_6,

Table 1
Effect of anesthesia on cardiovascular function in mice

Anesthesia/regimen	HR	MAP	SBP	References
Conscious telemetry	573 ± 14[a]	106 ± 5	–	(8)
Conscious telemetry	569 ± 9[a]	113 ± 1	–	(9)
Urethane + etomidate + morphine	607 ± 42	98 ± 12	118 ± 14	(10)
Avertin (Tribromoethanol)	518 ± 18	79 ± 1	–	(11)
Ketamine + xylazine	390 ± 15	105 ± 5	115 ± 7	(12)
Pentobarbital sodium	448 ± 20	82 ± 3	–	(13)
Isoflurane	581 ± 23	–	94 ± 2	(14)
Isoflurane	580 ± 25	81 ± 2	–	(15)

HR heart rate; *MAP* mean arterial pressure; *SBP* systolic blood pressure
[a]24 h averaged heart rate

increasingly becoming the anesthetic of choice because it is readily available and easy to titrate. Further, isoflurane provides rapid induction and recovery (17).

Technological advances have revolutionized cardiac phenotyping methods (18). Traditionally, heart function is evaluated by treadmill endurance assay and echocardiography (19, 20). New apparatus and computer software have been developed to study murine cardiac electrical activity and hemodynamics by electrocardiography (ECG) and left ventricular catheterization, respectively (16, 21). Here, we provide detailed methods for evaluating mouse treadmill endurance, ECG, and closed-chest left ventricular hemodynamics.

2. Materials

2.1. Treadmill Performance Regimen

1. Exe-3/6 open treadmill (Columbus Instruments, Columbus, OH, USA).
2. Timer.

2.2. 12-Lead ECG Assay

1. PowerLab 4/S data acquisition system (AD Instruments, Colorado Springs, CO, USA).
2. Single channel Bioamplifier model ML132 with subdermal needle electrodes (AD Instruments).
3. ECG lead selector (AD Instruments).

4. Chart software v5.5.6 (or higher) with ECG extension module (AD Instruments) (see Note 1).
5. Isoflurane portable anesthesia system (Summit Medical Equipment, Bend, OR, USA).
6. Isoflurane (VetOne, Median, ID, USA).
7. Oxygen tank containing 100% oxygen (Airgas National, Charlotte, NC, USA).
8. Thermophore heating pad (Medwing, Columbia, SC, USA).
9. Heating lamp (Tensor Lighting Company, Boston, MA, USA).
10. Vicks digital thermometer (Kaz, Hudson, NY, USA).
11. Thermolyne 589 rectal digital pyrometer (Barnstead International, Dubuque, IA, USA).
12. Straight serrated fine tip forceps (World Precision Instruments, Sarasota, FL, USA).

2.3. Closed-Chest Left Ventricular Catheterization Hemodynamic Assay

1. Millar MPVS-400 (Millar Instruments, Houston, TX, USA).
2. Millar ultraminiature pressure–volume (PV) catheter model SPR-839 (Millar Instruments).
3. Millar cuvette block (Millar Instruments).
4. Chart software v5.5.6 (AD Instruments).
5. Isoflurane portable anesthesia system (Summit Medical Equipment).
6. Isoflurane (VetOne).
7. Oxygen tank containing 100% oxygen (Airgas National).
8. Thermophore heating pad (Medwing).
9. Heating lamp (Tensor Lighting Company).
10. Vicks digital thermometer (Kaz).
11. Thermolyne 589 rectal digital pyrometer (Barnstead International).
12. Hair clippers (Wahl, Sterling, IL, USA)
13. Stereo microscope (Nikon, Melville, NY, USA).
14. Minivent mouse ventilation system type 845 (Hugo Sachs Elektronik, Hugstetten, Germany).
15. Tracheotomy cannula, 1.3 mm outer diameter (Harvard Apparatus, Holliston, MA, USA).
16. Surgical instruments: microsurgical spring scissors, straight serrated fine tip forceps, Dumont type or other fine tip straight and angled forceps, Kilner curved fine sharp point scissors, hemostats (World Precision Instruments), and Guthrie double hook retractor (Fine Science Tools, Foster City, CA, USA).

17. Bread silk suture # 4-0 (SofSilk USSC Sutures, Norwalk, CT, USA).
18. 25 μL 33G gas-tight Hamilton syringe and needle (Hamilton Company, Reno, NV, USA).
19. PE 10 polyethylene tubing (Clay Adams Division of Becton Dickinson and Company, Parsippany NJ, USA).
20. 30% hypertonic saline (Abbott Laboratories, North Chicago, IL, USA).
21. 0.9% isotonic saline (Abbott Laboratories).
22. 27G ½ in. and 30G ½ in. needles (Becton-Dickinson Medical Supply, Franklin Lakes, NJ, USA).
23. Cotton-tipped wooden applicators (Fisher Scientific, Pittsburgh, PA, USA).
24. Heparin multidose vial (Baxter Healthcare Corporation, Deerfield, IL, USA).
25. Dobutamine (Sigma, St. Louis, MO, USA).
26. 30G ½ cc insulin syringes (Becton-Dickinson Medical Supply).
27. PVAN data analysis software (Millar Instruments).

3. Methods

For all animal experiments, get approval from the Institute's Animal Care and Use Committee and follow NIH guidelines.

3.1. Treadmill Performance Regimen

1. Five-day training/treadmill acclimation (see Note 2). On day 1, place mouse on unmoving flat treadmill for 2 min. Place mouse on unmoving 7° inclined treadmill for 5 min (see Note 3). Next, run mouse at 5 m/min on 7° inclined treadmill for 15 min. Run mouse at 10 m/min for 5 min. On day 2, place mouse on unmoving flat treadmill for 2 min. Run mouse on 7° treadmill at 5 m/min for 5 min. Run mouse on 7° treadmill at 10 m/min for 15 min. Run mouse on 7° treadmill at 12 m/min for 5 min. On day 3, place mouse on unmoving flat treadmill for 2 min. Run mouse on 7° treadmill at 5 m/min for 5 min. Run mouse on 7° treadmill at 10 m/min for 15 min. Run mouse on 7° treadmill at 12 m/min for 10 min. On day 4, place mouse on unmoving flat treadmill for 2 min. Run mouse on 7° treadmill at 5 m/min for 5 min. Run mouse on 7° treadmill at 10 m/min for 20 min. Run mouse on 7° treadmill at 12 m/min for 5 min. Run mouse on 7° treadmill at 15 m/min for 5 min. On day 5, place mouse on unmoving flat treadmill for 2 min. Run mouse on 7° treadmill at 5 m/min for 5 min. Run mouse on 7° treadmill at 10 m/min for 20 min.

Run mouse on 7° treadmill at 12 m/min for 5 min. Run mouse on 7° treadmill at 15 m/min for 5 min.

2. Experimental assessment of treadmill performance on day 6. Place mouse on unmoving flat treadmill for 2 min. Run mouse on 7° treadmill at 5 m/min for 5 min. Increase speed by 1 m/min every 5 min (see Note 4). The maximal running capacity for each mouse is defined as the point where the mouse stays on the shocker for 10 s without attempting to reenter the treadmill. To calculate the total running distance, multiply the time spent at each speed setting by the speed of the treadmill and then add all of these distances together.

3.2. 12-Lead ECG Assay

1. Mouse anesthesia. Place the mouse into a new, empty cage alone for 5 min prior to anesthetizing (see Note 5). Gently transfer the mouse into a prewarmed induction chamber and anesthetize with 3% isoflurane at an oxygen flow rate of 2 L/min for 2.5 min. Promptly remove the mouse from the induction chamber. Place the mouse onto a prewarmed recording table with snout inserted into nose cone. Provide maintenance anesthesia of 1–1.5% isoflurane at an oxygen flow rate of 0.5–0.6 l/min (see Note 6). Secure mouse limbs to the recording table using tape placed over the paws. Leave one lower limb unsecured for monitoring the depth of anesthesia with toe pinch. Lubricate the rectal temperature probe with surgical lubricant and gently insert 2–3 mm into the rectum. Maintain mouse body temperature between 37 and 38°C during the procedure.

2. ECG needle electrode placement. Insert the sockets of the needle electrodes onto the pins in the ECG lead selector cable. Place the needles subdermally into their corresponding positions on the mouse. To place needles, gently lift the mouse skin with serrated forceps and insert the needle into the resulting skin tent. Limb leads should be placed parallel to the limb running distal to proximal. The chest electrode should always be inserted parallel to the sternum running in the direction of the head to the toe. For the V1 position, the needle should be ~1–2 mm to the right of the sternum. For the V2 position, the needle should be placed ~1–2 mm to the left of the sternum. V3 should be placed ~1–2 mm medially to the left midclavicular line. V4 should be placed at the left midclavicular line and slight further caudally. V5 should be located at the anterior axillary line slightly further caudally than V4. V6 should be placed at the midaxillary line slightly further caudally to V5.

3. ECG Recording. Record a 1-min-long rhythm strip from the lead II position once the mouse temperature has stabilized between 37 and 38°C. Then, sequentially record each limb lead

tracing for about 15–20 s. Then, sequentially record the chest lead tracings for about 15–20 s by moving the needle electrode to each placement listed above. Finally, record a second 1-min rhythm strip from the lead II position (see Note 7).

4. Mouse recovery. Immediately turn the maintenance isoflurane to 0% after the last lead is recorded. Allow the mouse to breath 100% oxygen at 0.5–0.6 L/min while it awakens. Carefully remove all electrodes, the rectal probe, and tape restraints. Once the mouse is awake and able to right itself, gently return it to the empty cage. Monitor the mouse carefully for the next 15 min before returning it to the original cage (see Note 8).
5. ECG analysis. Merge each lead tracing into a single signal averaged ECG (SAECG) by utilizing the block averaging function. Select the Mouse preset conditions and the Mitchell definition for corrected QT interval (QTc). After software analysis, review each of the SAECG to ensure that all the markings are placed appropriately (see Note 9).

3.3. Closed-Chest Left Ventricular Catheterization Hemodynamic Assay (See Note 10)

1. Mouse anesthesia. Place the mouse into a new, empty cage alone for 5 min prior to anesthetizing (see Note 5) (Fig. 1). Gently transfer the mouse into a prewarmed induction chamber and anesthetize with 3% isoflurane at an oxygen flow rate of 2 L/min for 2.5 min. Promptly remove the mouse from

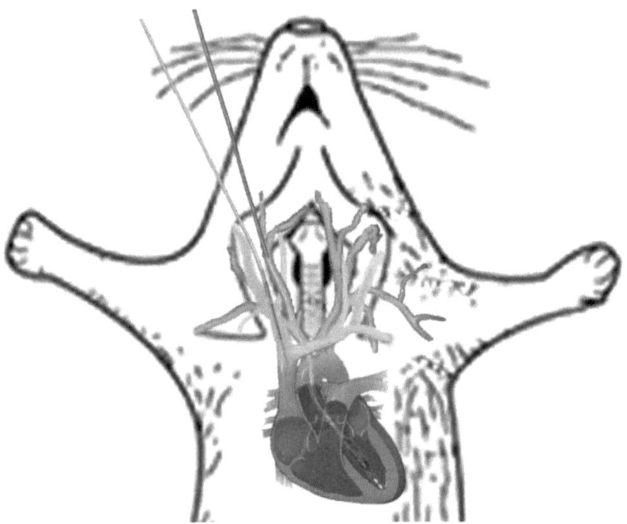

Fig. 1. Schematic diagram of closed-chest left ventricular catheterization. In the closed-chest approach, the PV catheter is introduced into the right internal carotid artery and advanced through the aortic valve into the left ventricle. The right external jugular vein is cannulated to allow delivery of fluids and hypertonic saline for volume calibration.

the induction chamber. Shave the neck and chest with clippers. Place the mouse onto a prewarmed recording table with snout inserted into the nose cone (see Note 6). Provide maintenance anesthesia of 1–1.5% isoflurane at oxygen flow rate of 0.5–0.6 L/min. Secure mouse limbs to the recording table using tape placed over the paws. Leave one lower limb unsecured to allow monitoring the depth of anesthesia with toe pinch. Lubricate the rectal temperature probe with surgical lubricant and gently insert 2–3 mm into the rectum. Maintain mouse body temperature between 37 and 38°C during entire procedure.

2. Tracheotomy and mouse ventilation. Make a skin incision at the anterior neck and separate the parotid glands and subcutaneous tissue overlying the trachea using blunt dissection. Under a stereo microscope, expose the trachea by cutting away the cricothyroid muscles. Make an incision in the trachea just above the cricoid cartilage (swollen region just under the cricothyroid muscle) using a 20G needle bent at a 90° angle. Enlarge the incision using microsurgical spring scissors. Remove the mouse snout from the nose cone. Place a looped suture around the teeth and secure above the head to stretch the mouse neck and create traction on the trachea. Cannulate the trachea using a 1.3 mm OD tracheotomy tube. Tie a suture around the trachea and the tracheotomy tube to hold in place. Connect the anesthesia supply tube from the nose cone to the intake tubing on the mouse ventilator and set the oxygen flow rate to 0.2–0.3 L/min. Ventilate the mouse with a tidal volume of 8–10 μL/g bodyweight at a respiratory rate of ~200 respirations per minute (22) (Fig. 2a).

3. Isolation of internal carotid artery. Locate the right carotid sheath running along the posteriolateral region of the right side of the trachea. Dissect away the right carotid sheath with blunt dissection. Separate the larger pulsating internal carotid artery from the internal jugular vein and vagus nerve. The vagus nerve should be carefully separated and left intact to prevent altering the autonomic innervation of the heart. Place looped sutures at the distal and proximal ends of the internal carotid artery and secure them with hemostats. Stretch and occlude the artery by pulling on the hemostats. Place a third looped suture between the outer sutures and leave it loosely tied (Fig. 2a).

4. Cannulation of internal carotid artery. Presoak the 1.4 F Millar PV catheter in 37°C normal saline for a few minutes. Open the Chart software to enable visualization of the pressure tracing while inserting the PV catheter. Stretch the carotid artery using the proximal and distal sutures to occlude blood flow. Grasp the end of the PV catheter with serrated

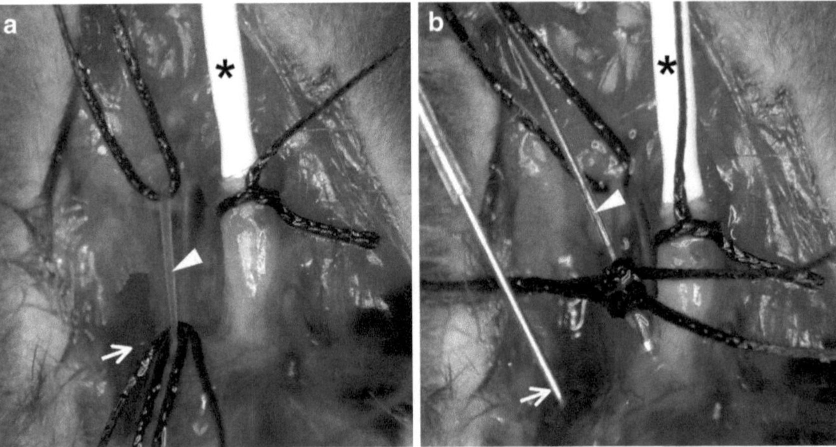

Fig. 2. Surgical isolation of trachea, carotid artery, and external jugular vein during left ventricular catheterization assay. (**a**) A photograph of the mouse neck under a stereoscopic dissection microscope following internal carotid artery isolation. *Arrowhead*, the internal carotid artery; *arrow*, the external jugular vein; *asterisk*, trachea. (**b**) A photograph of the mouse neck under a stereoscopic dissection microscope after PV catheter insertion and external jugular vein cannulation. *Arrowhead*, the PV catheter in the internal carotid artery; *arrow*, the 30G needle cannulating the external jugular vein; *asterisk*, trachea.

straight fine-toothed forceps which have the tips covered with PE 10 tubing. Incise the carotid artery with a 27G needle which has the tip bent at a 90 angle. While holding the 27G needle in the carotid artery, give gentle upward traction on the needle to open the incision and advance the PV catheter into the carotid artery underneath the 27G needle. Gently tie the middle suture around the carotid artery and the PV catheter. Release traction on the suture occluding the proximal end of the carotid artery. Advance the catheter further into the carotid artery until the proximal suture is above the volume electrodes on the PV catheter. Tie the proximal and middle sutures around the carotid artery and the catheter. The knots should be tight enough to prevent blood from leaking around the catheter, but loose enough to allow the catheter to slide within the artery (Fig. 2b).

5. Advancement of the PV catheter into the left ventricle. Carefully advance the PV catheter from the carotid artery into the aorta. Use the pressure tracing from the Chart software window to guide advancement. Confirm aortic location of the PV catheter by visualizing the pressure tracing oscillating with the dichrotic notch evident. Negotiate the PV catheter through the aortic valve with gentle in and out advancement. A drop in the diastolic pressure to near zero with the systolic pressure unchanged confirms the correct placement of the PV catheter in the left ventricle. Once inside the left ventricle, adjust the depth and angle until the

best change in relative volume units (RVU) is obtained (see Note 11).

6. Baseline data collection. After signal stabilization, record baseline PV loops at the steady state.
7. Locate inferior vena cava (IVC). Make a small subxiphoid incision and open up the peritoneum. Mobilize the liver away from the diaphragm by cutting the falciform ligament between the diaphragm and superior edge of the liver. Carefully spread the liver and diaphragm with a cotton-tipped applicator to visualize the IVC along the posterior abdominal wall to the right of the spinal column.
8. Transiently occlude the IVC. While visualizing the pressure–volume tracings in the Chart window, occlude the IVC with a cotton-tipped application for ~2 s. The pressure and volume tracings should exhibit a uniform 20–30% decrease during the occlusion.
9. Intraperitoneal dobutamine injection. Prepare 1 μg/μL solution of dobutamine in normal saline. Inject 5 μg/g body weight of dobutamine via intraperitoneal injection.
10. Monitor dobutamine response. At 5 min postdobutamine injection, record PV loops.
11. Hypertonic saline calibration. Carefully expose the right external jugular vein by bluntly dissecting the surface of the right neck along the midclavicular line (Fig. 2a). Cannulate the vein with a 30G needle attached to a 25 μL Hamilton syringe with PE 10 tubing (Fig. 2b). Quickly and evenly inject 5 μL of 30% hypertonic saline into the right external jugular vein while visualizing the PV tracings in the Chart window. During injection, the pressure tracing should remain unchanged, while the volume tracing exhibits a uniform rise (Fig. 3).
12. Blood cuvette calibration. At the end of PV recording, remove ~200 μL of blood from the mouse by either direct cardiac puncture or large vein sampling using a heparinized syringe. Fill each well in the Millar-supplied cuvette with heparinized blood and partially submerge the cuvette in a 37°C water bath. Place the PV catheter into each well and record the resulting RVU. Using a Microsoft Excel graph, plot the measured RVU for each well on the *x*-axis against the known well volumes on the *y*-axis (see Note 12). Determine the slope and *y*-intercept for the line.
13. After finishing these calibrations and other experimental manipulations, euthanize the mouse and save your Chart file for future analysis (see Note 13).
14. Determination of offset voltage (Vp). Select the region of the hypertonic saline injection beginning with the first volume

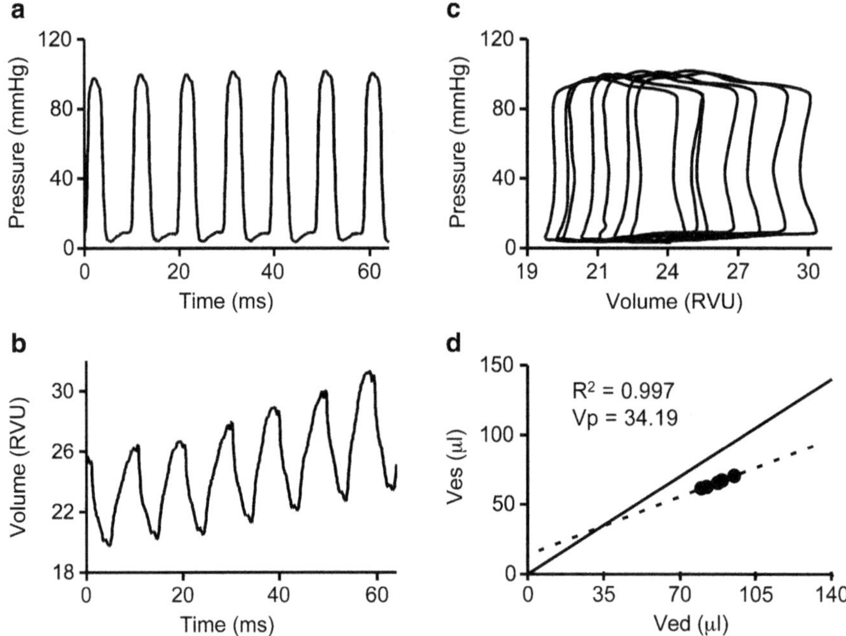

Fig. 3. A representative example of hypertonic saline calibration during left ventricular catheterization. (**a**) Pressure tracing during hypertonic saline calibration. (**b**) Volume tracing during hypertonic saline calibration. (**c**) PV loops from hypertonic saline calibration. *RVU* relative volume units. (**d**) Linear regression graph for determination of the offset voltage. In this particular example, *Ves*= 0.59 × *Ved*+ 13.90. *Ved* end-diastolic volume; *Ves* end-systolic volume; *Vp* offset voltage.

loop before injection to the maximum volume achieved during saline injection (Fig. 3b). Import this data into PVAN by clicking on the heart icon in the Chart window. Click on the steady state/saline icon and enter the slope and *y*-intercept values determined from the blood cuvette calibration. Perform the saline calibration analysis to construct a linear regression graph (Fig. 3d). Save the *V*p, offset voltage, value for use in determining the actual volume of the left ventricle.

15. Baseline and dobutamine stress analysis. Select the corresponding loops to be analyzed and click on the heart icon in the Chart window. Select steady state/saline icon and enter the slope, *y*-intercept, and *V*p from cuvette calibration and offset voltage determination. Select hemodynamics to complete the data analysis and view the left ventricular functional results. These data may be exported to Microsoft excel by saving the data file.

16. Preload reduction IVC occlusion analysis. Select the IVC occlusion loops to be analyzed beginning with the first loop before occlusion started and ending at the point where the pressure tracing is 20–30% decreased from baseline (Fig. 4a, b). Click on the heart icon in the Chart window and

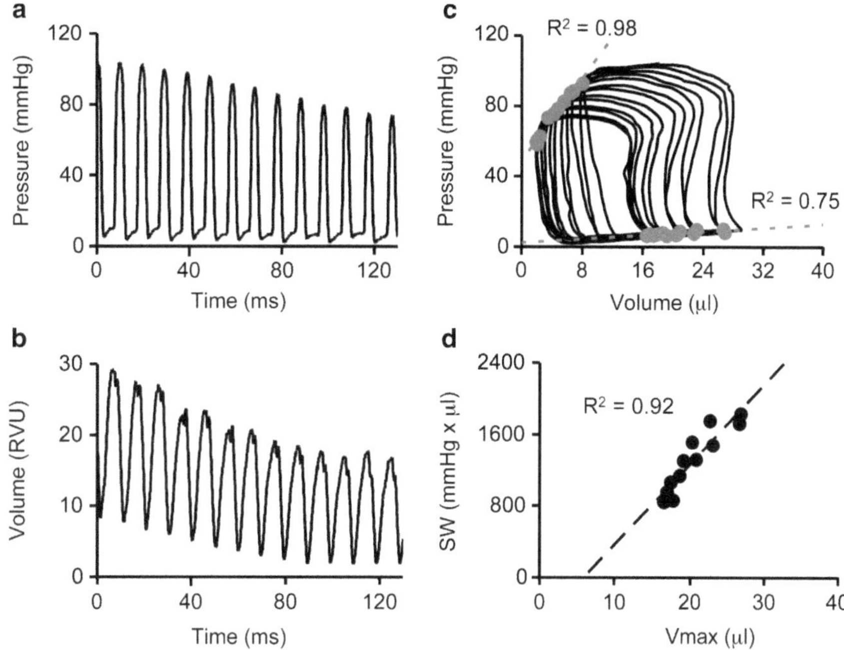

Fig. 4. A representative example of preload reduction via IVC occlusion during left ventricular catheterization. (**a**) Pressure tracing during IVC occlusion. (**b**) Volume tracing during IVC occlusion. (**c**) PV loops from IVC occlusion. The *top dotted line* can be used to determine ESPVR: *Pes*. In this example, ESPVR: *Pes*= 4.18 × *Ves*+ 11.95. The *bottom dotted line* can be used to determine EDPVR: *Ped*. In this example, EDPVR: *Ped*= 0.26 × *Ved*+ 2.23. *EDPVR* end-diastolic pressure volume reserve; *ESPVR* end-systolic pressure volume reserve; *Ped* end-diastolic pressure; *Pes* end-systolic pressure; *Ved* end-diastolic volume; *Ves* end-systolic volume. (**d**) Preload recruitable stroke work (SW) graph from IVC occlusion. In this example, *SW*= 89.63 × *Ved*–5.90. *V*max, maximum volume.

select occlusion. Enter the slope, *y*-intercept, and offset voltage for the mouse as determined above. Click on the contractility icon to determine the end-systolic pressure volume reserve (ESPVR) and end-diastolic pressure volume reserve (EDPVR) (Fig. 4c). Determine the preload recruitable stroke work by clicking on the icon to compare the stroke work to the maximum volume (Fig. 4d).

4. Notes

1. Chart software version 5 requires a PC capable of running Windows 2000 or higher or Mac running Mac OS X 10.1 or higher. Chart software versions 6 and 7 require either Windows XP SP2 or Mac OS X 10.4 or higher.
2. Treadmill endurance testing requires 6 consecutive days of treadmill running. Prior to the experimental testing of cardiovascular treadmill endurance on day 6, mice must be

subjected to 5 days of increasing treadmill exposure to train them for treadmill running.

3. To make a positive 7° inclined treadmill, set the crank at zero and place the incline rod into the hole on the distal end of the treadmill. Then, insert the pin into the first hole on the rod with the treadmill raised above the hole and lower the treadmill onto the pin.
4. A ruler or other flat instrument may be used to help mouse stay in the proper orientation on the running lanes and help mouse off the shocking grid. A more demanding protocol can be used for young adult mice (16).
5. Moving the mouse to a new empty cage prior to study ensures an equal activity level for all mice prior to study.
6. Mouse heart rate is strongly dependent on temperature and the mouse should be kept warm during all phases of the procedure. Induction chamber and operating tables should be monitored with a thermometer to ensure the temperature is kept between 36 and 38°C.
7. While switching between different leads, the recording of the ECG tracing should be switched to the monitoring mode. This will aid in differentiating between different leads during analysis and remove any noise from the tracings during switching. Certain heating pads may cause electrical interference during recording and should be switched off during recording.
8. Mice with muscle disease require extra care during anesthesia because of poor respiratory function. Isoflurane should be carefully titrated to prevent respiratory depression and maintain a respiratory rate above 80 per minute.
9. To thoroughly evaluate heart electrical activity, we recommend performing 12-lead ECG assay. Depending on the model used, one may notice abnormal changes in certain leads only. In the case of mdx mouse model of Duchenne muscular dystrophy, most of the characteristic changes are found in lead II except for the Q amp. We usually get Q amp data from lead I recording (16, 23). For additional information on ECG tracing analysis, please refer to AD Instrument's ECG Analysis Guide.
10. The left ventricular catheterization of mice has been described using both a closed-chest internal carotid approach and an open-chest apical stab approach (23, 24). We describe the closed-chest approach. This approach yields more physiologic hemodynamics than the open-chest approach (25). Further, the closed-chest approach can be performed without ventilation. However, if you are working with mouse models of muscular dystrophy, we recommend utilization of a ventilator

during left ventricular hemodynamic study because respiratory function is often impaired in these models.

11. Negotiation of the PV catheter into the left ventricle requires gentle in and out movement of the catheter especially when negotiating the aortic valve. Ideal placement of the PV catheter in the left ventricle may require adjusting the angle and rotation of the catheter.
12. Included with the Millar cuvette are the known volumes for each well to be used for extrapolating the left ventricular volume from the RVUs.
13. PV loop analysis is best performed off-line once the experiment has been completed.

Acknowledgments

This work is supported by grants from the National Institutes of Health HL-91883 and the Muscular Dystrophy Association (DD). We thank Mr. Nate Marschalk and Dr. Dejia Li for helpful discussion on treadmill function assay.

References

1. McNally, E.M., and MacLeod, H. (2005) Therapy Insight: cardiovascular complications associated with muscular dystrophies. *Nat Clin Pract Cardiovasc Med* 2, 301–308.
2. Yue, Y., Li, Z., Harper, S.Q., Davisson, R.L., Chamberlain, J.S., and Duan, D. (2003) Microdystrophin gene therapy of cardiomyopathy restores dystrophin-glycoprotein complex and improves sarcolemma integrity in the Mdx mouse heart. *Circulation* 108, 1626–1632.
3. Li, J., Wang, D., Qian, S., Chen, Z., Zhu, T., and Xiao, X. (2003) Efficient and long-term intracardiac gene transfer in [delta]-sarcoglycan-deficiency hamster by adeno-associated virus-2 vectors. *Gene Ther* 10, 1807–1813.
4. Duan, D. (2006) Challenges and opportunities in dystrophin-deficient cardiomyopathy gene therapy. *Hum Mol Genet* 15, R253–R261.
5. Allamand, V., and Campbell, K.P. (2000) Animal models for muscular dystrophy: valuable tools for the development of therapies. *Hum Mol Genet* 9, 2459–2467.
6. Kass, D.A., Hare, J.M., and Georgakopoulos, D. (1998) Murine cardiac function : a cautionary tail. *Circ Res* 82, 519–522.
7. Janssen, B.J.A., De Celle, T., Debets, J.J.M., Brouns, A.E., Callahan, M.F., and Smith, T.L. (2004) Effects of anesthetics on systemic hemodynamics in mice. *Am J Physiol Heart Circ Physiol* 287, H1618–H1624.
8. Butz, G.M., and Davisson, R.L. (2001) Long-term telemetric measurement of cardiovascular parameters in awake mice: a physiological genomics tool. *Physiol Genomics* 5, 89–97.
9. Carlson, S.H., and Wyss, J.M. (2000) Long-term telemetric recording of arterial pressure and heart rate in mice fed basal and high NaCl diets. *Hypertension* 35, E1–E5.
10. Segers, P., Georgakopoulos, D., Afanasyeva, M., Champion, H.C., Judge, D.P., Millar, H.D., et al. (2005) Conductance catheter-based assessment of arterial input impedance, arterial function, and ventricular-vascular interaction in mice. *Am J Physiol Heart Circ Physiol* 288, H1157–H1164.
11. Lin, M., Liu, R., Gozal, D., Wead, W.B., Chapleau, M.W., Wurster, R., et al. (2007) Chronic intermittent hypoxia impairs baroreflex control of heart rate but enhances heart rate responses to vagal efferent stimulation in anesthetized mice. *Am J Physiol Heart Circ Physiol* 293, H997–H1006.

12. Pacher, P., Liaudet, L., Bai, P., Mabley, J.G., Kaminski, P.M., Virag, L., et al. (2003) Potent metalloporphyrin peroxynitrite decomposition catalyst protects against the development of doxorubicin-induced cardiac dysfunction. *Circulation* 107, 896–904.
13. Pacher, P., Batkai, S., and Kunos, G. (2004) Haemodynamic profile and responsiveness to anandamide of TRPV1 receptor knock-out mice. *J Physiol* 558, 647–657.
14. Shioura, K.M., Geenen, D.L., and Goldspink, P.H. (2008) Sex-related changes in cardiac function following myocardial infarction in mice. *Am J Physiol Regul Integr Comp Physiol* 295, R528–534.
15. Zuurbier, C.J., Emons, V.M., and Ince, C. (2002) Hemodynamics of anesthetized ventilated mouse models: aspects of anesthetics, fluid support, and strain. *Am J Physiol Heart Circ Physiol* 282, H2099–H2105.
16. Bostick, B., Yue, Y., Long, C., Marschalk, N., Fine, D.M., Chen, J., et al. (2009) Cardiac expression of a mini-dystrophin that normalizes skeletal muscle force only partially restores heart function in aged Mdx mice. *Mol Ther* 17, 253–261.
17. Pacher, P., Nagayama, T., Mukhopadhyay, P., Batkai, S., and Kass, D.A. (2008) Measurement of cardiac function using pressure-volume conductance catheter technique in mice and rats. *Nat Protocols* 3, 1422–1434.
18. Hoit, B.D. (2004) Murine physiology: measuring the phenotype. *J Mol Cell Cardiol* 37, 377–387.
19. Fewell, J.G., Osinska, H., Klevitsky, R., Ng, W., Sfyris, G., Bahrehmand, F., et al. (1997) A treadmill exercise regimen for identifying cardiovascular phenotypes in transgenic mice. *Am J Physiol Heart Circ Physiol* 273, H1595–H1605.
20. Lerman, I., Harrison, B.C., Freeman, K., Hewett, T.E., Allen, D.L., Robbins, J., et al. (2002) Genetic variability in forced and voluntary endurance exercise performance in seven inbred mouse strains. *J Appl Physiol* 92, 2245–2255.
21. Branco, D., Wolf, C., Sherwood, M., Hammer, P., Kang, P., and Berul, C. (2007) Cardiac electrophysiological characteristics of the mdx 5cv mouse model of Duchenne muscular dystrophy. *J Interv Card Electr* 20, 1–7.
22. Schwarte, L.A., Zuurbier, C.J., and Ince, C. (2000) Mechanical ventilation of mice. *Basic Res Cardiol* 95, 510–520.
23. Bostick, B., Yue, Y., Long, C., and Duan, D. (2008) Prevention of dystrophin-deficient cardiomyopathy in twenty-one-month-old carrier mice by mosaic dystrophin expression or complementary dystrophin/utrophin expression. *Circ Res* 102, 121–130.
24. Georgakopoulos, D., Mitzner, W.A., Chen, C.-H., Byrne, B.J., Millar, H.D., Hare, J.M., et al. (1998) In vivo murine left ventricular pressure-volume relations by miniaturized conductance micromanometry. *Am J Physiol Heart Circ Physiol* 274, H1416–1422.
25. Hoit, B.D., Ball, N., and Walsh, R.A. (1997) Invasive hemodynamics and force-frequency relationships in open- versus closed-chest mice. *Am J Physiol Heart Circ Physiol* 273, H2528–H2533.

Chapter 7

Golden Retriever Muscular Dystrophy (GRMD): Developing and Maintaining a Colony and Physiological Functional Measurements

Joe N. Kornegay, Janet R. Bogan, Daniel J. Bogan, Martin K. Childers, and Robert W. Grange

Abstract

Studies of canine models of Duchenne muscular dystrophy (DMD) provide insight regarding disease pathogenesis and treatment efficacy. To take maximal advantage, colonies of affected dogs must be maintained and outcome parameters developed. In this chapter, we review our 25 years of experience with the golden retriever muscular dystrophy (GRMD) model. Key challenges in colony development (breeding, neonatal death, and the risk of inbreeding) and representative functional measurements (tibiotarsal joint angle and torque force; and eccentric contraction decrement) are discussed.

Key words: Duchenne muscular dystrophy, Golden retriever muscular dystrophy, Neonatal care, Inbreeding, Joint angle, Torque force

1. Introduction

Although cellular and gene therapies for Duchenne muscular dystrophy (DMD) are promising, key questions must first be addressed in relevant animal models. Spontaneous forms of X-linked muscular dystrophy due to dystrophin deficiency have been identified in mice (1, 2), multiple dog breeds (3–7), and cats (8, 9). Unlike the dystrophin-deficient mdx mouse, which remains relatively normal clinically, affected dogs develop progressive, fatal disease strikingly similar to the human condition. Accordingly, studies in the canine dystrophin-deficient models, such as golden retriever muscular dystrophy (GRMD) (3, 10), may be more likely than those in mdx mice to predict pathogenesis and outcome of treatment in DMD.

Dongsheng Duan (ed.), *Muscle Gene Therapy: Methods and Protocols*, Methods in Molecular Biology, vol. 709,
DOI 10.1007/978-1-61737-982-6_7,

Most DMD natural history studies have included measurements of muscle strength, joint contractures, and timed function tests. Results from these tests are used to track disease progression and offer insight into clinical milestones, such as the loss of ambulation and the need for ventilatory support. Both muscle weakness and joint contractures contribute to postural instability and ultimate loss of ambulation (11).

Contracture and muscle strength scores in DMD, for the most part, correlate and deteriorate synchronously over time (12). Contractures generally are attributed to shortening of involved muscles. Otherwise, distal joint contractures in DMD have been attributed to two mechanisms. On the one hand, there may be an imbalance of agonist and antagonist muscles operating around a joint. Weakness of antagonist extensor muscles correlates highly with flexor contracture severity in DMD (12). As opposing extensor muscles weaken, flexor contractures worsen. Alternatively, distal limb flexor contractures in DMD have been attributed to postural instability caused by selective sparing or tightness of proximal flexor muscles (11, 13).

Over the past 25 years, we have conducted extensive studies in dogs with GRMD (also termed canine X-linked muscular dystrophy, CXMD). This chapter outlines the strategies and precautions in establishing and maintaining the GRMD colony and functional evaluation of disease progression in GRMD by tibiotarsal joint (hock; ankle) angle and torque force, and eccentric contraction-induced force decline.

2. Materials

2.1. Colony Development and Maintenance

1. GRMD dogs (see Note 1).
2. Reagents used for measuring serum progesterone levels (Biovet–Canada, St-Hyacinthe, Québec, Canada).
3. Reagents used for semen collection (see Note 2): wAg-Tek Semen Collection Cones (Neogen Corp., Lansing, MI, USA), disposable centrifuge tube (sterile, polypropylene, 15 mL) (Thermo Fisher Scientific, Waltham, MA, USA).
4. Reagents used for artificial insemination (AI): Priority Care Nonspermicidal Sterile Lubricating Jelly (5 oz) (First Priority, Inc, Elgin, IL, USA), Pipette 9" w/ flex tip (Veterinary Concepts, Spring Valley, WI, USA), Norm-Ject 10 mL Airtite syringes (Agtech, Inc., Manhattan, KS, USA).
5. Software for calculation of the degree of inbreeding (CompuPed, Man's Best Friend Software, Franklin, IL, USA).
6. Precision Balance Toploading Scale (Model SB8000, Mettler Toledo, Columbus, OH, USA).

2.2. Physiological Functional Measurements

1. Preanesthetic agents: Atropine Sulfate Inj (Baxter Healthcare Corporation, Deerfield, IL, USA), Acepromazine Maleate Inj (Boehringer Ingelheim, St. Joseph, MO, USA), Butorphanol Tartrate (Fort Dodge Animal Health, Fort Dodge, IA, USA).
2. Anesthetic agents: Propofol (Abbott Animal Health, North Chicago, IL, USA), SevoFlo (sevoflurane, Abbott Laboratories).
3. Postanesthetic reversal agent: Naloxone HCl (Hospira, Lake Forest, IL, USA).
4. Pulse oximeter: Cardell Multiparameter Monitor 9405 (Minrad International, Orchard Park, NY, USA).
5. Goniometer (Smith & Nephew Roylan, Germantown, WI, USA).
6. Monopolar Electromyography (EMG) Needle Electrodes (Houston's Electrode Sales, Penn Valley, CA, USA).
7. TE42 EMG unit (Teca Corp, Pleasantville, NY, USA).
8. Force measuring system with servomotor (Aurora Scientific, Ontario, Canada).
9. Statistical analysis software (Sigma Stat, San Rafael, CA, USA).

3. Methods

3.1. Colony Development and Maintenance

1. For GRMD breeding (see Notes 3 and 4), check for the signs of proestrus in GRMD carriers 3 times a week (Monday, Wednesday, and Friday). At the first sign of proestrus, measure serum luteinizing hormone (LH) and progesterone levels. When LH is positive, ovulation has occurred.
2. Initiate breeding every other day by AI when progesterone levels are >3 ng/mL and vaginal cytology shows >60% cornified epithelial cells. Breeding is continued until progesterone levels are >10 ng/mL and white blood cells are seen on vaginal cytology, indicating the onset of diestrus.
3. Obtain semen from affected male. Uncap the centrifuge tube and place it inside the collection cone. Pull the centrifuge tube down into the cone until about 1 in. of tube is left inside the cone (making a seal with the tube so semen does not leak out). Fold down the top of cone once. Extend the penis from the prepuce. Grasp the bulbocarvernosus muscle of the penis. The penis will engorge and the dog ejaculates into the collection cone. Omit the first and the third fractions. Remove the cone and recap the tube.

4. Inseminate the bitch by AI. Add a syringe to the flextip part of the pipette. Insert the tip of the pipette in the tube just below the surface of the semen. Draw the sample up slowly without flexing the joint at the end of the pipette. Draw up extra air in the syringe to push the semen through the pipette and avoid leakage. Add nonspermicidal jelly to the tip of the pipette. Place a finger inside the vulva to guide the pipette. Direct the pipette dorsally through the vulva so as to avoid the recess of the vestibule and into the vagina. Attach the syringe to the flex tip of the pipette and infuse the semen. Remove the syringe and add air to force the full volume of semen into the vagina.
5. Evaluate ultrasound and measure relaxin levels 30 days from the onset of diestrus. If confirmed pregnant, the whelping date is calculated to be 27 days later. Body temperature is measured beginning a few days before this date to more precisely predict whelping. With this breeding protocol, conception rates are approximately 85%.
6. Carriers are not removed from the colony at a set age. Rather, we remove them based on the regularity of their estrus cycle, rate of pregnancy, litter size, and pup survival. With these factors in mind, most carriers continue to be bred until they are 5–7 years of age.
7. All bitches are observed at the anticipated time of whelping to ensure that appropriate neonatal care can be provided.
8. Affected pups are weak at birth and require nutritional supplementation for up to 4 weeks. We closely monitor pups during the first 24 h of life to ensure that colostrum is ingested. Colostrum from other dogs is harvested and frozen at –80°C so that it can be used as a supplement for weaker pups that are pushed aside by more vigorous littermates and/or do not have satisfactory weight gain.
9. Pups are weighed soon after birth and, on average, 4 times daily for the first 3 weeks (Table 1). Each pup's physical response to outside stimuli and hydration level are observed at weighing intervals. Newborn pups are evaluated on their overall responsiveness and strength. Righting response, whereby a pup is placed on its back and allowed to right itself, in addition to rooting, latching onto teats, and suckling ability of the pup when placed near the dam are observed. Any evidence of weakness of these responses or failure to thrive is noted. This would include the aforementioned physical responses and any evidence of respiratory distress, abdominal distension, and/or significant weight loss (i.e., sustained lack of weight gain or loss of 20 g. over a 24-h period).
10. In our hands, survival of GRMD dogs is typically 70–80%. Adequate numbers of dogs can be obtained for long-term

comparative studies (14–18). An average of 2.25 affected pups from each litter survive beyond 10 days of age. To provide an approximate number of affected dogs that can be produced each year by each carrier, the following formula is used: Affected pups/carrier each year = 2.25 (surviving pups/litter) × 0.83 (estrous cycles/year considering multiple factors) × 0.85 (conception rate) (see Note 4). Using this formula, each carrier produces only about 1.58 affected pups/annually. Thus, to produce 15 affected dogs each year, one must maintain ten carriers.

11. Based on results of studies to determine the effect of inbreeding on pup mortality (see Note 5 and Table 2), we introduce

Table 1
Schedule for recording dog weights

Age group	Frequency to record weight
Neonates (<24 h old)	1× soon after birth
Preweaning (1 day to 3 weeks old)	3–5× daily
Preweaning normal/carrier (3–6 weeks old)	1× weekly
Preweaning affected (3–6 weeks old)	2–3× daily
Postweaning normal/carrier (6 weeks to 3 months)	1× weekly
Postweaning affected (6 weeks to 3 months)	1× weekly
Juvenile/subadult normal/carrier (3–12 months)	1× monthly
Juvenile/subadult affected (3–12 months)	1× weekly
Adult normal/carrier (>12 months)	1× monthly
Adult affected (>12 months)	1× weekly
Any sternally recumbent dog (any age)	1× daily

Table 2
Mortality rate with respect to % inbreeding

Inbreeding (%)	Group size (*n*)	Number of litters	Number of pups died	Mortality rate (%)
0.00	84	12	4	4.76
4.69–12.89	101	14	13	12.87
15.63–19.53	74	10	14	18.92
20.31–24.22	94	14	21	22.34
25.78–40.63	106	16	34	32.08

new breeding stock unrelated to the colony, on average, at 5-year intervals. We continue to evaluate this strategy and plan to begin introducing new breeding stock at 3-year intervals.

12. Studies of the effects of various levels of postnatal care on pup mortality have guided methods used to manage pups during the first 2 weeks of life (see Note 6). Pups with partial postnatal care, defined as weighing pups and supplementing those experiencing difficulty with bitch's milk, milk replacer, or both using gastric tubes or bottles, tend to do better than those that receive no supplement care or are removed from the bitch entirely.
13. Studies of the effect of the GRMD trait on fetal (in utero) mortality have shown that GRMD pups are not at increased risk (see Note 7).
14. Numerous pups have been submitted for necropsy to determine cause of death. Respiratory and intestinal infections are the most common causes of neonatal death.

3.2. Physiological Functional Measurements

Most of our research using GRMD dogs has been focused on preclinical studies of potential treatments for DMD and has been done in collaboration with other scientists. To better utilize the GRMD model in therapeutic trials, we have developed various phenotypic tests to objectively characterize disease progression. Affected dogs have joint contractures (14) and demonstrate weakness of individual (15) and grouped (16) muscles. We have found a correlation between tibiotarsal joint torque and angle measurements (see Note 8). As with mdx mice, weakness is exacerbated by eccentric contractions (17). Importantly, by comparing serial measurements from treated and untreated dogs, one can document improvement or delayed progression of disease (18) (see Note 9). In our hands, results from dystrophin-deficient heterozygous males and homozygous females do not differ (see Note 10).

1. Measurement of the tibiotarsal joint angle. The tibiotarsal joint angle has been used as a surrogate biomarker for contracture severity in GRMD. Initially, we measured this angle in GRMD dogs undergoing serial peroneus longus force measurements (14, 15). Subsequent to this study, so as to be able to compare results with historical values, we have continued to use this basic approach without the need for surgery or immobilization of the limb. Others have described methods to measure the tibiotarsal joint angle at maximal flexion and extension in normal dogs (19, 20). Values for normal dogs using our method approximate, but are somewhat less than those recorded at maximal extension.

   To measure the tibiotarsal joint angle, first anesthetize the animal using the following agents and doses (also see materials). Preanesthetic agents 20–30 min prior to anesthesia induction: atropine sulfate (0.04 mg/kg, IM); acepromazine maleate (0.02 mg/kg, IM) for dogs weighing greater than 5 kg; and butorphanol tartrate (0.4 mg/kg, IM). Anesthetic induction agents: propofol (up to 3 mg/kg, IV – slowly!) for dogs weighing greater than 5 kg; for pups <5 kg, mask down using sevoflurane (to effect, inhaled), intubate, and maintain with sevoflurane. Monitor the dog by continuous recording of the following parameters with a pulse oximeter: ECG, heart and respiratory rate, blood pressure, end tidal (Et) CO_2, and saturation of hemoglobin by peripheral oxygen (SpO_2). Record these values as well as capillary refill time and anesthetic setting every 15 min.

Position the dog in dorsal recumbency and hold the pelvic limb so that the hip (coxofemoral) and knee (stifle) each forms a 90° angle (Fig. 1). The distal limb is supported with a finger placed just below the tibiotarsal joint (hock). The tibia should be parallel to the table. A goniometer is held so that the pivot point is centered over the lateral malleolus of the fibula, with one arm of the goniometer extending along the axis of the tibia/fibula and the other along the metatarsus. Note, the paw tends to deviate dorsally from the line of the metatarsus. The arm of the goniometer should be placed along the metatarsus and not aligned with the end of the paw.

Record the joint angle and repeat the measurement on the other limb.

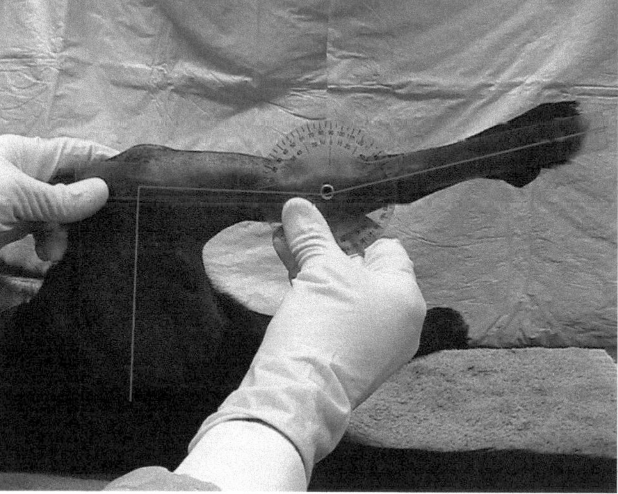

Fig. 1. Placement of the goniometer to measure the tibiotarsal joint angle. The dog is anesthetized and positioned in dorsal recumbency, with the stifle held at a 90° angle and the tibia/fibula parallel to the table. The pivot point of the goniometer is centered over the lateral malleolus of the fibula. One arm extends along the axis of the tibia/fibula, with the other along the metatarsus. The angle here is approximately 165°.

Tibiotarsal joint angles of GRMD and normal dogs are not statistically different at 3 months of age (14). However, by 6 months, GRMD angles are more acute, in keeping with the plantigrade posture and gait seen with more severely affected dogs.

Monitor affected and carrier dogs closely during anesthetic recovery until fully awake and in sternal recumbency. For rapid recovery, you may administer naloxone (up to 0.4 mg/kg, given in 0.5 dose increments, SQ). The first dose should be given while the dog is still intubated and breathing O_2. The second dose, if necessary, is given after the dog is extubated and/or if respiration drops below seven breaths per minute.

2. Measurement of tibiotarsal force and torque. The measurement of muscle contractile function is an important end-point for assessing the efficacy of therapies in DMD and the mdx mouse and GRMD dog models. Muscles in DMD and both the mdx mouse and GRMD dog are variably affected (21–23). In general, flexor muscles are affected initially, perhaps because of their predominant role in movement early in life. Extensors are affected later, due presumably to greater use in weight bearing and the damaging effects of lengthening (eccentric) contractions. Thus, ideally, a technique that allows evaluation of both flexor and extensor muscles is needed. Moreover, results of force measurements must be interpreted in light of this differential muscle involvement, especially taking into account the animal's age and the natural history of the disease (see Note 8). With this background information in mind, we were motivated to develop a measurement system to obtain in vivo force that satisfied two key characteristics including (1) minimally invasive, ideally not involving surgery, so that multiple measurements could be made over time, and (2) simultaneous evaluation of both extensor and flexor muscles to define the degree of differential muscle involvement. In the method described here, we measure isometric force/torque generated by extension and flexion of the tibiotarsal joint subsequent to percutaneous stimulation of the tibial and common peroneal nerves, respectively. Tests can be performed sequentially over time, thus allowing definition of the natural history of the disease. In the case of treatment trials, baseline measurements can be made at 2–3 months of age prior to the onset of treatment, so that each dog serves as its own control. Tetanic values are selectively evaluated because they are more consistent than those for the twitch. As with mdx mice, weakness is exaggerated by eccentric contractions (17) (see separate protocol below).

   Anesthetize and position the dog as described in the tibiotarsal joint angle measurement protocol. In a preliminary study, mean alveolar concentration (MAC) values for isoflurane did not significantly affect force measurement values (Schueler RO, Koch H, and Kornegay JN, unpublished data).

Calibrate the force transducer (see Note 11). Place one of the pelvic limbs in a stereotactic frame such that the hip (coxofemoral), stifle, and hock joints are all at 90° angles (Fig. 2). Suspend the tibiotarsal joint in a stirrup and immobilize the paw against a pedal using tape. With the paw secured to the pedal, the activation of either the flexors or the extensors results in generation of torque (see Note 12).

The common peroneal nerve (tibiotarsal joint flexion) can be palpated just distal to the head of the fibula as it courses across the bone. Paired stimulating and reference 27-gauge monopolar needle electrodes are placed subcutaneously on either side of the nerve. The electrodes are optimally positioned and the voltage is adjusted until twitch force (Pt) reaches a maximum. To maximally activate the muscles of interest, stimulation voltage is 50% greater than necessary to achieve Pt (generally, supramaximal pulses of 150 V are applied). An elec-

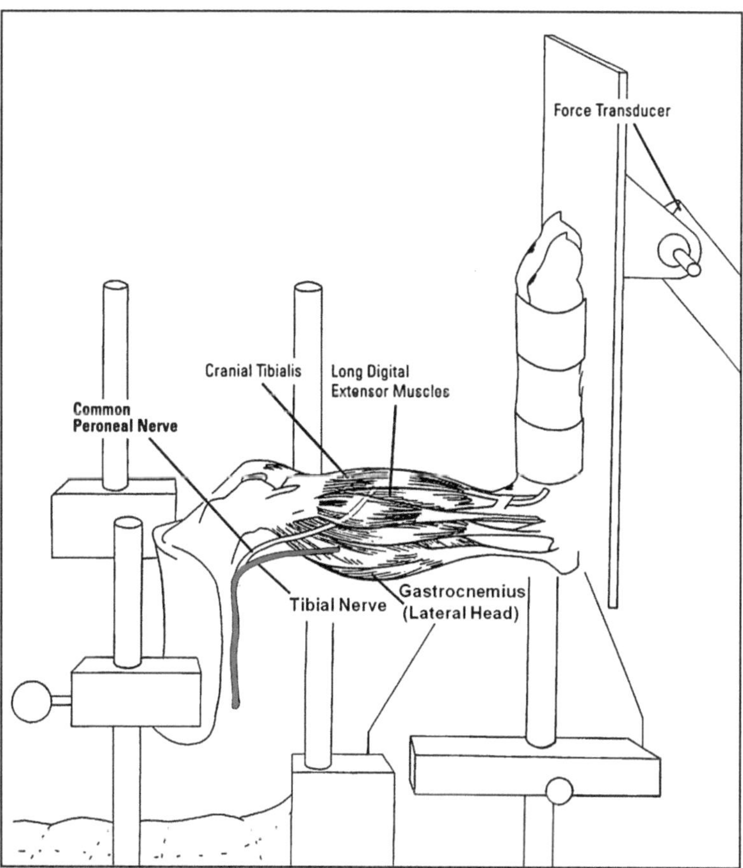

Fig. 2. Stereotactic frame and nerves/muscles involved in tibiotarsal joint torque measurements. Modified from Childers et al. (17).

tromyography (EMG) unit can be used to record electrical activity in tibiotarsal joint flexor muscles (cranial tibial and long digital extensor) to ensure proper needle placement. Once the maximal single twitch potential has been obtained, the nerve is stimulated with a tetanic run of 250 pulses (50/s) to obtain a tetanic mechanical potential. The potential should plateau, indicating that summation has occurred.

Force (torque) generated by tibiotarsal joint extension is measured using the same basic procedure. The tibial nerve is stimulated as it courses caudally and distally within the gastrocnemius muscle bellies. Paired stimulating and reference 27-gauge monopolar needle electrodes are placed deep within the musculature caudal and just proximal to the stifle. The approximate point of placement is along a line that bisects the 90° angle formed by the stifle (knee). The electrodes are optimally positioned and the voltage is adjusted until twitch force (Pt) reaches a maximum. Then follow the same protocol as described for the peroneal nerve.

Torque is recorded by the computer software. Because torque (generally recorded in Newton – meters) is the product of both force and distance of the lever arm, it is important to establish the length of the lever arm in each experiment. The lever arm for tibiotarsal joint flexion is determined by measuring the distance from the axis of rotation to the distal-most end of the tape surrounding the paw (Fig. 2). In our hands, this distance approximates 75% of the overall paw length measured from the tip of the digit to the point of the hock. For tibiotarsal extension, a pressure-sensitive film can be used to record the center of digital pressure. Data indicate that the distance measured from the axis to central pressure pad is 0.76% of the total paw length (Childers MK and Tegeler C, unpublished data). Force (Newtons) can be determined by dividing the torque (Newton – meters) by the lever arm length (meters).

For both flexion and extension measurements, passive force is subtracted from total force produced; only active force generated by muscles is measured. Because dogs vary in weight, the absolute tetanic force (Newtons) is divided by the body mass (kg) to obtain mass-corrected isometric force (Newtons/kg). Multiple tetanic recordings should be made, allowing 5 min for recovery between measurements. We typically reposition the electrodes once and make two recordings for both flexion and extension. The higher of the two values is recorded.

We have evaluated force and torque generated by tibiotarsal joint flexion and extension to determine both the natural history of GRMD (16) and the response of GRMD dogs to treatment (18) (see Note 13).

3. Tibiotarsal torque eccentric contraction decrement. Dystrophin serves to buttress the muscle cell membrane. In the absence of dystrophin, the membrane is prone to tearing during minimal exercise. Dystrophic muscles are particularly prone to injury subsequent to eccentric (lengthening) contractions (21). As an example, mdx mice have near-normal absolute force measurements, but demonstrate greater-than-normal force decrement after eccentric muscle contractions (24). We have previously shown that GRMD dogs also have a greater-than-normal force decrement with eccentric contractions (17). In an initial study, eccentric contractions were induced in flexor muscles of the cranial tibial compartment by stimulating the sciatic trunk in the midthigh area (17). This caused contraction of both the tibiotarsal joint flexors and extensors. Because the extensors are more powerful, eccentric (lengthening) contractions were induced in the flexors.

   More recently, we have developed a technique whereby the common peroneal nerve is stimulated while simultaneously extending the tibiotarsal joint. Briefly, as described above, the dog's paw is secured to the pedal of a servomotor–transducer system. The servomotor, under computer control, moves the pedal at a given rate and duration to extend (lengthen) the muscles of the cranial tibial compartment.

   Dogs are anesthetized using the protocol described above and placed in dorsal recumbency. Measure the tibiotarsal joint angle (see separate protocol). Calibrate the force transducer before each measurement. Depending on the system used, this may not be necessary. With the Aurora system, a calibration file that has been preset during manufacture is loaded at the beginning of the measurements. Place one of the pelvic limbs in a stereotactic frame such that the hip (coxofemoral), knee (stifle), and hock (tibiotarsal) joints are all at 90° angles (see tibiotarsal joint torque force protocol).

   Palpate the common peroneal nerve just distal to the head of the fibula as it courses across the bone. Paired stimulating and reference 27-gauge monopolar needle electrodes are placed subcutaneously on either side of the nerve. The electrodes are optimally positioned and the voltage is adjusted until twitch force (Pt) reaches a maximum. To maximally activate the muscles of interest, stimulation voltage is 50% greater than necessary to achieve Pt (generally, supramaximal pulses of 150 V are applied). An electromyography (EMG) unit can be used to record electrical activity in tibiotarsal joint flexor muscles (cranial tibial and long digital extensor) to ensure proper needle placement. Eccentric contractions are induced using square wave pulses of 100 μs duration in a tetanic run for 1 s at a frequency of 50 Hz. The contraction is held isometric at the optimal fiber length (L_o)

(see Note 12) for the first 900 ms. In the final 100 ms, the muscles of the cranial tibial compartment are stretched by the servomotor at a rate of 0.7 muscle lengths/s, such that the muscles are stretched to 107% of L_o. Stretches are performed 10 times, once every 5 s. Two additional series of ten stretches are performed, each separated by 4 min. Thus, the muscles of the cranial tibial compartment are stretched a total of 30 times over an ~10 min period. Torque is recorded by the computer software, and force is subsequently calculated by dividing the torque by the lever arm distance (~0.75 of paw length).

Contraction-induced injury is quantified by the force deficit (P_d), which is defined as the percent difference between maximum isometric force (P_o) before and after exercise (in this case, eccentric stretches). Force deficit is expressed as a percent change of the initial value of P_o, using the following equation: Force deficit $(P_d) = [P_o - P_{o\ \text{after stretch}} / P_o] \times 100$.

Based on our preliminary results, GRMD dogs have a decrement that varies from ~40 to 85%, while the decrement for normal and carrier dogs rarely exceeds 25%. However, data must be collected systematically from normal and dystrophic dogs and correlated with histologic markers such as Evans blue dye exclusion before definitive conclusions can be drawn.

4. Notes

1. An affected GRMD dog studied by our group until 40 months of age (3) is the common sire of all dogs in colonies in the US (the University of North Carolina at Chapel Hill [UNC-CH], the University of Missouri and the Fred Hutchinson Cancer Center in Seattle) and worldwide (Brazil, France, Japan, and the Netherlands). The original founder dog had a spontaneous mutation in the dystrophin gene whereby an RNA processing error occurred due to a single base change in the 3′ consensus splice site of intron 6 (10). Exon 7 is consequently skipped during RNA processing. The resulting transcript predicts that the dystrophin reading frame is terminated within its N-terminal domain in exon 8. A truncated, apparently unstable dystrophin molecule is produced.

   The first colony of affected GRMD/CXMD dogs was established at Cornell University by Barry Cooper using one of the dogs evaluated by our group. We used dogs from Cornell to establish a second colony at North Carolina State University in the late 1980s. This colony was moved to the University of Missouri-Columbia in 1994 and to UNC-CH in 2007. The original founder golden retriever dog was outbred to beagles and other mixed breed dogs. Thus, we have used the term GRMD to designate the specific splice site mutation in affected dogs rather than a disease of purebred golden retrievers.

2. Some have had success with the use of frozen semen (25). However, we use strictly fresh semen.
3. Conception rate and gestation in GRMD and other canine models are not impaired. Breeding affected males to obligate carriers produces an expected 1:1 ratio of affected to unaffected dogs (25% affected males, 25% affected homozygous females, 25% obligate female carriers, and 25% normal males). We use homozygous females and affected males interchangeably on studies, as there is no evidence that the severity of disease varies with gender (see Note 10). The mean litter size is seven pups, with a range of 1–13. This compares favorably with the average litter size of 7.6 seen in golden retrievers (26).
4. Dogs cycle in estrus on an average 10-month interval. We typically do not breed on the first estrous cycle or on the one following parturition. With these variables in mind, each carrier can be bred on average 0.83 times each year.
5. To determine the effect of inbreeding and postnatal care on pup mortality and thus guide overall colony management, we have systematically analyzed data from 66 litters with a total of 459 pups born over a 10-year period. As discussed in Note 1, all dogs in the colony originated from a single founder male. Ten additional founding dogs (three males, seven females) were added to the colony over the course of the 10 years reviewed. These dogs were assumed to be unrelated to the existing population and to each other since the exact relationships of these individuals were not known. Inbreeding has ranged from mating of cousins to continual line breeding (sire × daughters/granddaughters). All litters were produced from three different types of matings: (a) normal females × affected males, (b) carrier females × normal males, and (c) carrier females × affected males. Demographic data collected from each litter included inbreeding values for parents and offspring, initial number of offspring, fraction of offspring surviving until 14 days of age, sex and muscular dystrophy genotype, age of dam, parity (primiparous or multiparous dams), pup weights, pup condition, and postnatal care. The muscular dystrophy genotype of each individual was determined by elevated serum creatine kinase values or by restriction fragment length polymorphism (RFLP) analysis. Pup morality was defined as any death within the first 14 days of life. Inbreeding coefficients were calculated using the computer program CompuPed based on Wright's formula.

Models of the relationship of mortality and percent inbreeding were created using logistic regression. Hosmer–Lemeshow goodness of fit tests for the logistical models were done. There was evidence of a positive relationship between inbreeding and mortality rate ($p = 0.6950$, where $p > 0.05$ indicates a good fit). Mortality correlated positively with inbreeding for the overall colony, including normal, carrier, and affected dogs (Table 2). A total of 375 dogs (81.7%) had some calculated degree of inbreeding, and 81 (21.6%) of these died within the 14-day neonatal period. The coefficient for the indicator variable "affected" was 2.6458. This value must be exponentiated for interpretation. We found $e^{2.6458} = 14.094$ ("odds ratio"), implying that the odds of dying, at any level of percent inbreeding, are about 14 times greater for affected pups than for normal/carrier pups.

6. To determine the effect of postnatal care on pup mortality, we have evaluated the effects of care classified as none, partial, or complete. No intervention was defined as monitoring weights only. Partial intervention included weighing pups and supplementing those experiencing difficulty with bitch's milk, milk replacer, or both using gastric tubes or bottles. Complete intervention involved removing pups experiencing difficulty from the dam entirely and feeding as with the partial group. Litters were grouped according to their level of postnatal care and mortality data were compared for evidence of increased pup survival in any given group. Pups with partial postnatal care tended to do better than those with either no or complete care.
7. We have evaluated the effect of the GRMD trait on in utero mortality by comparing the expected genotypic ratios with the actual ratios of pups born in each litter. An overall evaluation of these data was made in relation to percent inbreeding and mortality rate, as well as an individual comparison by genotype. The Chi-square statistic for each litter was plotted against the percent inbreeding and the plot seemed random with no obvious pattern. Consistent with the plot, the correlation coefficient (specifically Spearman's Rho) was not significantly different from 0. This suggests that inbreeding does not affect prenatal mortality in our colony. Parity (primiparous or multiparous) also did not affect pup survival.
8. Consistent with the association between unbalanced muscle strength and contractures in DMD (above), tibiotarsal joint extension force and the extension/flexion ratio (Fig. 3a) correlate positively with joint angle in GRMD dogs at 6 months of age. Dogs with weaker extension force have more acute joint angles. Interestingly, flexion values correlate negatively, suggesting that flexor muscle functional hypertrophy can

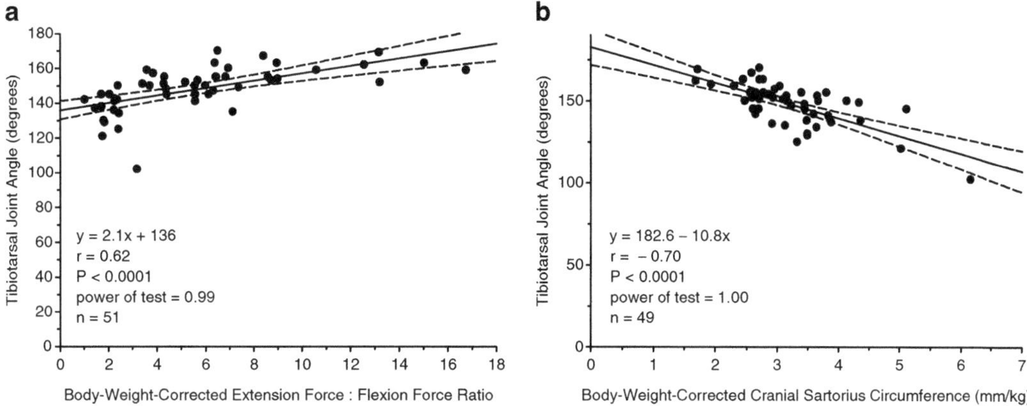

Fig. 3. Correlations among phenotypic tests in GRMD dogs. Scattergrams with regression lines drawn to show correlations between tibiotarsal joint angle (vertical axis in both) and the body-weight-corrected tibiotarsal extension/flexion force ratio (**a**) and cranial sartorius circumference (**b**) (horizontal axis in both) in GRMD dogs at 6 months of age. Joint angle correlates strongly ($p < 0.0001$) with both parameters, with the correlation being direct (*r* is positive) for the extension/flexion force ratio and inverse (*r* is negative) for cranial sartorius circumference.

contribute to contractures. In this sense, an increase in flexor muscle force could, paradoxically, indicate a more severe phenotype. Presumably, flexor muscles in more severely affected dogs undergo early necrosis and subsequent functional hypertrophy. Consistent with this hypothesis, we showed that GRMD dogs treated with prednisone (2 mg/kg beginning at 2 weeks) had increased tibiotarsal joint isometric extension force and a paradoxical decrease in flexion force at 6 months of age (18). The decrease in flexion force may have occurred because prednisone reduced early flexor muscle necrosis and subsequent functional hypertrophy. As with DMD, there is also an association between proximal flexor muscle function/mass and distal joint angles in GRMD. Specifically, tibiotarsal joint angle correlates negatively with cranial sartorius muscle weight (27) and circumference (Fig. 3b).

9. GRMD is variably expressed in affected dogs, with some having a much more severe phenotype than others (14–17). Phenotypic variation may confound statistical analysis, necessitating larger group sizes to achieve significance. In the case of localized cell or genetic treatments, the contralateral limb may serve as a control, with values compared between treated and untreated limbs. With systemic treatments, this problem can be at least partially offset if baseline measurements are done prior to the initiation of treatment. In this way, each dog serves as its own control. Accordingly, outcome is determined based on the difference in values at two or more time points in each treated animal, rather than at a single point in a group of animals.

10. Studies have suggested that gender affects phenotype in both the mdx mouse (28) and GRMD dog (29). Homozygous females reportedly have less severe clinical signs, due presumably to various factors, including effects of estrogen on muscle regeneration and inflammation. Such a gender effect could influence interpretation of preclinical studies in which both male and female animals are used. We have previously used heterozygous males and homozygous females interchangeably in our studies. Thus, in the context of the phenotypic studies discussed here, we sought to define whether gender influenced outcome. Results from these tests did not confirm an effect of gender on phenotype in GRMD dogs at 6 months of age (Table 3).
11. Depending on their range, different force transducers may have to be used for the flexion and extension recordings. Calibration should be repeated before each measurement.
12. Torque is the application of a force at some distance (the moment or lever arm) that is perpendicular to the axis of rotation (e.g., the tibiotarsal joint) and is calculated from the product of the force and the moment arm distance. The lever arm for our overall measurement system and calculation of torque has been described in Subheading 3.2. For individual muscle fibers, the length–tension relationship dictates that force generation is highest when there is maximal overlap of actin and myosin filaments and this is termed the optimal fiber length (L_0) (30). The force component of the torque generated at the tibiotarsal joint will, therefore, be influenced by the optimal fiber length of the multiple muscles that are

Table 3
Effect of gender on test results in GRMD dogs at 6 months

Test	Males ($n=27$)	Females ($n=24$)	p value
TTJ angle (°)	150.6 ± 10.86	145.3 ± 14.47	0.14
TTJ-corrected flexion (N/kg)	0.44 ± 0.11	0.45 ± 0.15	0.79
TTJ-corrected extension (N/kg)	2.11 ± 1.07	2.17 ± 0.72	0.604
Extension/flexion ratio	5.71 ± 4.24	5.64 ± 3.14	0.95
CS circumference (mm/kg)	3.04 ± 0.69	3.40 ± 0.96	0.134

TTJ tibiotarsal joint; *Corrected* body weight corrected; *CS* cranial sartorius

activated to contract. In addition, other factors such as the lever arm of the individual muscles as they cross the target joint and tendon slack length will influence the force component (31). The angle at which maximal joint torque is generated during isometric contractions has been termed the optimal joint angle (32).

We conducted studies to determine the optimal joint angle for this torque measurement (16). In a preliminary study of two normal dogs, values measured at 90° exceeded those made at 45° and 135° for both flexion and extension. Accordingly, the tibiotarsal joint angle was positioned at a 90° angle for subsequent studies. To more critically evaluate the tibiotarsal joint angle at which L_0 would be achieved, measurements were subsequently evaluated at 10° increments between 60° and 120° in seven GRMD and four normal dogs. For both groups, flexion plateaued at approximately 100°, while extension was minimal at about 70° (Childers MK, Kornegay JN, unpublished data). The length–tension relationship was not shifted for normal vs. GRMD dogs.

13. In our initial study of force and torque generated by tibiotarsal joint flexion and extension, force values were measured at 3, 4.5, 6, and 12 months of age. Absolute and body-mass-corrected GRMD twitch and tetanic force values were lower than normal at all ages ($p < 0.01$ for most). However, tarsal flexion and extension were differentially affected. Flexion values were especially low at 3 months, whereas extension was affected more at later ages. Several other GRMD findings differed from normal. The twitch/tetany ratio was generally lower; posttetanic potentiation for flexion values was less marked; and extension relaxation and contraction times were longer. The consistency of GRMD values was studied to determine which measurements would be most useful in evaluating treatment outcome. Standard deviation was proportionally greater for GRMD vs. normal recordings. More consistent values were seen for tetany vs. twitch and for flexion vs. extension. Left and right limb tetanic flexion values did not differ in GRMD; extension values were more variable. These results suggested that measurement of tarsal tetanic flexion force should be most useful to document therapeutic benefit in GRMD dogs (note, see discussion of the paradoxical decrease of tibiotarsal flexion values seen in GRMD dogs treated with prednisone in Note 8). Groups of 15 and five would be necessary to demonstrate differences of 0.2 and 0.4 in the means of treated and untreated GRMD dogs at 6 months of age, with associated powers of 0.824 and 0.856, respectively (16).

References

1. Bulfield, G., Siller, W.G., Wight, P. A. L., and Moore, K. J. (1984) X chromosome-linked muscular dystrophy (mdx) in the mouse. *Proc Natl Acad Sci USA* 81, 1189–1192.
2. Gillis, J. M. (1999) Understanding dystrophinopathies: an inventory of the structural and functional consequences of the absence of dystrophin in muscles of the mdx mouse. *J Muscle Res Cell Motil* 20, 605–625.
3. Kornegay, J. N., Tuler, S. M., Miller, D. M., and Levesque, D. C. (1988) Muscular dystrophy in a litter of golden retriever dogs. *Muscle Nerve* 11, 1056–1064.
4. Cooper, B. J., Winand, W. N., Stedman, H., Valentine B. A., Hoffman, E. P., Kunkel, L. M., et al. (1988) The homologue of the Duchenne locus is defective in X-linked muscular dystrophy of dogs. *Nature* 334, 154–156.
5. Schatzberg, S. J., Olby, N. J., Breen, M., Anderson, L. V. B., Langford, C. F., Dickens H. F., et al. (1999) Molecular analysis of a spontaneous dystrophin "knockout" dog. *Neuromuscul Disord* 9, 289–295.
6. Jones, B. R., Brennan, S., Mooney, C. T., Callanan, J. J., McAllister, H., Guo, L. T. et al. (2004) Muscular dystrophy with truncated dystrophin in a family of Japanese Spitz dogs. *J Neurol Sci* 217, 143–149.
7. Baltzer, W. I., Calise, D. V., Levine, J. M., Shelton, G. D., Edwards, J. F., and Steiner, J. M. (2007) Dystrophin-deficient muscular dystrophy in a Weimaraner. *J Am Anim Hosp Assoc* 43, 227–232.
8. Gaschen, F. P., Hoffman, E. P., Gorospe, J. R., Uhl, E. W., Senior, D. F., Cardinet, G. H. 3rd, et al. *(1992)* Dystrophin deficiency causes lethal muscle hypertrophy in cats. *J Neurol Sci* 110, 149–159.
9. Winand, N. J., Edwards, M., Pradhan, D., Berian, C. A., and Cooper, B. J. (1994) Deletion of the dystrophin muscle promoter in feline muscular dystrophy. *Neuromuscul Disord* 4, 433–445.
10. Sharp, N. J. H., Kornegay, J. N., Van Camp, S. D., Herbstreith, M. H., Secore, S. L., Kettle, S., et al. (1991) An error in dystrophin mRNA processing in golden retriever muscular dystrophy, an animal homologue of Duchenne muscular dystrophy. *Genomics* 13, 115–121.
11. Vignos, P. J., Jr., Spencer, G. E., Jr, and Archibald, J. C. (1963) Management of muscular dystrophy of childhood. *JAMA* 184, 89–96.
12. Brooke, M. H., Fenichel, G. M., Griggs, R. C., Mendell, J. R., Moxley, R., Miller, J. P., et al. (1983) Clinical Investigation in Duchenne dystrophy: 2. Determination of the "power" of therapeutic trials based on the natural history. *Muscle Nerve* 6, 91–103.
13. Siegel, I. M., Miller, J. E., and Ray, R. D. (1968) Subcutaneous lower limb tenotomy in the treatment of pseudohypertrophic muscular dystrophy. J Bone Joint Surg 50A, 1437–1443.
14. Kornegay, J. N., Sharp, N. J. H., Schueler, R. O., and Betts, C. W. (1994) Tarsal joint contracture in dogs with golden retriever muscular dystrophy. *Lab Anim Sci* 44, 331–333.
15. Kornegay, J. N., Sharp, N. J. H., Bogan, D. J., Van Camp, S. D., Metcalf, J. R., and Schueler, R. O. (1994) Contraction tension and kinetics of the peroneus longus muscle in golden retriever muscular dystrophy. *J Neurol Sci* 123, 100–107.
16. Kornegay, J. N., Bogan, D. J., Bogan, J. R., Childers, M. K., Cundiff, D. D., Petroski, G. F., et al. (1999) Contraction force generated by tarsal joint flexion and extension in dogs with golden retriever muscular dystrophy. *J Neurol Sci* 166, 115–121.
17. Childers, M. K., Okamura, C. S., Bogan, D. J., Bogan, J. R., Petroski, G. F., McDonald K., et al. (2002) Eccentric contraction injury in dystrophic canine muscle. *Arch Phys Med Rehabil* 83, 1572–1578.
18. Liu, J. M. K., Okamura, C. S., Bogan, D. J., Bogan, J. R, Childers, M. K., and Kornegay, J. N. (2004) Effects of prednisone in canine muscular dystrophy. *Muscle Nerve* 30, 767–773.
19. Jaegger, G., Marcellin-Little, D. J., and Levine, D. (2002) Reliability of goniometry in Labrador retrievers. *Am J Vet Res* 63, 979–986.
20. Nicholson, H. L., Osmotherly, P. G., Smith, B.A., and McGowan, C. M. (2007) Determinants of passive hip range of motion in adult Greyhounds. *Aust Vet J* 85, 217–221.
21. Edwards, R. H. T., Jones, D. A., Newham, D. J., and Chapman, S. J. (1984) Role of mechanical damage in the pathogenesis of proximal myopathy in man. *Lancet* 1, 548–552.
22. Dangain, J. and Vrbova, G. (1984) Muscle development in mdx mutant mice. *Muscle Nerve* 7, 700–704.
23. Valentine, B. A. and Cooper, B. J. (1991) Canine X-linked muscular dystrophy: selective involvement of muscles in neonatal dogs. *Neuromuscul Disord* 1, 31–38.

24. Moens, P., Baatsen, P. H. W. W., and Marechal, G. (1993) Increased susceptibility of EDL muscles from mdx mice to damage induced by contractions with stretch. *J Mus Res Cell Motil* 14, 446–451.
25. Thomassen, R., Sanson, G., Krogenæs, A., Fougnerb, J. A., Andersen Berg, A., and Farstad, W. (2006) Artificial insemination with frozen semen in dogs: a retrospective study of 10 years using a non-surgical approach. *Theriogenology* 66, 1645–1650.
26. Kelley, R. L. (2001) Factors influencing canine reproduction and nutritional management of the pregnant bitch. Canine Reproduction and Nutritional Health – Tufts Animal Expo, pp 9–14.
27. Kornegay, J. N., Cundiff, D. D., Bogan, D. J., Bogan, J. R., and Okamura, C. S. (2003) The cranial sartorius muscle undergoes true hypertrophy in dogs with golden retriever muscular dystrophy. *Neuromuscul Disord* 13, 493–500.
28. Salimena, M. C., Lagrota-Candido, J., and Quírico-Santos, T. (2004) Gender dimorphism influences extracellular matrix expression and regeneration of muscular tissue in mdx dystrophic mice. *Histochem Cell Biol* 122, 535–544.
29. Valentine, B. A., Cooper, B. J., de Lahunta, A, O'Quinn, R., and Blue, J. T. (1988) Canine X-linked muscular dystrophy. An animal model of Duchenne muscular dystrophy: clinical studies. *J Neurol Sci* 88, 69–81.
30. Gordon, A. M., Huxley, A. F., and Julian, F. J. (1966) The variation in isometric tension with sarcomere length in vertebrate muscle fibres. *J Physiol* 184, 170–192.
31. Hoy, M. G., Zajac, F. E., and Gordon, M. E. (1990) A musculoskeletal model of the human lower extremity: the effect of muscle, tendon, and moment arm on the moment-angle relationship of musculotendon actuators at the hip, knee, and ankle. *J Biomechanics* 23, 157–169.
32. Lieber, R. L. and Boakes, J. L. (1988) Sarcomere length and joint kinematics during torque production in frog hindlimb. *Am J Physiol* 254, C759–C768.

Part II

New Technology in Muscle Gene Therapy

Chapter 8

Directed Evolution of Adeno-Associated Virus (AAV) as Vector for Muscle Gene Therapy

Lin Yang, Juan Li, and Xiao Xiao

Abstract

Adeno-associated virus (AAV) is emerging as a vector of choice for muscle gene therapy because of its effective and stable transduction in striated muscles. AAV naturally evolve into multiple serotypes with diverse capsid gene sequences that are apparently the determinants of their tissue tropism and infectivity. Certain AAV serotypes show robust gene transfer upon direct intramuscular injection, while others are effective in crossing the endothelial barrier to reach muscle when delivered intravenously. Muscular dystrophy gene therapy requires efficient body-wide muscle gene transfer. However, preferential liver transduction by nearly all natural AAV serotypes could be an undesirable feature for muscle-directed applications, especially by means of systemic gene delivery. Here we describe a method of in vitro evolution and in vivo selection of AAV capsids that target striated muscles and detarget the liver. Using DNA shuffling technology, we have generated a capsid gene library by in vitro scrambling and shuffling the capsid genes of natural AAV1 to AAV9. To minimize the bias and limitation of in vitro screening on culture cells, we performed direct in vivo panning in adult mice after intravenous injection of the shuffled capsid library that packaged their own coding sequences. The AAV variants enriched in the heart and muscle are retrieved by capsid gene PCR and subsequently characterized for their tissue tropisms. This directed evolution and in vivo selection method should be useful in generating novel gene therapy vectors for muscle and heart and other tissues.

Key words: AAV, Transduction, DNA shuffling, In vivo selection, Tissue tropism

1. Introduction

Adeno-associated virus (AAV) is a promising vector for muscle gene therapy because of its efficient muscle transduction (1) and persistent transgene expression in myocytes (2). The AAV vectors show good safety profile in clinical trials (3). Importantly, some new AAV serotypes were successfully characterized recently with greatly improved heart and muscle transduction efficiency in systemic

Dongsheng Duan (ed.), *Muscle Gene Therapy: Methods and Protocols*, Methods in Molecular Biology, vol. 709, DOI 10.1007/978-1-61737-982-6_8,

gene delivery. This discovery sheds light on gene therapy of major muscular diseases such as congestive heart failure and muscular dystrophies where transduction of a majority of cardiomyocytes or multiple groups of striated muscles is required (4, 5).

Besides identification of new serotypes, genetic modification might further improve the AAV vector for its clinical application. One issue in utilizing the AAV vector is its broad tissue tropism, which is underscored in systemic gene transfer since transduction of the nontarget organs, primarily the liver, will dramatically decrease muscle transduction efficiency and pose potential safety concerns on ectopic gene expression. To improve tissue-specificity and targeting capacity of the AAV viral particles, random mutagenesis on the capsid genes, in addition to rationale engineering, is one of the most powerful ways to enhance capsids diversity and to alter capsids tropisms for intended tissues. Directed evolution is one of the recently developed approaches for genetic engineering. It involves the creation and expansion of enormous genetic information based on a gene pool and high-throughput screening for desirable genetic attributes on a particular biological platform (6). In this chapter, we describe a strategy where the *cap* genes of AAV serotypes are diversified by in vitro recombination using DNA shuffling. The resulted virus libraries containing a broad permutation of mutants and variants are panned (screened) directly in vivo in mice, but not on cultured cells, for enrichment and selection of muscle-tropic novel AAV capsids, as an example (7).

DNA shuffling is a method of random fragmentation followed by PCR reassembly to introduce or exchange new genetic mutations into the genes by patchy homologous recombination (6). AAV naturally evolves into different serotypes with variable tissue tropism or immunogenicity (8). The AAV *cap* genes of these serotypes are also the ideal templates for generation of novel recombinants by shuffling their DNA and select for the functionally viable mutants and variants. The modified AAV *cap* genes are cloned into suitable plasmid vector for production of chimeric virus library with the highly diversified biological properties and for further characterization of their tissue tropism.

A number of in vitro systems based on cultured cell lines have been used for the panning of virus vectors after DNA shuffling (9, 10). However, in gene therapy application AAV vector is destined to confront more complicated physiological environment in human or animal bodies, e.g., the serum proteins, the endothelial barriers, the extracellular matrix barrier, and the infectivity to the targeted cells, etc. Considering those barriers together, we chose to screen the AAV library directly in the C57BL/6J mice to mimic the realistic environment in vivo in preclinical or clinical situations.

2. Materials

2.1. In Vitro Recombination of AAV Cap Genes by DNA Shuffling

1. Starting AAV plasmid series: the AAV *cap* genes of serotypes 1, 2, 3B, 4, 6, 7, 8, 9 were respectively cloned into the packaging plasmid that contains the *rep* gene of AAV2 for the further molecular manipulation. The capsid genes will be used as the template for PCR amplification.
2. Primers for amplification and sequencing of AAV *cap* genes were listed in Table 1.
3. Pfu DNA polymerase and supplied buffer (Stratagene, Cedar Creek, TX, USA).
4. 100 mM dNTPs (Invitrogen, Grand Island, NY, USA).
5. Qiaquick gel extraction kit (Qiagen, Valencia, CA, USA).
6. DNase I (Sigma-Aldrich, St Louis, MO, USA).
7. 1M magnesium chloride solution (Sigma-Aldrich).
8. PTC-200 PCR thermocycler (MJ Research, Waltham, MA, USA).

2.2. Construction of Plasmid Library and Generation of Virus Library

1. Restriction endonucleases *Hind* III, *Not* I, *Hae* III, *Pst* I, and *Taq* I (New England BioLab, Ipswich, MA, USA).
2. Alkaline phosphatase, calf intestinal (New England BioLab).
3. T4 DNA ligase with supplied buffer (New England BioLab).
4. *E. coli* ElectroMAX DH10B™ cells (Invitrogen).
5. MicroPulser electroporator (Bio-Rad, Heracules, CA, USA).
6. Plasmid XX2 encoding AAV2 *rep* and *cap* genes.
7. pHelper plasmid containing five genes from adenovirus (Stratagene).

Table 1
Primer sequences for molecular manipulation of AAV *cap* genes

Primer	Sequence	Function
A	5′-CCCAAGCTTCGATCAACTACGCAGACAGGTA CCAA-3′	Binds *Hind* III site in AAV *rep* gene
B	5′-ATAAGAATGCGGCCGCAGAGACCAAAGTTCAACT GAAACGA-3′	Binds *Not* I site before polyA tail following AAV *cap* gene
C	5′-TATAAGTGAGCCCAAACGGGTG-3′	Binds AAV *rep* gene upstream *Hind* III site
D	5′-AAGGCCTACGACCAGCAG-3′	Binds AAV *cap* gene
E	5′-GCTAATAACCTTACCAGCAC-3′	Binds AAV *cap* gene

8. Adenovirus serotype 5 (ATCC, Manassas, VA, USA).
9. HEK 293 cells (ATCC).
10. Benzonase® nuclease (Sigma-Aldrich).
11. Sodium deoxycholate (Sigma-Aldrich).
12. Proteinase K (Roche, Nutley, NJ, USA).
13. Glycogen (Sigma-Aldrich).
14. Optima™ L-80 XP preparative ultracentrifuge (Beckman Coulter, Fullerton, CA, USA).
15. Slide-A-Lyzer dialysis cassette (Pierce, Rockford, IL, USA).

2.3. Screening of Virus Library in Animal Model

1. C57BL/6J mice (The Jackson Laboratory).
2. Phenol/chloroform/isoamyl alcohol (25:24:1, pH 6.7/8.0) (Fisher Scientific, Waltham, MA, USA).
3. RNase A (Qiagen).
4. iProof™ high-fidelity DNA polymerase (Bio-Rad) (see Note 1).
5. DNA sequencing service (see Note 2).

2.4. Production and Characterization of Recombinant AAVs

1. Packaging Plasmids AAV1, AAV2, AAV3B, AAV4, AAV6, AAV7, AAV8, and AAV9 all contain the *rep* gene of AAV2 and *cap* gene from each specific serotype.
2. Vector plasmid AAV-luc containing CMV promoter-driven firefly luciferase reporter gene.
3. Tissue tearor (Misc Equipment, Midland, ON, Canada).
4. Luciferase assay system (Promega; Madison, WI, USA).
5. Coomassie (Bradford) protein assay kit (Fisher Scientific).
6. DNeasy blood and tissue kit (Qiagen).
7. TaqMan® MGB probes and universal PCR master mix (Applied Biosystems, Foster City, CA, USA).
8. 7,300 Real-Time PCR system (Applied Biosystems).

3. Methods

The basic design of the AAV *cap* gene shuffling and in vivo panning is briefly described below. First, to shuffle the DNA, a collection of capsid genes of different AAV serotypes or genotypes is mixed and partially digested into small fragments by DNase I. The pool of the small fragments is then denatured into single-stranded DNA by boiling and cooling down, so that those ssDNA fragments with patchy homologies will reanneal with one another. A portion of those random fragments will reconstitute the full-length *cap* genes, consisting of random pieces from different serotypes of AAV. The single-stranded DNA gaps between those imperfectly annealed fragments will be repaired and subsequently amplified by primer-less PCR reaction (the random DNA fragments with homologies will serve as "primers" in the PCR reaction). As a result, novel capsid genes of full-length are generated and purified by size fractionation with gel electrophoresis. The gel purified full-length *cap* genes will be cloned into a plasmid that contains two inverted terminal repeats (ITR) and the *rep* gene of AAV2. After inserting the *cap* gene behind the *rep* gene, the plasmid is now in theory "infectious" and ready for infectious AAV particle production. Thus, a pool of infectious AAV plasmids containing the shuffled novel AAV *cap* genes will be used in the next step to produce a library of AAV vial particles, which contains novel capsids encapsidating the corresponding *cap* genes. The library will be directly injected intravenously in mice for in vivo panning. To retrieve novel AAV variants displaying improved tissue tropism and targeting capability, the AAV *cap* genes will be recovered by PCR using the total DNA isolated from the desirable tissues, e.g., skeletal muscle or heart, etc. The retrieved genes encoding novel capsid variants, which are enriched in the given tissue, will be recloned into the infectious plasmids and the viral particles are panned again in vivo in the same tissue for further enrichment and characterization.

3.1. DNA Shuffling

1. Use AAV *cap* genes of eight serotypes 1, 2, 3B, 4, 6, 7, 8, 9 as the PCR template for *cap* gene amplification (see Note 3).
2. Amplify a 2.65-kb *cap* gene fragment from these eight parental plasmids by the same primers C and B (Table 1) and Pfu DNA polymerase for high-fidelity PCR. Purify the amplified DNA fragments by Qiaquick gel extraction kit and mix DNA fragments in the same molar ratio to a total quantity of 4 μg.
3. Digest the DNA mixture by 0.04 U of DNase I in 100-μL system containing 50 mM Tris–HCl (pH 7.5) and 10 mM $MnCl_2$ at 15°C for 2 min. Add 20 μL of 0.5M EDTA (pH 8.0) to stop the reaction. Inactivate DNase I by treatment at 75°C for 10 min.
4. Separate the digested DNA in 2% agarose gel by running at 5 V/cm for 1.5–2 h (see Note 4). Exercise DNA fragments of 300–500 nucleotides from the gel. Purify DNA fragments into 20 μL TE 8.0.
5. Use 300 ng of the DNA fragments (300~500 bp) as templates for AAV *cap* gene reconstruction. Perform a 20-μL PCR reaction using 2 μL 10× buffer, 200 μM of dNTP, and 1 U of Pfu DNA polymerase. Run the PCR program at 94°C for 5 min and then 22 cycles of 94°C for 1 min, 50~55°C for

1 min (+0.1°C per cycle), and 72°C for 1–4.5 min (+5 s/cycle). Program 72°C for 10 min at the end of PCR reaction.

6. Augment the reassembled AAV *cap* gene for further cloning manipulation. Use 1 μL of the primer-less PCR product as template for the secondary amplification in a 50-μL system containing 5 μL 10× buffer, 200 μM of dNTP, 0.5 μM of primers A and B (Table 1), and 2.5 U of Pfu DNA polymerase. Run the PCR program at 94°C for 5 min, 75°C for 5 min, 60°C for 1 min, and 72°C for 4.5 min, followed by 29 cycles of 94°C for 1 min, 60°C for 1 min, 72°C for 4.5 min, and finally by 72°C for 10 min.

3.2. Construction of Shuffled Plasmid Library

1. Clone the shuffled AAV *cap* genes. Digest 8 μg of pAAV2-ITRs plasmid (an "infectious" plasmid containing the AAV2 *rep* gene and *cap* gene and ITRs) with *Hind* III and *Not* I followed by alkaline phosphatase treatment for 1 h to remove the AAV2 *cap* gene. Concentrate 2 μg of the shuffled *cap* genes by ethanol precipitation, dissolve DNA in 20 μL of TE 8.0, and completely digest by *Hind* III and *Not* I. Purify the 6.0-kb fragment of pAAV2-ITRs plasmid (without the *cap* gene) and the 2.6-kb pooled shuffled AAV *cap* genes from agarose gel and dissolved in 20 μL of TE 8.0.
2. Set up the ligation at the vector and insert molar ratio of 1:3 to shuffle the AAV *cap* genes into the pAAV2-ITRs vector (see Note 5). Set up multiple parallel ligation reactions with each 10-μL system containing 1 μL 10× buffer, 80 ng of vector, 120 ng of insert, and 0.5 μL of T4 DNA ligase. Incubate at 16°C overnight. Heat inactivate ligase at 65°C for 15 min.
3. Transform the ligation reaction into the DH10B competent cells using a MicroPulser electroporator. Specifically, add 1 μL of the ligation mix into 30 μL of DH10B cells. Incubate on ice for 5 min. Electro-shock the bacteria at 2.5 kV (see Note 6). Add 300 μL terrific broth medium to the cuvette. Transfer the bacteria to a 1.5-mL microfuge tube and incubate at 37°C for 40 min for recovery. Repeat this procedure 30–60 times. Spread the DH10B *E. coli* cells from each tube onto two 15-cm LB plates containing ampicillin.
4. Grow at 37°C for 14 h. Record the colony numbers on all the plates.
5. Randomly pick 40 colonies for overnight culture and perform miniprep. Use these as templates for PCR amplification with primers A and B (Table 1). Perform restriction digestion for sampling analysis of the inserted *cap* genes. Use following enzymes for diagnosis analysis including *Hae* III, *Pst* I, and *Taq* I. These enzymes yield distinguishable digestion patterns between AAV serotypes.

6. Verify the plasmid clones for the presence of the chimeric AAV *cap* gene. Pick ten clones used for midiprep. Cotransfect 3 µg of the midiprep DNA and 3 µg of pHelper plasmid into 293 cells in the six-well plate. The cells are collected 48 h later; collect cells and freeze and thaw cells 3 times. Detect the viral particles by dot-blot using the *rep-2* gene probe. Assess the viability of shuffled AAV *cap* genes for viral DNA packaging (see Note 7).
7. Record the approximate numbers of the *E. coli* colonies. Collect the colonies by adding 3 mL of TB medium to each plate, and gently wash and scrap the E. coli off the plates with a sterile L-shaped glass rod (see Note 8). Pool the colonies from all the plates and thoroughly mix them. Add sterile glycerol to a final concentration of 15% and divide the bacterial stock into 10–15 aliquots for storage at –80°C.
8. Use one aliquot of the bacterial library for propagation in 2 L of TB medium containing ampicillin at 37°C for 12 h. Use this bacterial culture for large-scale purification of plasmid DNA library by equilibrium centrifugation in CsCl-ethidium bromide gradients (11).

3.3. Generation of Chimeric Virus Library

1. Convert the shuffled chimeric AAV plasmid DNA library into viral particle libraries for screening and panning in vivo. To ensure that each virus particle is made of the specific capsid encoded by its own encapsidated *cap* gene, first produce an intermediate shuttle library from the plasmid library of *cap* genes (12). In detail, transfect 12.5 µg of the plasmid library, 12.5 µg of XX2, and 25 µg of pHelper into four 15-cm plates of 293 cells in 80% confluence. XX2 is used to improve packaging efficiency of the plasmid library since it is not likely that every shuffled AAV *cap* gene produces sufficient capsids to package its own DNA. Two days later, harvest the cells for virus extraction by freeze–thaw cycles. Determine the titer of the viruses contained in the supernatant by dot blot using *rep-2* probe.
2. Infect 293 cells with the AAV shuttle library to produce the competent virus library displaying shuffled AAV *cap* genes. Use twenty 15-cm plates of 293 cells and an infection multiplicity of 1,000 AAV particles for virus production to enable each 293 cell to be infected by only one viable AAV virion. Coinfect with adenovirus 5 at 100–200 m.o.i. (multiplicity of infection) to provide the necessary viral elements for AAV replication. Collect cells at 36 h later by spin at 1,300 × *g* for 10 min and resuspend the pellet in 10 mL of fresh DMEM medium.
3. Subject the cellular pellet to three freeze–thaw cycles in 15 mL tube. Then incubate at 55°C for 60 min to inactivate adenovirus.

4. Add 3 μL of Benzonase® nuclease to the cell lysate. Mix and then incubate at 37°C for 1 h and vortex every 15 min.
5. Add 850 μL of 10% deoxycholate solution to above lysate. Mix and then incubate at 37°C for another 30 min and swirl every 10 min.
6. Centrifuge the cell lysate at 2,000 × *g* for 15 min. Transfer the supernatant to a new 15-mL tube.
7. Add 7.2 g cesium chloride and vortex until completely dissolved. Split the mixture into two 14-mL SW41centrifuge tubes. Add 1.5 g/mL cesium chloride/PBS solution carefully to form a cushion at the bottom. Spin the gradient at 210,000 × *g* for 24 h at 12°C in an SW41 rotor in a Beckman ultracentrifuge.
8. Collect 4.5 mL of banded viral particles from above the 1.5 g/mL cesium chloride cushion in each tube via a 10-mL syringe and 18-gauge needle. Filter the fraction through a 0.45-μm filter membrane into a new 15-mL centrifuge tube and add 1.5 g/mL cesium chloride solution cushion as described. Centrifuge the gradient at 210,000 × *g* for 48 h at 12°C in an SW41 rotor.
9. Collect 0.5 mL fractions starting immediately above the cesium chloride cushion and identify the virus-rich fraction by quantitative DNA dot-blot. Use a dialysis cassette to desalt the virus-rich fraction by three cycles of dilution with 500 mL of dialysis buffer (PBS supplemented with 12% mannitol and 6 mM magnesium chloride) (see Note 9).
10. Determine the titer of the virus library by dot-blot using the *rep-2* probe.

3.4. Biopanning of Virus Library in Mouse Model

1. Our aim for modification of AAV *cap* gene is to improve its tissue tropism for muscle gene therapy. An ideal screening platform of the virus library should be an in vivo animal model. Considering the prospective application of the modified vectors, we chose the C57BL/6J mouse strain for this purpose because of its clear genetic background and wide usage in transgenic mouse and gene therapy studies.
2. Inject 5×10^{11} v.g. of the AAV library in 200 μL PBS to one adult male and one adult female C57BL/6J via the tail vein.
3. Sacrifice the mice 72 h later to collect the heart, hind limb muscles, and liver. Ground each tissue (200–400 mg) into powder under liquid nitrogen using a mortar and pestle. Isolate total DNA using the phenol/chloroform extraction method (11).
4. Determine the DNA concentration by UV spectrophotometer. Use ~300 ng of the DNA as template for PCR amplification

with the iProof DNA polymerase. The 50-μL reaction system includes 10 μL 5× buffer, 200 μM dNTP, 0.5 μM of primers A and B each, and 0.5 μL of the iProof DNA polymerase. The PCR program is 98°C for 60 s; 75°C for 5 min (add enzyme for hot start during this time), 68°C for 20 s, and 72°C for 80 s; followed by 29 cycles of 98°C for 10 s, 68°C for 20 s, and 72°C for 80 s; and finally 72°C for 10 min.

5. Run 10 μL of the PCR reaction on 1% agarose gel for detection of the retrieved AAV *cap* genes. Set up multiple parallel PCR reactions and purify the 2.6-kb retrieved *cap* genes by using the Qiaquick Gel Extraction Kit (Qiagen).
6. Ligate the purified AAV *cap* genes from the heart or muscle into the pAAV2-ITRs vector for reconstruction of the plasmid libraries.
7. Pick ~300 clones from the heart and muscle plasmid libraries, respectively, to accelerate the screening course. Perform amplification and restriction analysis of the *cap* genes using three enzymes *Hae* III, *Pst* I, and *Taq* I.
8. Identify multiple AAV *cap* gene variants from the heart or muscle plasmid libraries by the restriction analysis. Mix their correspondent plasmids in the same molar ratio for packaging of the secondary viral particle libraries respectively by cotransfecting with pHelper plasmid into 293 cells as described in Subheading 3.3.
9. Perform additional rounds of screening of the virus libraries to further enrich the AAV variants with heart or muscle tropism. Inject the heart or muscle-enriched AAV libraries (5×10^{11} v.g.) into C57BL/6J mice via tail vein. Retrieve the AAV *cap* genes from mouse tissues using the similar methods as described above. Use these *cap* genes to reconstruct the plasmid libraries by ligation into the pAAV2-ITRs vector.
10. Select respectively from the heart, muscle, and liver plasmid libraries 100, 100, and 50 clones for sequence analysis of their AAV *cap* genes using primers B, C, D, and E (Table 1). Align the complete *cap* gene sequences using the Clustal X program (13) to identify different *cap* genotypes and calculate their gene frequencies among different tissues. Those AAV *cap* gene variants with higher gene frequency in heart or muscle but scarce in liver are predicted to have enhanced muscle targeting and infection capability (see Note 10).

3.5. Characterization of Tissue Tropism of AAV Variants

1. For the AAV *cap* genes enriched in heart or muscle identified and confirmed by sequence analysis, excise them from their original infectious plasmid (ITR-containing) backbone by *Hind* III/ *Not* I digestion and clone into the AAV2 packaging vector plasmid (without ITRs). The resultant AAV packaging

plasmids contain the *rep-2* gene and the chimeric *cap* genes, ready for packaging of reporter AAV vectors for tissue tropism characterization.

2. Purify the selected packaging plasmid and the AAV-luc vector containing luciferase reporter gene by CsCl-ethidium bromide gradient centrifugation. Use these plasmids to produce recombinant AAV vector packaged by chimeric *cap* gene. Perform cotransfection using 6.25 μg specific packaging plasmid, 18.75 μg AAV-luc reporter vector plasmid, and 25 μg pHelper into the 293 cells in 15-cm plate by calcium phosphate transfection.
3. Harvest and lyse the cells for purification of AAV vector as described in Subheading 3.3. Omit the 55°C heat inactivation treatment after the freeze–thaw cycle because no adenovirus is used in this stage of the vector production. Determine titers of the recombinant AAVs by quantitative dot-blot using the CMV probe.
4. Inject 3×10^{11} v.g. of the recombinant AAV-luc into 8-week-old male C57BL/6J mice in 200 μL PBS via tail vein. Treat four to five mice in each group for each novel AAV capsid variant. Administer the same dose of recombinant AAV9 vector (produced from the AAV2/9 plasmid) into a different mouse group as a positive control or reference standard.
5. Two weeks later, sacrifice the mice and collect the muscle and nonmuscle organs including the heart, abdomin, diaphragm, tibialis anterior, gastrocnemius, quadricep, forelimb, masseter, tongue, liver, spleen, pancreas, kidney, testis, lung, brain, etc. Freeze in liquid nitrogen and store in –80°C before use.
6. For analyzing lucifase gene expression in those tissues, homogenize100 mg of the frozen tissue in 0.8 mL lysis buffer with the mechanic tearor on ice (see Note 11). After vortex, centrifuge the sample at $14{,}000 \times g$ for 3 min at 4°C and transfer the supernatant into a new tube (see Note 12).
7. Prepare the luciferase reagent by dissolving the substrate in the buffer provided by the manufacturer 30 min before experiment. Mix 20 μL of tissue lysate and 50 μL of luciferase substrate into the microplate for luminescence detection. After shaking for 5 s, record the luminescent signal for 10 s at 340 nm using a microplate reader.
8. Determine protein concentration of the tissue samples by the Bradford protein assay. Briefly, mix 10 μL of the protein samples or the BSA standards with 200 μL of the 1:30 diluted dye reagent in the microtiter plate. After 5 min, obtain the protein concentration at 595 nm using a microplate reader. Express the normalized luciferase activity as relative light unit per mg protein.

9. Determine the AAV vector genome in mouse tissues by quantitative real-time PCR. Extract the genomic DNA from 10~25 mg of mouse tissue using DNeasy blood and tissue kit. Use 10 ng of this DNA as template for quantitative PCR in a 25-μL reaction system containing 12.5 μL 2× Taqman Universal PCR Master Mix, 0.5 μM of the forward and reverse primers each, and 0.15 μM of the Taqman MGB probe. The primers and probe for quantitative PCR detection of CMV promoter sequence in the vector DNA are listed in Table 1. Use Applied Biosystems 7,300 Real-Time PCR System for monitoring and recording of the PCR reaction.
10. Use the mouse glucagon gene (a single-copy gene per haploid genome) as the internal cell genome copy number control for real-time quantitative PCR. Its primers and probe are also listed in Table 1. Use serial dilutions of given copy numbers of reference plasmids, which contain either the CMV promoter sequence or the glucagon gene sequence as the DNA copy number standards in the qPCR analyses. Report the AAV vector biodistributions in different tissues as vector genome copies per diploid cell by dividing the copy number of vector DNA by one half the mouse glucagon gene copy numbers in the same amount of total DNA for the qPCR reactions. For example, if the vector copy number in a muscle sample determined by qPCR is 5,000 and the cellular DNA glucagons gene copy number of the same sample is 10,000, then the vector copy number is defined as one copy per diploid cell, or per nucleus in muscle.

4. Notes

1. The Pfu DNA polymerase was competent in amplification of the chimeric AAV capsid genes after the initial DNA shuffling. However, the yields of capsid genes were low when this enzyme was used for PCR amplification using the tissue DNA from the mice as the template. The iProof DNA polymerase was used in this step and produced higher yields of AAV capsid genes from mouse tissues for further cloning and screening work.
2. Sequencing work was done using the 96-well sample handling method for large number of clones after first round of capsids DNA recovery from the tissues.
3. We initially recognized a *Hind* III site within the AAV3B capsid gene sequence. To avoid its interference with the successive molecular cloning of chimeric capsid genes, we silenced

this enzyme site by mutagenesis before using this gene as template for DNA shuffling.

4. Dyes were not contained in the DNA loading buffer in this step to prevent its disturbance in observation of the DNA bands under UV light.
5. Ligation efficiency was crucial for construction of a plasmid library of abundant diversity. We have observed only tens of bacterial colonies or thousands of ones from 1 μL of different ligation products. The key factor for a successful ligation was the quality of the DNA vector and insert. The cloning backbone plasmid should be thoroughly digested by *Hind* III/ *Not* I in a 200-μL reaction overnight and then dephosphoralyted completely. With respect to the insert, a large volume of capsid genes should be first concentrated by ethanol precipitation and then digested thoroughly by *Hind* III/ *Not* I. An empty backbone plasmid ligation should be used to serve as the negative control side by side with the chimeric capsid gene ligation. The negative control should yield no colonies or barely any colonies.
6. Cautions should be taken to make sure no air bubble existed in the competent bacterial cells in the cuvette, in order to avoid overcurrent pop during electroporation.
7. We usually use 1-h incubation at 37°C to digest the contaminated DNA in AAV samples purified with CsCl cushion. However, for the AAV samples from crude cell lysate, 25–100 μL of samples were digested in a 200-μL reaction with 10 U of DNase I at 37°C for 4 h to completely remove the DNA contamination.
8. Adding another 1.5 mL of medium for the second washing will ensure collection of most of the bacterial colonies from the plate.
9. The dialysis buffer had to be stirred to avoid aggregation and precipitation of AAV virions.
10. The AAV7 packaging plasmid was included in production and in vivo selection of the secondary AAV library. It served as an internal control to study the biodistribution of AAV variants among mouse tissues.
11. The tissues could be first cut with scissors in the lysis buffer to facilitate homogenization using the mechanic tearor.
12. In a preliminary experiment, we observed rapid decrease of luciferase activities in tissue extracts when they were placed under room temperature. Thus, it was essential to always keep the tissue lysates or supertanants on ice except noted in the luciferase assay to accurately measure the luciferase activity in different mouse tissues.

Acknowledgments

This work was supported by National Institutes of Health grants AR 45967 and AR 50595.

References

1. Xiao, X., Li, J., and Samulski, R. J. (1996) Efficient long-term gene transfer into muscle tissue of immunocompetent mice by adeno-associated virus vector. *J Virol* 70, 8098–8108.
2. Duan, D., Sharma, P., Yang, J., Yue, Y., Dudus, L., Zhang, Y., et al. (1998) Circular intermediates of recombinant adeno-associated virus have defined structural characteristics responsible for long-term episomal persistence in muscle tissue. *J Virol* 72, 8568–8577.
3. Kay, M. A., Manno, C. S., Ragni, M. V., Larson, P. J., Couto, L. B., McClelland, A., et al. (2000). Evidence for gene transfer and expression of factor IX in haemophilia B patients treated with an AAV vector. *Nat Genet* 24, 257–261.
4. Wang, Z., Zhu, T., Qiao, C. P., Zhou, L. Q., Wang, B., Zhang, J., et al. (2005) Adeno-associated virus serotype 8 efficiently delivers genes to muscle and heart. *Nat Biotech* 23, 321–328.
5. Inagaki, K., Fuess, S., Storm, T. A., Gibson, G. A., Mctiernan, C. F., Kay, M. A., et al. (2006) Robust systemic transduction with AAV9 vectors in mice: Efficient global cardiac gene transfer superior to that of AAV8. *Mol Ther* 14, 45–53.
6. Stemmer, W. P. (1994) DNA shuffling by random fragmentation and reassembly: In vitro recombination for molecular evolution. *Proc Natl Acad Sci USA* 91, 10747–10751.
7. Yang, L., Jiang, J., Drouin, L. M., Agbandje-McKenna, M., Chen, C., Qiao, C., Pu, D., Hu, X., Wang, D. Z., Li, J., Xiao, X. (2009) A myocardium tropic adeno-associated virus (AAV) evolved by DNA shuffling and in vivo selection. *Proc Natl Acad Sci USA* 106, 3946–3951.
8. Gao, G., Alvira, M. R., Somanathan, S., Lu, Y., Vandenberghe, L. H., Rux, J. J., Calcedo, R., et al. (2003) Adeno-associated viruses undergo substantial evolution in primates during natural infections. *Proc Natl Acad Sci USA* 100, 6081–6086.
9. Soong, N. W., Nomura, L., Pekrun, K., Reed, M., Sheppard, L., Dawes, G., et al. (2000) Molecular breeding of viruses. *Nat Genet* 25, 436–439.
10. Powell, S. K., Kaloss, M. A., Pinkstaff, A., McKee, R., Burimski, I., Pensiero, M., et al. (2000) Breeding of retroviruses by DNA shuffling for improved stability and processing yields. *Nat Biotech* 18, 1279–1282.
11. Sambrook, J. and Russell, D. W. (2001) Molecular Cloning: A Laboratory Manual. Cold Spring Harbor Laboratory Press. Third Edition New York.
12. Müller, O. J., Kaul, F., Weitzman, M.D., Pasqualini, R., Arap, W., Kleinschmidt, J.A., et al. (2003) Random peptide libraries displayed on adeno-associated virus to select for targeted gene therapy vectors. *Nat Biotech* 21, 1040–1046.
13. Larkin, M. A., Blackshields, G., Brown, N. P., Chenna, R., McGettigan, P. A., McWilliam, H., et al. (2007) Clustal W and Clustal X version 2.0. *Bioinformatics* 23, 2947–2948.

Chapter 9

Systemic Gene Transfer to Skeletal Muscle Using Reengineered AAV Vectors

Jana L. Phillips, Julia Hegge, Jon A. Wolff, R. Jude Samulski, and Aravind Asokan

Abstract

Gene therapy of musculoskeletal disorders warrants efficient gene transfer to a wide range of muscle groups. Reengineered adeno-associated viral (AAV) vectors that selectively transduce muscle tissue following systemic administration are attractive candidates for such applications. Here we provide examples of several lab-derived AAV vectors that display systemic tissue tropism in mice. Methods to evaluate the efficiency of gene transfer to skeletal muscle following intravenous or isolated limb infusion of AAV vectors in mice are discussed in detail.

Key words: AAV, Reengineering, Capsid, Tropism, Muscle, Gene transfer, Isolated limb infusion, Gene therapy

1. Introduction

Musculoskeletal disorders include a broad spectrum of diseases that often share clinical symptoms involving muscle weakness, damage, and degeneration. For instance, the muscular dystrophies (MD) are a group of inherited muscle disorders characterized by progressive muscular degeneration. Diseases including Duchenne, Becker, limb girdle, myotonic, and others are included within this disease group (1, 2). A wide variety of neuromuscular diseases affecting the central nervous system such as cerebral palsy, spinal muscular atrophy, and congenital myasthenia gravis also share similarities with musculoskeletal complications seen in MD (3, 4). Another prominent example is lysosomal storage disease, a heterogeneous group of inherited metabolic diseases resulting from defects in the degradation or transport of several distinct by-products of cellular turnover (5, 6). The various

Dongsheng Duan (ed.), *Muscle Gene Therapy: Methods and Protocols*, Methods in Molecular Biology, vol. 709, DOI 10.1007/978-1-61737-982-6_9, © Springer Science+Business Media, LLC 2011

subtypes are characterized by multisystemic involvement with neurodegenerative features and musculoskeletal complications often seen in most severe forms of the disease.

Gene therapy of musculoskeletal disorders is a promising strategy, wherein corrective gene replacement can abrogate disease symptoms. For instance, promising preclinical studies using truncated forms of the dystrophin gene in mouse models of muscular dystrophy have been reported (7–10). Correction of glycogen storage disorders such as Pompe disease by efficient expression of enzymes such as acid alpha-glucosidase in skeletal muscle has also been shown (11, 12). However, a significant hurdle towards achieving efficient gene transfer in skeletal muscle is the paucity of vectors that can selectively transduce a broad spectrum of muscle groups following systemic administration.

We recently generated a panel of synthetic adeno-associated viral (AAV) vectors by reengineering a previously identified heparan sulfate receptor footprint on AAV serotype 2 (13). This approach yielded several chimeric AAV capsids displaying systemic muscle tropism following intravenous administration in mice (Table 1). In particular, an AAV2/AAV8 chimeric (AAV2i8) with a unique capsid surface footprint readily traversed the blood

Table 1
Summary of tissue tropisms and relative transduction efficiencies of reengineered AAV vectors[a]

AAV mutant(s)	Amino acid sequence	Transduction efficiency (muscle)	Transduction efficiency (liver)	Tropism
AAV2	585-RGNRQA-590	+	+++	Liver
AAV8	585-QQNTAP-590	++++	++++	Systemic
2i1; 2i6	585-SSSTDP-590	–	–	N/A
2i3a/3b	585-SSNTAP-590	–	–	N/A
2i4	585-SNSNLP-590	–	–	N/A
2i5	585-SSTTAP-590	–	–	N/A
2i9	585-SAQAQA-590	–	–	N/A
2i8; 2irh43; 2irh49-53,57,58,64	585-QQNTAP-590	++++	+	Systemic muscle
2i10; 2irh40	585-QANTGP-590	++	+	Systemic muscle
2irh38	585-QTNTGP-590	++	+	Systemic muscle
2irh2	585-QTNGAP-590	++	+	Systemic muscle
2i11; 2i12; 2irh32-34	585-NATTAP-590	++	+	Systemic muscle

[a]Transduction profiles were derived from live animal imaging, luciferase expression, and vector genome biodistribution data; (+) one relative log unit

vasculature and selectively transduced cardiac and whole body skeletal muscle tissues with high efficiency. Unlike other AAV serotypes, which are preferentially sequestered by the liver, AAV2i8 displays reduced hepatic tropism. The unique pharmacokinetic and pharmacodynamic features of AAV2i8 might be particularly relevant for systemic gene delivery in the treatment of musculoskeletal disorders. The current report describes methods for evaluating the systemic muscle tropism of AAV vectors following intravenous or isolated limb infusion in mice.

2. Materials

2.1. Generating Reengineered AAV2 Vectors

2.1.1. Mutagenesis

1. Plasmid containing AAV2 capsid gene coding for VP1, VP2, and VP3 proteins (see Note 1).
2. Primers for site-directed mutagenesis (Table 2) (IDT, Coralville, IA, USA).
3. Quikchange II site-directed mutagenesis kit® (Stratagene, La Jolla, CA, USA).
4. Molecular biology/PCR grade water.
5. Biometra Thermal Gradient Cycler (or other equivalent).
6. Electroporation-competent *E. coli* DH10B (Stratagene) (see Note 2).
7. Ampicillin LB agar plates for selection.
8. Plasmid DNA Miniprep and Maxiprep kits (Qiagen, Valencia, CA, USA).

2.1.2. AAV Vector Production

Several detailed protocols for the production of recombinant AAV vectors packaging different transgene cassettes have been reported (14, 15). Specific reagents for generation of reengineered AAV2 vectors relevant to the current report include:

1. Plasmid containing reengineered AAV2 capsid genes.
2. Plasmid containing AAV inverted terminal repeats (ITRs) flanking the chicken beta-actin (CBA) promoter-driven firefly luciferase (Luc) cassette.

2.2. Evaluating Systemic Gene Transfer to Skeletal Muscle Using AAV Vectors

2.2.1. Intravenous (Tail Vein) Administration of AAV Vectors

1. Balb/c male or female mice (see Note 3).
2. 28–30G insulin syringes (BD, Franklin Lakes, NJ, USA).
3. AAV Vector suspended in 1× phosphate-buffered saline (PBS).
4. 70% isopropyl alcohol to cleanse tail vein injection site.
5. Heat lamp or warm water (~48°C) to increase circulation and dilate the tail vein.
6. Plastic mouse restrainer for tail vein injection (Braintree Scientific, MA, USA).

Table 2
Nucleotide sequences of primers utilized in site-directed mutagenesis for reengineering AAV2 capsids

Name	Primer sequence
2i1; 2i6	5′-ctaccaacctccagagtagcagcacagatccagctaccgcagatg-3′
2i3b	5′-ctaccaacctccagagtagcaacacagcaccagctaccgcagatg-3′
2i4	5′-ctaccaacctccagagtaacagcaacctaccagctaccgcagatg-3′
2i5	5′-ctaccaacctccagagtagcaccacagcaccagctaccgcagatg-3′
2i7	5′-ctaccaacctccaggcagccaacacagcagcagctaccgcagatg-3′
2i8; 2irh43, 2irh49-53, 57,58, 64	5-′ctaccaacctccagcaacagaacacagcaccagctaccgcagatg-3′
2i9	5′-ctaccaacctccagagtgcccaggcacaagcagctaccgcagatg-3′
2i10, rh40	5′-tctaccaacctccagcaggcaaacacgggtcctgctaccgcagat-3′
2i11, 12, rh32-34	5′-tctaccaacctccagaacgccaccactgcccccgctaccgcagat-3′
2irh2	5′-tctaccaacctccagcagacaaacggggctcctgctaccgcagat-3′
2irh38	5′-tctaccaacctccagcagacaaacacgggtcctgctaccgcagat-3′

2.2.2. Isolated Limb Infusion of AAV Vectors

1. Balb/c male or female mice (see Note 4).
2. AAV suspended in 1× PBS.
3. Polyethylene PE-10 tubing (BD).
4. 30G needles (Thermofisher, Waltham, MA, USA).
5. High-pressure syringe pump (Harvard Apparatus).
6. Isofluorane.
7. Heat pad (Braintree Scientific).
8. Mineral oil or other eye lubricant.
9. Betadine and 70% isopropyl alcohol.
10. Sterile scissors and forceps (Braintree Scientific).
11. Sterile cotton swabs.
12. 5-0 silk suture (Braintree Scientific).
13. Small rubber-bands.
14. Clippers with #40 blade (Braintree Scientific).

2.2.3. Live Animal Bioluminescence Imaging (See Note 5)

1. Balb/c male or female mice from Subheadings 2.2.1 or 2.2.2.
2. Xenogen Lumina® live animal bioluminescence imaging system and software (Caliper Life Sciences, CA, USA).
3. Isofluorane.
4. Luciferin substrate (Nanolight, Pinetop, AZ, USA or Caliper Lifesciences, CA, USA).
5. 28G insulin syringes (BD).

2.2.4. Harvesting Different Muscle Groups

1. Balb/c male or female mice from above sections.
2. Dissecting scissors and forceps.
3. 25G needles (1/2 in.).
4. 20–60 mL luer-lock syringes.
5. 1× PBS.
6. 3.8% sodium citrate or 0.5 mM EDTA (anticoagulant).
7. 2 mL microcentrifuge tube.
8. 2.5% Avertin or anesthetic of choice (Sigma, St. Louis, MO, USA).

2.2.5. Tissue Lysate Luciferase Expression Assay

1. TissueLyser II (Qiagen) (see Note 6).
2. 96-Well flat bottom, white polystyrene assay plates.
3. 96-Well flat bottom, clear polystyrene assay plates.
4. Assay-grade Luciferin substrate (Promega, Madison, WI, USA).
5. 2× Passive Lysis Buffer (Promega).
6. Victor 2® Luminometer (Perkin Elmer, Waltham, MA, USA).
7. Bradford protein assay reagent (Bio-Rad, Hercules, CA, USA).
8. Eppendorf 5424 microcentrifuge.

3. Methods

3.1. Generating Reengineered AAV2 Vectors

3.1.1. Mutagenesis (See Note 7)

1. Setup reactions for site-directed mutagenesis as outlined in the Instruction Manual for the Quikchange II site-directed mutagenesis kit.
2. Setup thermal cycling parameters as follows: denaturation at 95°C for 30 s; annealing at 55°C for 1 min; extension at 68°C for 7 min for a total of 25–30 cycles.
3. Upon completion of thermal cycling, cool the reaction at 4°C followed by treatment with *DpnI* to digest plasmid template.
4. Transform using electroporation-competent DH10B followed by plating of transformed cells onto LB Amp agar plates and incubate overnight at 37°C.
5. Subject individual clones amplified overnight to plasmid DNA isolation using a miniprep kit and submit for DNA sequencing to validate modified capsid gene sequence.

3.1.2. AAV Vector Production

Several detailed protocols for the production of recombinant AAV vectors packaging different transgene cassettes have been reported. No modifications to published methods are required for the generation of reengineered AAV2 vectors (see Note 8).

3.2. Evaluating Systemic Gene Transfer to Skeletal Muscle Using AAV Vectors

3.2.1. Intravenous (Tail Vein) Administration of AAV Vectors

1. Allow mice to sit under heat lamp for 1–3 min, or submerge tail in warm water for 30 s to 1 min to allow the tail vein to dilate.
2. Restrain the mouse with a plastic restrainer and wipe tail with 70% ethanol.
3. Insert needle almost parallel to the left or right lateral tail vein.
4. Ensure that the bevel of the needle is facing upward and inject the vector dose slowly.
5. Injection quality is considered good if the vein clears. However, if injection fluid accumulates subcutaneously, remove needle and reinject at a proximal site on the tail vein.

3.2.2. Isolated Limb Infusion of AAV Vectors

1. Remove hub from one of the 30G needles and insert into the end of the PE-10 tubing. This constitutes the injection end. Use another 30G needle and insert to the opposite end (leaving hub intact for syringe attachment). The length of tubing can vary based on space between the test animal and syringe pump setup.
2. Load the syringe with 200, 500 μL, or 1 mL of vector dose ranging from 10^9 to 10^{11} particles per animal.
3. Anesthetize mice using 2.5% isofluorane delivered through a nose-cone apparatus. Alternatively, intraperitoneal injection of low-dose Avertin or preferred anesthetic can be used.
4. Once mice are completely anesthetized, apply rubber band around the base of the hind limb (Fig. 1) to occlude blood flow.
5. Tape foot down to surgical board to secure and shave hair around incision site using clippers.
6. Sterilize with three rotations of Betadine and 70% isopropyl alcohol using sterile swabs.
7. Make a 1 cm incision to expose the medial saphenous vein (Fig. 1).
8. Insert needle until bevel is inside of the vein.
9. Infuse at 1 mL/min.
10. After injection is complete, leave needle in for another minute until pressure has decreased and entire dose has been delivered (see Note 9).
11. Suture skin together and apply antibiotic ointment.
12. Allow mice to recover on heat pad until awake and moving around.

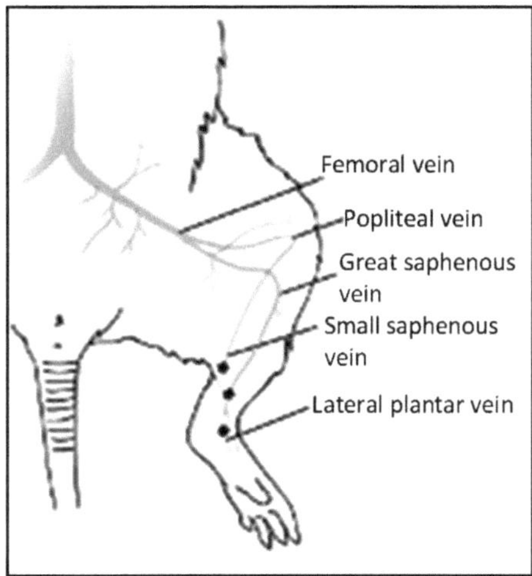

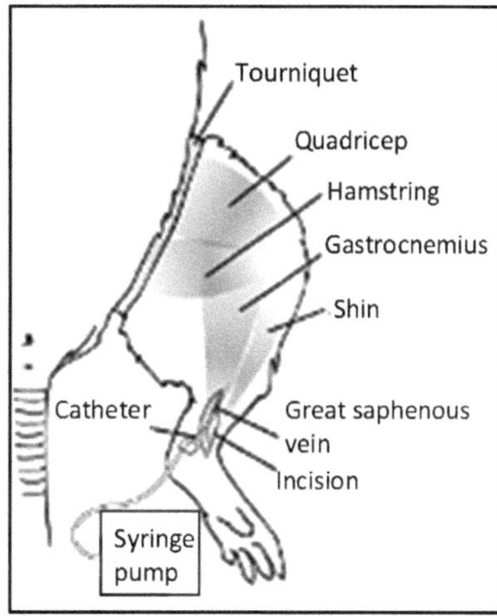

Fig. 1. Schematic of isolated limb infusion procedure for evaluation of systemic gene transfer to skeletal muscle using AAV vectors. (*Left*) Cartoon shows the venous structure of the hind limb. (*Right*) The great saphenous vein is utilized for vector administration. The catheter tip is inserted into the distal portion of the great saphenous vein. As the syringe pump injects the virus, the entire saphenous vein will clear, including all branching veins that feed from different hind limb muscle groups. The tourniquet will prevent any leakage from the saphenous vein into the vena cava, which will in turn preclude systemic organ transduction. This schematic was previously published in ref (20) and has been reproduced in accordance with the rights and permissions policy of Nature Publishing Group.

3.2.3. Live Animal Bioluminescence Imaging (See Note 5)

1. Anesthetize mice using 2.5% isofluorane using Xenogen anesthesia setup.
2. Once completely anesthetized, inject 120 μL of a 25 mg/mL luciferin substrate per 20 g body weight through intraperitoneal injection and wait 5 min.
3. Place animals in individual nose-cones within imaging chamber and reduce isofluorane to 1.5%.
4. After 5 min, acquire image and obtain overlaid bioluminescent and grayscale mouse images at 5 s, 1, and 5 min exposure time.
5. Allow mice to recover on a heat pad until fully awake.

3.2.4. Harvesting Different Muscle Groups

1. Anesthetize mice with 2.5% Avertin (or anesthetic of choice).
2. Once mice are completely anesthetized, make a midline incision from the bottom (between the hind limbs) all the way up to the second rib.
3. Separate muscle and skin with needles or retractors, and cut diaphragm to expose the heart.
4. Once heart is exposed, insert 25G needle (attached to a 20–60 mL leur-lock syringe containing 20 mL 1× PBS with 100 μL of 3.8% sodium citrate or 0.5 mM EDTA) into the left ventricle. Begin perfusion followed by incision of right atrium of the heart to allow outflow of perfusate.
5. Upon completion of perfusion, euthanize the mouse via cervical dislocation and remove specific muscle groups (five sets

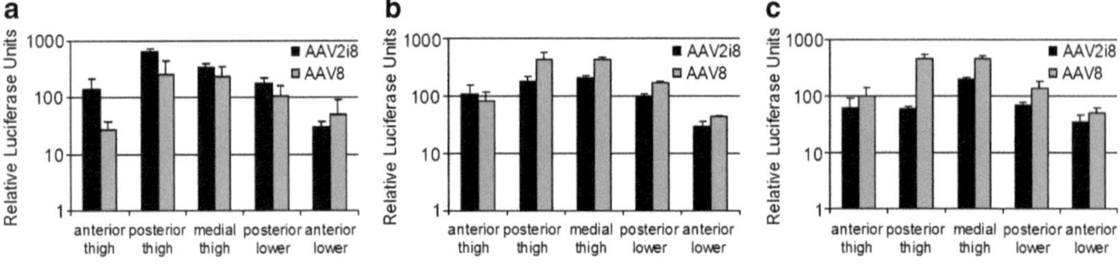

Fig. 2. Relative luciferase expression levels in different skeletal muscle groups following isolated limb infusion of AAV2i8 (*black bars*) and AAV8 (*gray bars*) vectors in mice. Different volumes of injection, 200 μL (**a**, low), 500 μL (**b**, medium), and 1 mL (**c**, high), were administered into the saphenous vein as described in the main text. All studies were carried out in triplicate. *Error bars* indicate one standard deviation. Note that AAV2i8 vectors display increased gene transfer efficiency at low volume of injection, while AAV8 vectors transduce certain skeletal muscle groups with higher efficiency at high injection volume.

of fore and hind limb skeletal muscle, diaphragm, abdominal, intercostal muscles, and other organs, if warranted) (Fig. 2). Be sure to rinse with 1× PBS before storage in 2 mL microcentrifuge tubes.

6. Tissues may be prepared for luciferase assay immediately, or frozen at –80°C for up to 6 months.

3.2.5. Luciferase Expression Assay with Tissue Lysates

1. For use with TissueLyser II – add sterile, stainless steel beads to 2 mL microcentrifuge tubes and run at 10 Hz for 2 intervals of 10 s each (see Note 6).
2. Take approximately 50 μL (~15–25 mg of tissue homogenate) and place in a separate tube for viral DNA isolation, if required.
3. Add 150 μL of 2× passive lysis buffer to new sample and mix by pipetting several times.
4. Allow sample to sit on ice for 5 min, then mix by pipetting.
5. Centrifuge samples at 2,350 × *g* for 3 min.
6. Take 50 μL of supernatant and place in the 96-well, flat bottom, white polystyrene plates. Tissue lysates from untreated mice are used as background controls (see Note 10).
7. Add 100 μL of assay-grade luciferin substrate to each sample well and obtain readout using the Victor 2® luminometer within 10 min of substrate addition.
8. To determine protein concentration of supernatant, add 1 μg to a 96 well, flat bottom, clear polystyrene assay plate. Add 200 μL of Bio-rad protein assay reagent (diluted 1:5 using sterile dH_20) to each well, mix by pipetting, and incubate at room temperature for 5 min. Obtain protein concentrations using a standard curve generated using different concentrations of BSA (Sigma, St. Louis, MO) and normalize tissue lysate luciferase expression levels to relative light units (RLU)/μg of cellular protein (Fig. 2).

4. Notes

1. Different plasmid reagents have been established for production of naturally occurring AAV isolates as well as lab-derived strains. For this specific application, the pXR2 plasmid backbone or pACG backbones described earlier are appropriate templates (16, 17). Alternative templates containing capsid gene encoding for VP1, VP2, VP3 proteins of AAV serotype 2 can also be utilized for site-directed mutagenesis. It is important to note that primer sequences might vary accordingly.
2. Stratagene recommends utilization of Electroten® blue electroporation-competent *E. coli* for transformation following site-directed mutagenesis. We have successfully obtained mutant clones at a high frequency (50% or more) using the DH10B strain. Chemically competent *E. coli* may also be utilized, although the transformation efficiency might be lower.
3. Previous studies have demonstrated gender differences in transduction efficiency of different AAV vectors in mice (18, 19). In particular, certain AAV serotypes tend to show higher transduction levels in the male liver following intravenous administration. Other tissue types do not appear to demonstrate marked differences in transduction efficiency.
4. Gender differences may be seen as noted earlier. In addition, for isolated limb infusion studies, any mouse strain such as C57/Bl6 or other relevant strain may be used. Balb/c mice (white coat) are typically utilized for applications involving live animal imaging.
5. Although the Xenogen live animal imaging system allows monitoring kinetics and dynamics of gene expression, quantitative analysis of luciferase expression in harvested tissue/organs is a well-established alternative to this approach. Accordingly, we have included procedures for quantitation of bioluminescence in tissue lysates as an alternative for live animal imaging.
6. The Qiagen TissueLyserII® is ideally suited for processing tissue and rapid generation of tissue lysates. Alternatively, tissues can be minced/ground using dissecting scissors or homogenized using a small mortar and pestle, respectively.
7. We have optimized PCR conditions for carrying out mutagenesis of different AAV capsid genes. However, reaction conditions for site-directed mutagenesis are also outlined by the manufacturer (Stratagene Quikchange® kit). Primer design and reaction conditions might require further optimization for different genes.

8. It should be noted that reengineered AAV2 vectors, relevant to the current study, do not bind heparan sulfate (13). Therefore, heparin affinity column chromatography is not suitable for purification of aforementioned AAV vectors. Standard purification procedures involving cesium chloride ultracentrifugation have been described (14, 15).
9. For high-pressure isolated limb infusions, it is important to leave the catheter tip in vein for another minute after injection is complete to minimize leakage/back-flow. For high volume injections, fluid leakage from the injection site is likely. Such can be minimized by optimizing injection conditions and intravenous residence time of the catheter for differently sized mice (20).
10. For luciferase assays, presence of minor amounts of tissue debris in the supernatant will not interfere with the assay. It is critical to skip rows and columns (i.e., place samples in alternating wells) to avoid overlap in signal between wells. This step is critical to avoid false positives.

References

1. Guglieri, M., Straub, V., Bushby, K., Lochmüller, H. (2008) Limb-girdle muscular dystrophies. *Curr Opin Neurol* 21, 576–584.
2. Deconinck, N., Dan, B. (2007) Pathophysiology of duchenne muscular dystrophy: current hypotheses. *Pediatr Neurol* 36, 1–7.
3. Farrar, M.A., Johnston, H.M., Grattan-Smith, P., Turner, A., Kiernan, M.C. (2009) Spinal muscular atrophy: molecular mechanisms. *Curr Mol Med* 9, 851–862.
4. Meriggioli, M.N., Sanders, D.B. (2009) Autoimmune myasthenia gravis: emerging clinical and biological heterogeneity. *Lancet Neurol* 8, 475–490.
5. Aldenhoven, M., Sakkers, R.J., Boelens, J., de Koning, T.J., Wulffraat, N.M. (2009) Musculoskeletal manifestations of lysosomal storage disorders. *Ann Rheum Dis* 68, 1659–1665.
6. Mah, C., Cresawn, K.O., Fraites, T.J. Jr, Pacak, C.A., Lewis, M.A., Zolotukhin, I., Byrne, B.J. (2005) Sustained correction of glycogen storage disease type II using adeno-associated virus serotype 1 vectors. *Gene Ther* 12, 1405–1409.
7. Duan, D. (2008) Myodys, a full-length dystrophin plasmid vector for Duchenne and Becker muscular dystrophy gene therapy. *Curr Opin Mol Ther* 10, 86–94.
8. Muir, L.A., Chamberlain, J.S. (2009) Emerging strategies for cell and gene therapy of the muscular dystrophies. *Expert Rev Mol Med* 25, e18.
9. Zhu, T., Zhou, L., Mori, S., Wang, Z., McTiernan, C.F., Qiao, C., Chen,C., Wang, D.W., Li, J., Xiao, X. (2005) Sustained whole-body functional rescue in congestive heart failure and muscular dystrophy hamsters by systemic gene transfer. *Circulation* 112, 2650–2659.
10. Tang, Y., Cummins, J., Huard, J., Wang, B. (2010) AAV-directed muscular dystrophy gene therapy. *Expert Opin Biol Ther* 10, 395–408.
11. Pastores, G.M. (2008) Musculoskeletal complications encountered in the lysosomal storage disorders. *Best Pract Res Clin Rheumatol* 22, 937–947.
12. Mah, C., Pacak, C.A., Cresawn, K.O., Deruisseau, L.R., Germain, S., Lewis, M.A., Cloutier, D.A., Fuller, D.D., Byrne, B.J. (2007) Physiological correction of Pompe disease by systemic delivery of adeno-associated virus serotypes 1 vectors. *Mol Ther* 15, 501–507.
13. Asokan, A., Conway, J.C., Phillips, J.L., Li, C., Hegge, J., Sinnott, R., Yadav, S., DiPrimio, N., Nam, H.J., Agbandje-McKenna, M., McPhee, S., Wolff. J., Samulski, R.J. (2010) Reengineering a receptor footprint of adeno-associated virus enables selective and systemic

gene transfer to muscle. *Nat Biotechnol* 28, 79–82.

14. Choi, V.W., Asokan, A., Haberman, R.A., Samulski, R.J. (2007) Production of recombinant adeno-associated viral vectors for in vitro and in vivo use. *Curr Protoc Mol Biol* Chapter 16 Unit 16.25.
15. Grieger, J.C., Choi, V.W., Samulski, R.J. (2006) Production and characterization of adeno-associated viral vectors. *Nat Protoc* 1, 1412–1428.
16. Zolotukhin, S., Potter, M., Zolotukhin, I., Sakai, Y., Loiler, S., Fraites, T.J. Jr, Chiodo, V.A., Phillipsberg, T., Muzyczka, N., Hauswirth, W.W., Flotte, T.R., Byrne, B.J., Snyder, R.O. (2002) Production and purification of serotype 1, 2, and 5 recombinant adeno-associated viral vectors. *Methods* 28, 158–167.
17. Rabinowitz, J.E., Rolling, F., Li, C., Conrath, H., Xiao,W., Xiao, X., Samulski, R.J. (2002) Cross-packaging of a single adeno-associated virus (AAV) type 2 vector genome into multiple AAV serotypes enables transduction with broad specificity. *J Virol* 76, 791–801.
18. Pañeda, A., Vanrell, L., Mauleon, I., Crettaz, J.S., Berraondo, P., Timmermans, E.J., Beattie, S.G,, Twisk, J., van Deventer, S., Prieto, J., Fontanellas, A., Rodriguez-Pena, M.S., Gonzalez-Aseguinolaza, G. (2009) Effect of adeno-associated virus serotype and genomic structure on liver transduction and biodistribution in mice of both genders. *Hum Gene Ther* 20, 908–917.
19. Davidoff, A.M., Ng, C.Y., Zhou, J., Spence, Y., Nathwani, A.C. (2003) Sex significantly influences transduction of murine liver by recombinant adeno-associated viral vectors through an androgen-dependent pathway. *Blood* 102, 480–488.
20. Hagstrom, J.E., Hegge, J., Zhang, G., Noble, M., Budker, V., Lewis, D.L., Herweijer, H., Wolff, J.A. (2004) A facile nonviral method for delivering genes and siRNAs to skeletal muscle of mammalian limbs. *Mol Ther* 10, 386–398.

Chapter 10

Bioinformatic and Functional Optimization of Antisense Phosphorodiamidate Morpholino Oligomers (PMOs) for Therapeutic Modulation of RNA Splicing in Muscle

Linda J. Popplewell, Ian R. Graham, Alberto Malerba, and George Dickson

Abstract

Duchenne muscular dystrophy (DMD) is caused by mutations that disrupt the reading frame of the human *DMD* gene. Selective removal of exons flanking an out-of-frame DMD mutation can result in an in-frame mRNA transcript that may be translated into an internally deleted, Becker muscular dystrophy (BMD)-like, but functionally active dystrophin protein with therapeutic activity. Antisense oligonucleotides (AOs) can be designed to bind to complementary sequences in the targeted mRNA and modify pre-mRNA splicing to correct the reading frame of a mutated transcript so that gene expression is restored. AO-induced exon skipping producing functional truncated dystrophin exon has been demonstrated in animal models of DMD both in vitro and in vivo, and in DMD patient cells in vitro in culture, and in DMD muscle explants. More recently, AO-mediated exon skipping has been confirmed in DMD patients in Phase I clinical trials. However, it should be noted that personalized molecular medicine may be necessary, since the various reading frame-disrupting mutations are spread across the *DMD* gene. The different deletions that cause DMD would require skipping of different exons, which would require the optimization and clinical trial workup of many specific AOs. This chapter describes the methodologies available for the optimization of AOs, and in particular phosphorodiamidate morpholino oligomers (PMOs), for the targeted skipping of specific exons on the *DMD* gene.

Key words: Optimization, Exon skipping, DMD, Antisense oligonucleotides, Bioinformatics, Functional analysis

1. Introduction

Duchenne muscular dystrophy (DMD) is a severe muscle-wasting disease, affecting 1:3,500 live male births, caused by the lack of functional dystrophin protein in skeletal muscles, as a result of

Dongsheng Duan (ed.), *Muscle Gene Therapy: Methods and Protocols*, Methods in Molecular Biology, vol. 709, DOI 10.1007/978-1-61737-982-6_10, © Springer Science+Business Media, LLC 2011

frame-disrupting deletions or duplications or, less commonly, nonsense or missense mutations in the *DMD* gene (1). Mutations that maintain the reading frame of the gene and allow expression of semi-functional, but internally deleted dystrophin are generally associated with the less severe Becker muscular dystrophy BMD) (1, 2).

Transforming an out-of-frame DMD mutation into its in-frame BMD counterpart with antisense oligonucleotides (AOs) is the basis of the potentially exciting exon skipping therapy for DMD (3). The hybridization of AOs to specific RNA sequence motifs prevents assembly of the spliceosome, so that it is unable to recognize the target exons) in the pre-mRNA and include them in the mature gene transcript (4, 5). AOs have been used to induce skipping of specific exons such that the reading frame is restored and truncated dystrophin expressed in vitro in DMD patient cells (6–9) and in animal models of the disease in vivo (10–14).

Initial proof-of-principle clinical trials, using two different AO chemistries (phosphorothioate-linked 2′-*O*-methyl modified bases (2′OMePS) (15) and phosphorodiamidate morpholino oligomer PMO) (16)) for the targeted skipping of exon 51 of the *DMD* gene after intramuscular injection, have been performed recently with encouraging results. While both chemistries have excellent safety profiles (17, 18), PMOs appear to produce more consistent and sustained exon skipping in the *mdx* mouse model of DMD (19–21), in human muscle explants (22), and dystrophic canine muscle cells in vitro (23). However, for some human exons, 2′OMePS and PMO AONs performed equally well (18). Since the mutations that cause DMD are so diverse, skipping of exon 51 would have the potential to treat only 13% of DMD patients with genomic deletions on the Leiden DMD database (24). The continued development and analysis of AOs for the targeting of other *DMD* exons is therefore vital. There have been a number of studies published describing screens of large numbers of AOs of various chemistries for the targeted skipping of certain exons of the *DMD* gene (25–28). These studies used a number of tools in the design of AOs. In general terms, bioactive AOs target certain serine/arginine (SR)-rich protein-binding motifs, bind to their target more strongly, either as a result of being longer or by being able to access their target site more easily, have their target sites nearer to the splice acceptor site (3′ splice site), and have their target sites in areas of open conformation. However, it should be noted that designing an AO to have all of these properties will not necessarily guarantee bioactivity. The empirical analysis of designed AOs is still essential.

In this protocol, we outline the bioinformatic tools available to aid the design of PMOs for the targeted skipping of exons on the human *DMD* gene, and the functional analysis used to assess bioactivity in normal human skeletal muscle cells.

2. Materials

2.1. Design Process

1. The *DMD* gene sequence and access to the following bioinformatics tools including ESEfinder (http://rulai.cshl.edu/tools/ESE/), RescueESE (http://genes.mit.edu/burgelab/rescue-ese/), PESX (http://cubweb.biology.columbia.edu/pesx/) and splicefinder (http://www.splicefinder.net/indexm.php), mfold (http://mfold.bioinfo.rpi.edu/), and binding energy programme (http://rna.urmc.rochester.edu/rnastructure.html).
2. Source of genomic DNA from the species of interest. This will ideally be in the form of a YAC or BAC clone, but total genomic DNA may be used.
3. Gentra Puregene Cell kit (Qiagen Ltd., Crawley, West Sussex, UK).
4. PCR primers designed to amplify genomic DNA across the exon of interest, together with ~500 nucleotide (nt) of upstream and downstream introns. The "forward" (+strand) primer used for amplification should incorporate a minimal T7 RNA polymerase promoter sequence (5′-TAATACGACTCACTATAGG-3′) so that PCR products can be used as templates for production of pre-mRNA by in vitro transcription.
5. 2× PCR Master Mix with cresol red (GeneSys Ltd., Camberley, Surrey, UK).
6. A PCR machine.
7. Agarose gel electrophoresis equipment.
8. QIAquick gel extraction Kit (Qiagen Ltd.).
9. The mMESSAGE mMACHINE kit for T7 in vitro transcription (Ambion, Austin, TX, USA).
10. TURBO DNase (Ambion).
11. MEGAclear kit (Ambion).
12. A hexamer hybridization array screen (Nyrion, Edinburgh, Midlothian, UK).
13. Hybridization reaction buffer: 50 mM Tris–HCl, 5 mM $MgCl_2$, 50 mM KCl, 5 mM DTT (pH 8.5).
14. ExpandRT RNA-dependent DNA polymerase (Roche Diagnostics, Burgess Hill, West Sussex, UK).
15. Prepare hybridization wash buffer containing 100 mM NaCl, 0.1% SDS.
16. Access to a primer design software, such as VectorNTI (Invitrogen Ltd., Paisley, Renfrewshire, UK).

2.2. Functional Analysis

1. 1 mM PMO (Gene Tools, Philomath, OR, USA). Store in aliquots at –20°C.
2. 200 μM complementary leash. Store in aliquots at –20°C.
3. RNase-, DNase-free sterile H_2O.
4. Sterile 10× PBS.
5. A PCR machine.
6. Horizontal gel electrophoresis equipment.
7. UltraPure™ agarose (Invitrogen Ltd.).
8. 5× TBE buffer (SigmaAldrich, St. Louis, MO, USA). Store at room temperature (RT).
9. 10 mg/mL ethidium bromide (SigmaAldrich). Store at RT. Please note ethidium bromide is a mutagen/carcinogen.
10. 5× DNA loading buffer (Bioline UK Ltd., London, UK). Store at 4°C.
11. DNA Hyperladder V (Bioline UK Ltd.). Store at 4°C.
12. Basic cell-culturing equipment such as a 37°C, 5% CO_2 incubator, a Class 2 microbiological safety cabinet, a 37°C water bath, a microscope, and a haemocytometer.
13. Normal human skeletal muscle primary cells (TCS Cellworks, Buckingham, Bucks, UK). 500,000 cells/vial. Store in vapour phase of a liquid nitrogen storage facility.
14. Skeletal muscle cell growth medium (PromoCell GmbH, Heidelberg, Germany). Store at 4°C.
15. Skeletal muscle cell growth medium supplement mix (PromoCell GmbH). Store at –20°C. Final concentrations of growth factors in the complete medium are 5% foetal calf serum, 50 μg/mL fetuin, 1 ng/mL basic fibroblast factor, 10 ng/mL epidermal growth factor, 10 μg/mL insulin, 50 ng/mL amphotericin B, 50 μg/mL gentamicin, and 0.4 μg/mL dexamethasone.
16. Skeletal muscle cell differentiation medium (PromoCell GmbH). Store at 4°C.
17. Skeletal muscle cell differentiation medium supplement mix (PromoCell GmbH). Store at –20°C. The complete medium contains 10 μg/mL insulin, 50 ng/mL amphotericin B, and 50 μg/mL gentamicin (final concentration).
18. Dulbecco's modified Eagle's medium (SigmaAldrich). Store at 4°C.
19. Sterile-filtered, cell cultured tested 200 mM L-glutamine (SigmaAldrich). Store frozen at –20°C in 5.5 mL aliquots.
20. Foetal bovine serum, certified heat-inactivated (Invitrogen Ltd.). Store at –20°C in 50 mL aliquots.

21. ECM gel from Holm and Swarm murine sarcoma (SigmaAldrich). Defrost overnight at 4°C, and then aliquot into 1 mL aliquots and store at −20°C.
22. 0.25% Trypsin/EDTA (SigmaAldrich). Store frozen at −20°C in 5 mL aliquots.
23. Mr Frosty cryo-freezing container (Fisher Scientific, Loughborough, Leics., UK) or similar, which, when filled with isopropanol, allows a cooling rate of 1°C min^{-1} in a −80°C freezer.
24. Lipofectin transfection reagent (Invitrogen Ltd.). Store at 4°C.
25. QIAshredder kit (Qiagen Ltd.). Store at RT.
26. RNeasy Mini Kit (Qiagen Ltd.). Store at RT.
27. GeneScript RT-PCR system kit (GeneSys Ltd., Camberley, Surrey, UK).
28. 2× PCR Master Mix with cresol red (GeneSys Ltd.).
29. Nested PCR primers designed over exon–exon junctions to ensure not amplifying genomic DNA.
30. Total RNA harvested from cultured cells (see above).
31. DNA Hyperladder IV (Bioline UK Ltd., London, UK). Store at 4°C.
32. GeneTools software (Syngene, Cambridge, Cambridgeshire, UK).
33. QIAquick gel extraction Kit (Qiagen Ltd.).
34. Electrophoretically separated skipped and full-length products on ethidium bromide agarose gels.
35. Access to Nanodrop to measure RNA concentration spectrophotometrically.
36. Access to a sequencing facility to confirm correct skip. We use a commercial company Eurofins MWG Operon (Ebersberg, Germany).

3. Methods

3.1. Design Process

There have been a number of studies describing the screening of large arrays of AOs for the targeted skipping of certain *DMD* exons (25–28). These studies generally used a number of standard tools to aid the design of AOs. When optimizing a PMO for the targeted skipping of one particular *DMD* exon, it is important to use all the information available from the exon sequence itself through interrogation of the sequence with various softwares. This, together with previously published data, should provide a good starting point in the design of overlapping arrays of PMOs at 3–5 nt intervals over the regions of interest.

1. Exonic splicing enhancer/suppressor sequence prediction. There are a number of software programmes available for the prediction of putative SR protein-binding motifs (SF2/ASF, SC35, SRp40, SRp55, Tra2β, and 9G8). These include ESEfinder, RescueESE, PESX, and splicefinder. Obtain graphical representation of putative exonic splicing enhancer (ESE) sequences (i.e. SR protein-binding domains) and exonic splicing suppressor sequences within the exon of interest by entering the *DMD* exon sequence of interest. According to the output, design overlapping arrays of PMOs over the exon sequences predicted to be involved in recognition by ESE elements, or have an indication of being a strong SR protein-binding domain. Figure 1 shows the output for exon 46 of the human *DMD* gene. It depicts the target sites of PMOs we designed according to the method mentioned above. In theory, blocking of ESE sequences with PMOs should lead to non-recognition of the exon as exon and result in the targeted exon being spliced out with neighbouring introns (see Notes 1 and 2).
2. Predict the secondary structure of pre-mRNA. Published studies have suggested that AOs whose target sites overlap areas of open conformation or have their ends in open loop structures may be more bioactive. Pre-mRNA secondary structure can be predicted using the mfold RNA folding software by inputting the exon sequence of interest together with 50 nt of upstream and downstream introns, and with a maximum base-pairing distance of 100 nt. Figure 2 depicts the predicted pre-mRNA secondary structure for exon 46 and the target sites of a number of PMOs designed to induce exon 46 skipping.
3. Predict PMO-target binding energy. The thermodynamic strength of binding of 2′-OMe AOs (27) and PMOs (28) to their target sites has been reported to show strong significance to bioactivity. The stronger an AO binds to its target, the more bioactive the AO is. The theoretical strength of binding of designed PMOs can be calculated using the software http://rna.urmc.rochester.edu/rnastructure.html. The information required to calculate the theoretical binding energy is: (1) free energy of the exon itself (self-folding), (2) free energy of the PMO (self-folding), (3) bimolecular PMO–PMO binding energy, and (4) bimolecular PMO–exon binding energy. Then the theoretical binding energy of PMO to exonic target can be calculated using the following formula:

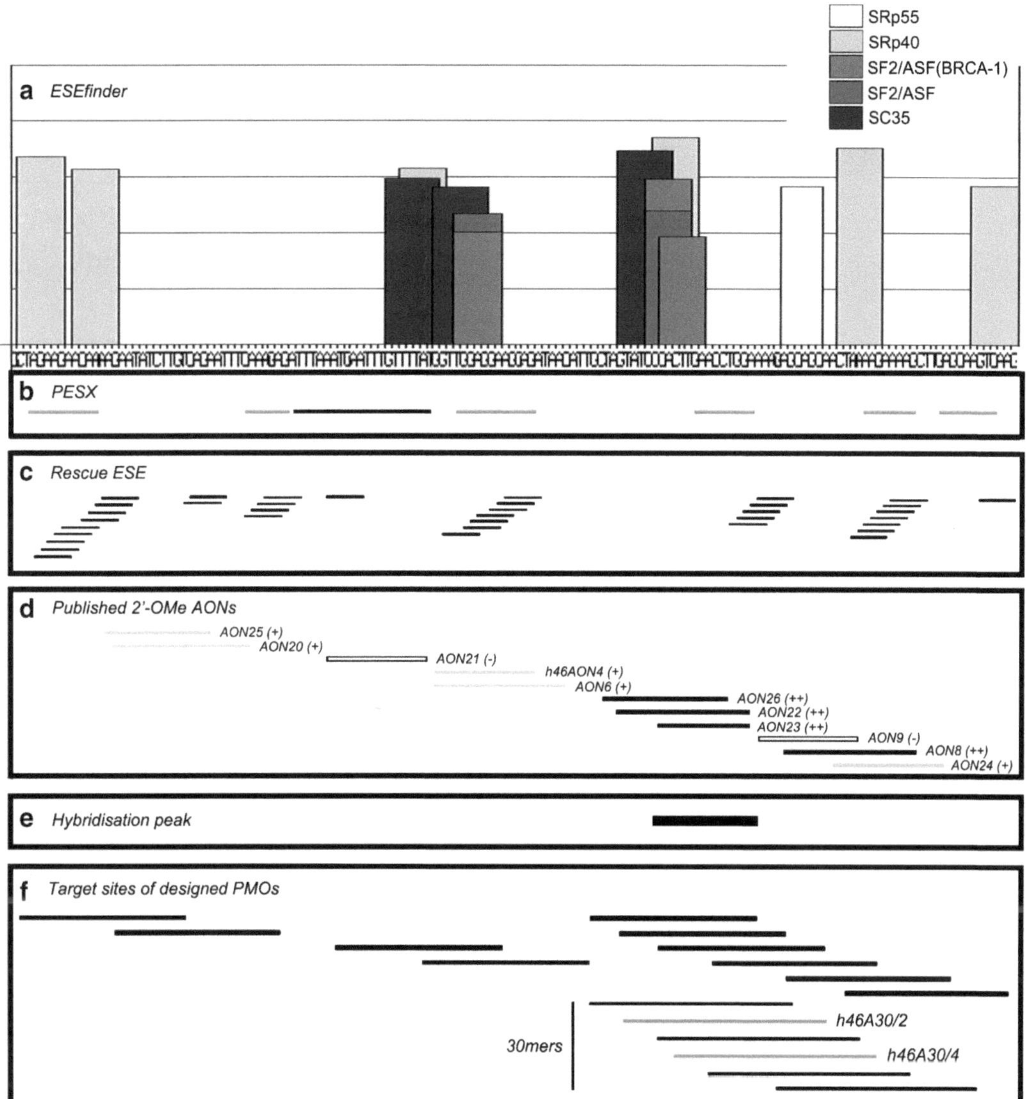

Fig. 1. Scheme summary of the tools used in the design of phosphorodiamidate morpholino oligomers (PMOs) to exon 46. (**a**) Results of ESEfinder analysis, showing the locations of SRp55, SRp40, SF2/ASF (BRCA1), SF2/ASF, and SC35 binding motifs above the established threshold value for each motif. (**b**) Output of PESX analysis showing the locations of exonic splicing enhancers (ESEs) as *grey lines*, and exonic splicing silencers as *black lines*. (**c**) RescueESE analysis showing predicted ESEs by *lines*, and where they overlap as a *ladder of lines*. (**d**) The positions of target sites of 2′OMePS AOs studied previously (25) are shown as *lines* and are colour-coded according to levels of bioactivity observed (*white* no exon skipping detected; *grey* exon skipping observed in up to 25% of transcripts; *black* exon skipping observed in over 25% of transcripts in normal control myotube cultures). (**e**) Location of the hybridization peak, as revealed by AccessMapper analysis of hybridization screen of synthetic pre-mRNA against random hexamer array, is shown by the *black line*. (**f**) The location of the target sites for all 25mer (*thick lines*) and 30mer (*thin lines*) PMOs used by ourselves in the study of exon 46 skipping are indicated by *lines*. The position of the most bioactive PMOs (h46A30/2 and h46A30/4) tested by ourselves for the targeted skipping of exon 46 are shown in *grey*.

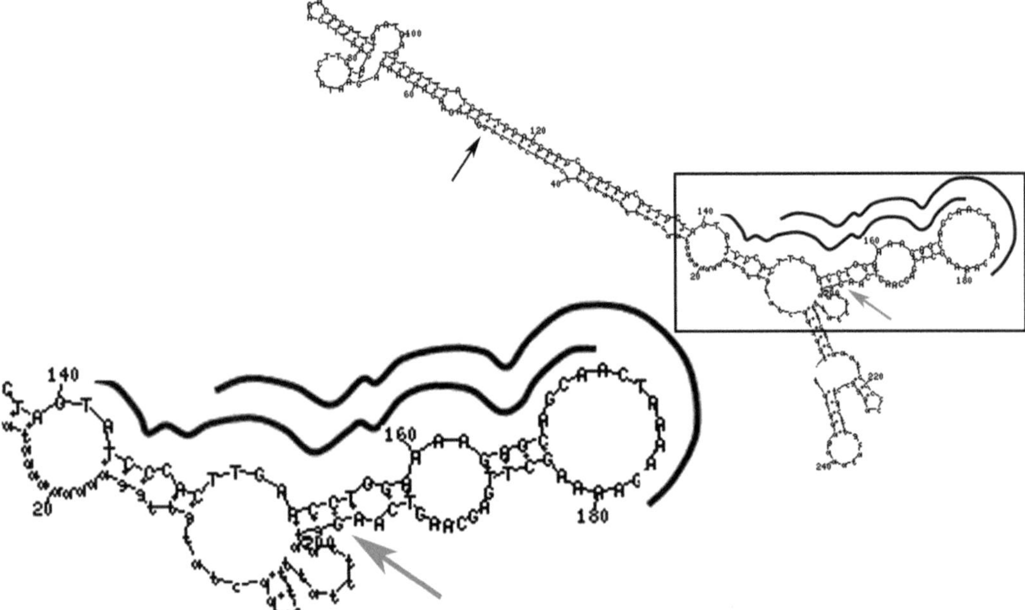

Fig. 2. Mfold secondary structure prediction for exon 46 of the human *DMD* gene. Mfold analysis was preformed using exon 46 plus 50 nucleotides (nt) of the upstream and downstream introns, and with a maximum base-pairing distance of 100 nt. The splice sites are indicated by the *grey* (splice acceptor site) and *black* (splice donor site) *arrows*. *Black lines* indicate the positions of the target sites of the two most bioactive PMOs (h46A30/2 and h46A30/4) for the targeted skipping of exon 46.

(PMO – exon binding energy) – (free energy of exon) –
(free energy of PMO) – (PMO – PMO binding energy)

4. Hybridization array analysis. Prepare the template for mRNA production. Using suitable primer design software, design PCR primers that are situated in the genomic DNA approximately 500 bp upstream and downstream of the exon of interest. The "forward" primer, located upstream of the target exon, should have a minimal T7 RNA polymerase promoter sequence added to its 5′ end. This sequence has been defined as 5′-TAATACGACTCACTATAGG-3′ and means that once the PCR product has been generated and purified, it can be used as a template for in vitro transcription.

   Set up the PCR reaction using 100–250 ng of genomic DNA. This would ideally be in the form of a YAC or BAC clone, which is now available for a wide variety of species from a number of sources. If this is not possible, extract total genomic DNA from cultured cells from the species of interest. Purify the PCR product (approximately 1.2 kb in length) from an agarose gel following electrophoresis.

   Prepare pre-mRNA using the mMESSAGE mMACHINE kit. Specifically, (1) thaw the frozen reagents of the kit; (2) add reagents to a microcentrifuge tube at RT in the order of

20 µL nuclease-free water, 10 µL 2× NTP/CAP, 2 µL 10× reaction buffer, 0.1–0.2 µg PCR-product template, and 2 µL enzyme mix; (3) flick the tube or pipette the mixture up and down gently, and then microfuge tube briefly to collect the reaction mixture at the bottom of the tube; (4) incubate at 37°C for 1 h. A 2 h incubation may be necessary to achieve maximum yields; (5) add 1 µL TURBO DNase, mix well and incubate at 37°C for 15 min. This will remove the DNA template from the reaction; (6) use Ambion MEGAclear kit to purify the RNA transcript from unincorporated nucleotides, DNase degradation products, enzymes, and salts; (7) quantify yield of RNA by reading absorbance at 260 nm on a Nanodrop; and (8) store RNA at –80°C until use.

Incubate the synthetic pre-mRNA in the hybridization reaction buffer on an immobilized array of oligonucleotides that comprises of all possible sequences of 6 nt in length (4,096 oligonucleotides in total). Detect pre-mRNA hybridization by the incorporation of dye-labelled ddNTPs using the ExpandRT RNA-dependent DNA polymerase. RNA, enzyme, and any excess labelled ddNTPs are washed off using the hybridization wash buffer. The oligonucleotides to which labelled ddNTPs have been added can then be detected by fluorescence. From the position on the array of the labelled oligonucleotides, the complimentary accessible sequences on the applied RNA are inferred by the computer software in the system (see Note 3).

5. Design leash. Due to the lack of charge, PMOs are unable to enter cells in vitro. Charge can be introduced by annealing the PMOs to the complementary phosphorothioate-capped oligodeoxynucleotide leashes (19). Figure 3 shows a leashed PMO designed by ourselves to induce exon 46 skipping.
6. Design primers for nested RT PCR. As the levels of skipped transcript are generally very small relative to the full-length transcript, it is necessary to perform nested RT-PCR analysis

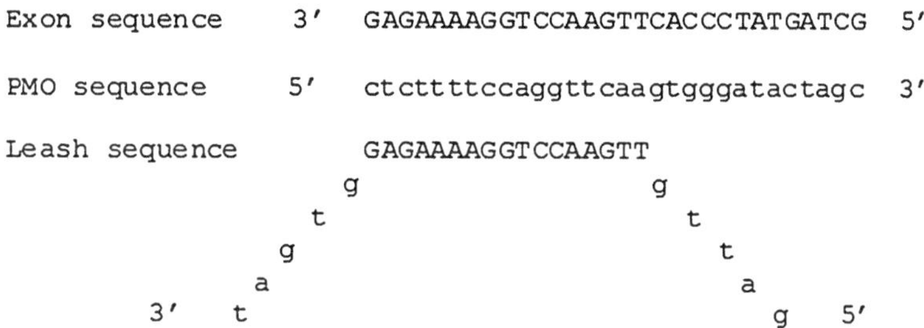

Fig. 3. Leash design strategy shown for a PMO targeting exon 46 of the human *DMD* gene. The complementary sequence of the PMO is 17 bases long, irrespective of whether a 25mer or 30mer PMO is being used, and always starts at the 5′ end of the PMO. The tails of the leash are always of the sequence "gattg" (5′ to 3′) at the 5′ end of the PMO, and "gtgat" (5′ to 3′) at the 3′ end of the PMO.

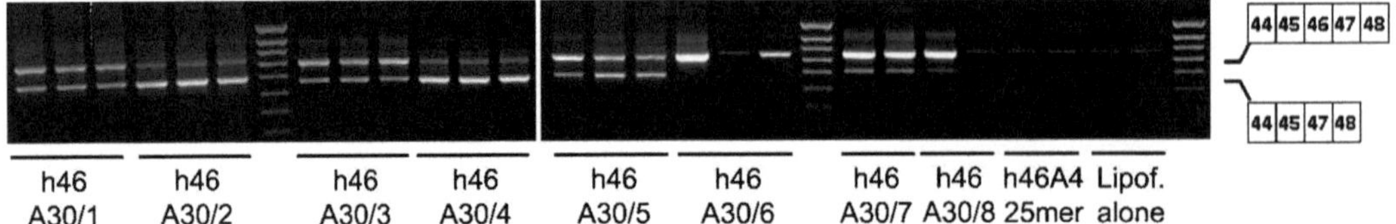

Fig. 4. Comparison of bioactivity of PMOs targeted to exon 46 in normal hSkMCs. Normal human skeletal myoblasts were transfected with PMOs indicated at 500 nM using lipofectin (1:4). RNA was harvested after 24 h and subjected to nested RT-PCR and products visualized by agarose gel electrophoresis. Samples were loaded in duplicate or in triplicate. The position of full-length product (exons 44, 45, 46, 47, and 48) and skipped product (exons 44, 45, 47, and 48) are indicated, differing in size by 156 bp (i.e. size of exon 46). The higher bioactivity of h46A30/2 and h46A30/4, relative to the other PMOs tested, and the complete inactivity of h46A30/6 and h46A4 (25mer) are clearly evident.

on the harvested RNA, to assess exon skipping induced by the PMOs. This is easy to achieve when the PCR primers are carefully designed so that they have specificity for either the full-length or skipped transcripts and all the primers have similar T_ms. Design primers using the VectorNTI software so that in each round either the forward or reverse primer is over an exon/exon junction. This will enable discrimination between amplified contaminating genomic DNA and amplified transcribed cDNA. The PCR products derived from genomic DNA will be much larger than the intronless mRNA-derived products. The primers should be designed so that both full-length and skipped amplicons are amplified by the same primer pair, but will result in different sized products. For example, for exon 46 skipping, the first round forward primer is designed to exon 43/exon 44 boundary and the reverse primer is designed to exon 49. The second round forward primer is designed to exon 44 and the reverse primer is designed to exon 48/exon 49 boundary. This will lead to amplification of a full-length product containing exons 44, 45, 46, 47, and 48 and amplification of an exon 46-skipped product containing only exons 44, 45, 47, and 48. The full-length product is 156 bp longer (156 bp is the size of exon 46). It is easily distinguishable on agarose gel electrophoresis (see Fig. 4). The following guidelines should be considered when designing the primers: (1) design the primers to have a C or A at the 3′ end and a GC content of ~50–60%, (2) to have similar T_ms for primer pairs, (3) to have a T_m greater than 55°C as calculated by G + C rule, and (4) to avoid any internal secondary structure and potential primer–dimer formation.

3.2. Functional Analysis

1. Annealing PMOs and leashes. PMOs are unable to enter cells in vitro due to their lack of charge. Introduce charge by annealing the PMOs to complementary phosphorothioate-capped oligodeoxynucleotide leashes (23). Prepare annealed leash/PMO stocks at 100 μM in 50 μL aliquots in PCR tube by adding together 12.5 μL 10× PBS, 7.5 μL RNase-, DNase-free

H_2O, 25 μL leash (200 μM), and 5 μL PMO (1 mM) and mix gently. Run on PCR programme as follows:

95°C for 5 min, 85°C for 1 min, 75°C for 1 min, 65°C for 5 min, 55°C for 1 min, 45°C for 1 min, 35°C for 5 min, 25°C for 1 min, and then hold at 15°C. Store leashed PMOs at 4°C for up to 6 weeks.

2. Verifying PMO has annealed to leash. Pipette 1 μL leashed PMO (100 μM) into one tube, 1 μL of 2× dilution of leash stock (200 μM) into another tube, and 1 μL of 10× dilution of PMO stock (1 mM) into a third tube. Add 2 μL 1× PBS followed by 1 μL H_2O to each tube and incubate at 37°C for 30 min. Add 1 μL of 5× DNA loading buffer. Load total volume onto 3% agarose gel in TBE with 0.5 μg/mL ethidium bromide (see below for details of agarose gel electrophoresis) and run samples against 7.5 μL of Hyperladder V. Visualize products with UV light. An increase in size should be evident in the leashed PMO lanes compared to leash alone if the PMO and leash have hybridized effectively (Fig. 5).
3. Culturing of normal human skeletal muscle primary cells. Prepare growth and differentiation media. Thaw the growth/differentiation medium supplement mix between 15 and 25°C. Remove the safety seal and open the basal growth/differentiation medium in a microbiological safety cabinet. Carefully open the screwtop of the supplement and transfer the contents with a 25 mL sterile pipette to the medium. Ensure the contents disperse and dissolve immediately into the medium. Do not touch the sides of the bottle neck with the pipette tip as this could lead to microbial contamination. Note the date of addition of the supplement mix on the bottle. Close the bottle and swirl gently until a homogenous mixture is formed. The complete medium should be stored in the dark

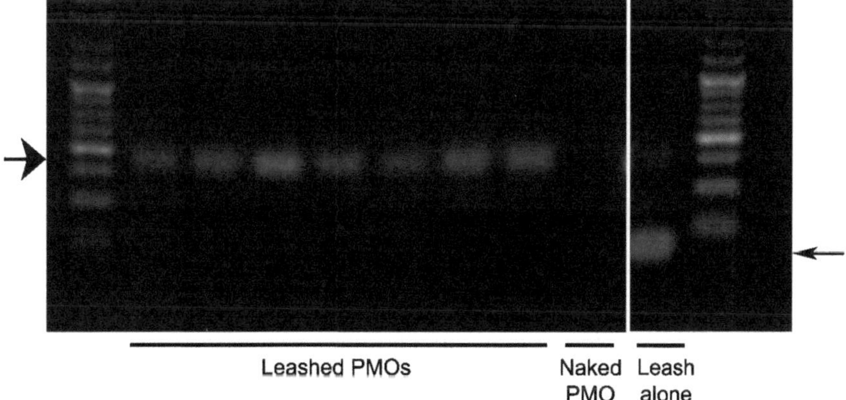

Fig. 5. Verification of annealing of leash to PMO. The larger size of the leashed PMOs (as indicated by *thick arrow*) relative to the size of the leash alone (as indicated by *thin arrow*) indicates successful annealing of complementary leash to PMO. It is interesting to note that PMO alone, being uncharged, does not electrophorize into the agarose gel.

at 4°C for up to 6 weeks; only the volume required should be pre-equilibrated before use rather than the whole bottle.

Initiate proliferating cultures from cryopreserved cells. Prior to thawing cells, supplement medium as described above. Using a microbiological safety cabinet and standard cell culture techniques, pre-equilibrate a 75 cm^2 culture flask with 14 mL of growth medium at 37°C in a 5% CO_2 incubator. Transfer the cryovial containing ~5×10^5 normal human skeletal muscle cells in a 1 mL volume quickly from liquid nitrogen to a 37°C water bath, ensuring the cap is not fully submerged. NB: Wear appropriate safety equipment, i.e. goggles, gloves, and lab coat. Swirl the vial gently but rapidly for 1–2 min until only a small piece of ice (grain of rice size) is left. Wipe the vial dry and transfer the vial on ice to the microbiological safety cabinet. Rinse the vial with alcohol and wipe to remove the excess. Open the vial and gently pipette the cell suspension up and down to evenly suspend the cells. Pipette the cells into the prepared culture flask in an arc on the surface of the medium and gently swirl the medium to disperse the cells for even growth. Examine the cells microscopically to check even distribution of the cells in the flask and transfer to the 5% CO_2 incubator. Do not disturb the culture for the next 16 h to allow cell attachment. After a maximum of 24 h, examine the culture microscopically to check if seeding has been successful, and to remove the DMSO present in the freezing mix, replace the medium with 15 mL of pre-equilibrated fresh medium. Be careful to run the medium over a cell-free surface of the flask and never over the cell layer as this may dislodge the cells. Return the cells to the incubator, re-feed with fresh media every 48 h, and subculture when the cells reach 60–80% confluence, while they are still actively dividing.

Subculture proliferating cultures. Once the cells have reached 80% confluence, passage in a biological safety cabinet as described below, to provide cells for transfection in 6-well plates, proliferating culture in a 75 cm^2 flask, and one vial of cells for cryopreservation. Prepare one 75 cm^2 flask by equilibrating 14 mL fresh supplemented growth medium for at least 30 min in the 5% CO_2 incubator. Pre-warm a 50 mL of DMEM supplemented with l-glutamine, 50 mL supplemented growth medium, and 20 mL supplemented growth medium with 10% FCS which acts as trypsin blocking solution (i.e. add 2.5 mL FCS to 47.5 mL supplemented growth medium) in a 37°C water bath. Aliquot out 18 mL ice-cold DMEM supplemented with l-glutamine onto ice. Quickly defrost 2×1 mL aliquots of ECM gel in a 37°C water bath for 1–2 min until only a small piece of ice (grain of rice size) is left. Add the ECM gel to ice-cold DMEM, pipette to mix, and then aliquot out onto 4×6-well plates. Incubate in 5% CO_2 incubator for at least 45 min before use.

To passage the cells, aspirate the old media and rinse the cell layer gently with 15 mL pre-warmed DMEM supplemented with l-glutamine to remove traces of FCS. Add 1 mL trypsin/EDTA (defrosted to RT), ensuring that the entire surface of the cell sheet is covered. Monitor trypsinization microscopically, tapping the flask sharply to detach the cells once they have rounded up. This should take around a minute. Prolonged exposure or incubation with higher concentrations of trypsin may cause irreversible changes to the cells. Quickly add 15 mL pre-warmed trypsin blocking solution (i.e. supplemented growth medium with 10% FCS) and transfer the cells to a centrifuge tube. Rinse the flask with 5 mL DMEM supplemented with l-glutamine to collect residual cells and add to the centrifuge tube. Centrifuge for 5 min at 220 × *g*. Discard the supernatant and gently resuspend the cell pellet in 2 mL pre-warmed supplemented growth medium with 10% FCS. Perform a cell count with a haemocytometer and adjust cell volume to give 500,000 cells/mL. Typically, this will give a final volume of cell suspension of 6 mL.

Remove 0.9 mL of the cell suspension into a cryovial, add 100 μL DMSO (to give freezing mix with 10% FCS and 10% DMSO), and pipette to mix. Freeze cryovials overnight in a suitable cryo-freezing container and then transfer to the vapour phase of a liquid nitrogen storage facility. Remove 1 mL of the cell suspension into the 75 cm^2 flask containing 14 mL of pre-equilibrated supplemented growth medium, and gently swirl to mix cells into the media. Check flask microscopically for even seeding of cells and transfer to 5% CO_2 incubator. Feed culture every 48 h with fresh growth media and repeat subculture procedure when they reach 60–80% confluence. It should be noted that normal human cells have a limited life span in vitro; TCS CellWorks guarantee their cryopreserved hSkMCs for ≥15 doublings. In our hands, this equates to seven passages. Remove the 4 × 6-well plates from the 5% incubator and carefully aspirate the ECM gel. Aliquot 1 mL of pre-warmed supplemented growth media into each well. Dilute the remaining cell suspension to a total volume of 24 mL with pre-warmed growth media. Mix thoroughly by pipetting and aliquot carefully 1 mL of cell suspension into each well. Gently rock the plates to ensure mixing of cell with the media. Check plates microscopically for even seeding of the cells and transfer to 5% CO_2 incubator. The cells are ready for transfection when they reach 80% confluence (typically 48 h after plating out). Feed every 48 h if required (see Notes 4–6).

4. Perform transfection assays. Once the cells in the 6-well plates have reached 80% confluence, they are ready for transfection. If the cells are allowed to pass 80% confluence, transfection

success is compromised. The method described here is for transfections with lipofectin as transfection reagent. Lipofectin is used in the absence of antibiotics to avoid cell death and it performs optimally in media without serum. As a negative control, leashed PMOs are used without lipofectin and, to assess toxicity of the transfection reagent, diluted lipofectin should be used alone, and blanks controls of DMEM alone included for reference. The protocol given is for the transfection of 24 wells (i.e. four 6-well plates). Adjust volumes given accordingly for bigger or smaller experiments.

Pre-warm 50 mL of supplemented differentiation media and 100 mL DMEM supplemented with l-glutamine. Aspirate growth media from cells and replace with 2 mL of differentiation media and transfer to a 5% incubator for 1 h. Dilute the leashed PMO in DMEM supplemented with l-glutamine to give the required final concentration in the transfection mix, e.g. for a final concentration of 500 nmol, use 15.5 μL of 100 μM of stock leashed PMO and 294.5 μL of DMEM to give enough for three wells. Mix the lipofectin before use and then dilute to give a final ratio to DNA used of 1:4, i.e. for 500 nmol of leashed PMO final concentration and to have sufficient for 24 wells, use 252 μL lipofectin with 2.016 mL DMEM. Let stand for 30 min at RT. Adjust the dilution of lipofection to maintain the ratio at 1:4, in relation to the final concentration of leashed PMOs used. Combine the diluted lipofectin with the diluted leashed PMO, i.e. to 310 μL of diluted leashed PMO, add 310 μL of diluted lipofectin. Mix gently by pipetting and incubate for 10–15 min at RT. Remove the differentiation medium from the cells and wash once with 2 mL of pre-warmed DMEM supplemented with l-glutamine. Remove the wash medium. Add 0.8 mL pre-warmed DMEM supplemented with l-glutamine to each well. Add 200 μL of leashed PMO/lipofectin mix to respective wells, typically setting up three reps of each, and mix gently by rocking the plates.

Incubate the cells at 37°C in a 5% CO_2 incubator for 4 h. Replace the transfection mix with 2 mL of pre-warmed supplemented differentiation media. Incubate the cells at 37°C in a 5% CO_2 incubator and harvest RNA at set time points following transfection. This would typically be 24 h after transfection, but can be extended to anything up to 14 days, with feeding of the cells every 48 h with 2 mL of fresh pre-warmed supplemented differentiation media (see Notes 7–10).

5. Extract RNA. Extract RNA from the cells using the RNeasy mini kit and QIAshredder columns at pre-determined time points after transfection of the cells with leashed PMOs. Typically, this is 24 h after transfection, but for studies examining the persistence of exon skipping, RNA should be harvested at day 1, 2, 3, 7, 10, and 14. The cultures should

be fed every 2 days to ensure their survival. It should be noted that to prevent damage and possible contamination of harvested RNA, all steps should be carried out carefully, but with speed. Gloves should be worn at all times.

Carefully remove cell culture medium supernatant from cells by aspiration and ensure that all supernatant is removed. The media will inhibit cell lysis and dilute the lysate and thereby affect binding of the RNA to RNeasy silica-gel membrane, so affecting RNA yield. Add 350 μL of buffer RLT to each well of the 6-well plate that is to be harvested. Carefully pipette the buffer a number of times over the surface of the well to ensure all cells are lysed. To ensure complete homogenization of the cells so that the highest yield possible of RNA is achieved and clogging of the RNeasy mini column is prevented, pipette the cell lysate directly onto a QIAshredder spin column placed in a 2 mL collection tube and centrifuge for 2 min at maximum speed. Remove QIAshredder column and discard. Homogenized cells lysates in buffer RLT can be stored at −70°C for several months. Simply cap, label, and freeze. To process frozen lysates, thaw samples at 37°C for 15–20 min to dissolve salts. If any insoluble material is still visible, centrifuge for 5 min at 3,000–5,000 × *g*. Transfer supernatant to a new sterile microcentrifuge tube and continue as normal.

Add 350 μL of 70% ethanol to homogenized lysate (either fresh or from frozen), and mix well by pipetting. Apply up to 700 μL of the sample, including any precipitate that may have formed to an RNeasy mini column placed in a 2 mL collection tube. Close the tube gently and spin for 15 s at ≥ 8,000 × *g*. Discard the flowthrough. Add 700 μL Buffer RW1 to the RNeasy column. Close the tube gently and centrifuge for 15 s at ≥ 8,000 × *g*. Discard the flowthrough. Pipette 500 μL Buffer RPE onto the RNeasy column. Buffer RPE is supplied as a concentrate. Before using for the first time, add 4 volumes of ethanol (96–100%) (i.e. 44 mL of ethanol to 11 mL of Buffer RPE concentrate) to obtain a working solution. Close the tube gently and centrifuge for 15 s at ≥8,000 × *g* to wash the column. Discard the flowthrough. Add another 500 μL Buffer RPE to the RNeasy column. Close the tube gently and centrifuge for 2 min at ≥ 8,000 × *g* to dry the RNeasy silica-gel membrane.

To elute the RNA from the column, carefully transfer the RNeasy column, ensuring no contact of the column with the flowthrough, to a new 1.5 mL collection tube. Pipette 30 μL RNase-free H_2O directly onto the RNeasy silica-gel membrane. Close the tube gently and centrifuge for 1 min at ≥8,000 × *g* to elute. Assess the concentration of RNA spectrophotometrically at 260 nm on a Nanodrop and adjust the concentration of the RNA to 0.5 μg/μL with nuclease-free sterile H_2O. Store RNA samples at −70°C (see Notes 11 and 12).

6. Perform nested RT-PCR analysis. As the levels of skipped transcript are generally very small relative to the full-length transcript, it is necessary to perform nested RT-PCR analysis on the harvested RNA. This is easy to achieve when the PCR primers are carefully designed so that they have specificity for either the full-length or skipped transcripts and all the primers have similar T_ms. Using a sample that should only contain full-length transcript (e.g. lipofectin alone), and a sample that, on the basis of previously published work and strength of PMO design tools used, should contain skipped transcript along with full-length transcript, ascertains the optimal temperature for the PCR by performing a temperature gradient PCR for both the first round and the second round. This will be the temperature that gives a clean full-length product in the lipofectin alone RNA sample and no skipped product, and a full-length and skipped product in the PMO-transfected RNA sample. However, for reassurance, you may wish to order from GeneArt RNA sequences that correspond to the full-length and, perhaps most importantly, to the skipped transcript. If there is no facility for performing temperature gradient, perform the nested RT-PCR on a full-length and skipped sample at an annealing temperature just below the T_m for the primers used. Adjust the annealing temperature subsequently if required. The methodology described is the one used by us for analysing RNA samples harvested from transfected cells on four 6-well plates (i.e. 24 samples) for the targeted skipping of exon 46.

   Perform the first round RT-PCR. Thaw all components of the GeneScript RT-PCR kit and keep on ice. Wear gloves throughout to avoid any contamination with RNases. Briefly vortex and microfuge all reagents before preparing the reactions. Set up two master mixes, prepared in two separate RNase- and DNase-free microcentrifuge tubes kept on ice (Tables 1 and 2).

   Carefully pipette 2 μL of each RNA to be tested (containing 1 μg of RNA) into the bottom of a 0.2 mL thin-wall PCR tube on ice, using a fresh tube for each sample. It is important to include a tube with 2 μL of RNAse-free sterile H_2O to rule out the possibility of cross-contamination of samples (25 tubes in total). Vortex master mix 1 and carefully pipette 10 μL into each PCR tube on ice, using a fresh tip for each tube to ensure no cross-contamination of samples. Vortex master mix 2 and carefully pipette 13 μL into each PCR tube on ice, using a fresh tip for each tube to ensure no cross-contamination of samples. Place the tubes in a thermal cycler at 45°C and cycle as follows:

   Synthesize the first strand cDNA. Perform Reverse transcription at 45°C for 30 min. Perform RT inactivation at 92°C for 5 min. Perform RNA/cDNA/primer denaturation.

Table 1
Composition of the master mixture 1 for the first round PCR (Subheading 3.2, step 6)

Master mix 1	Volume for 1 tube (25 μL final)	Volume for 24 tube stock	Final concentration (μM)
Nuclease-free sterile H_2O	Up to 10 μL	Up to 260 μL	
5 mM dNTP mix	1 μL	26 μL	200
100 μM Forward outer primer	0.25 μL	6.5 μL	1
100 μM Reverse outer primer	0.25 μL	6.5 μL	1

Table 2
Composition of the master mixture 2 for the first round PCR (Subheading 3.2, step 6)

Master mix 2	Volume for 1 tube (25 μL final)	Volume for 24 tube stock	Final concentration
Nuclease-free sterile H_2O	Up to 13 μL	Up to 338 μL	
5× RT-PCR buffer (includes 7.5 mM $MgSO_4$)	5 μL	130 μL	1.5 mM $MgSO_4$
GeneScript enzyme mix (blended mix of Accurase DNA polymerase and MMuLV reverse transcriptase)	0.25 μL	6.5 μL	0.625 U

Synthesize the second strand cDNA and perform PCR. Carry out 20 cycles PCR reaction. Each cycle includes denaturation at 92°C for 30 s, annealing at 64°C for 30 s, and extension at 68°C for 90 s. Carry out one more cycle with extension at 68°C for 10 min only. Remove the samples for the PCR machine, allow to cool to RT, and either set up second round or freeze samples at –20°C.

Perform the second round PCR. Defrost a tube of 2× PCR Master Mix (containing cresol red), vortex, microfuge briefly, and then keep on ice. Make up a stock in a DNase and RNase-free microfuge tube on ice (Table 3). Mix well and aliquot 23 μL of stock into 0.2 mL thin-walled PCR tubes

Table 3
Composition of the master mixture for the second round PCR (Subheading 3.2, step 6)

Stock	Volume for 1 tube (final volume 25 μL)	Volume for 24 tube stock (μL)	Final concentration
Nuclease-free sterile H_2O	10 μL	260	
100 μM forward inner primer	0.25 μL	6.5	1 μM
100 μM reverse inner primer	0.25 μL	6.5	1 μM
2× PCR MasterMix	12.5 μL	325	0.625 U Taq polymerase, 200 μM dNTPs, 1.5 nM $MgCl_2$

on ice (25 in total = 24 samples + H_2O control). Carefully remove cap from first round PCR tubes, ensuring gloves are changed frequently to avoid cross-contamination of samples. Aliquot 2 μL of first round product into respective second round PCR tubes, making sure that the tip is put right into the bottom of the PCR tube before dispensing from tip. Cap and place in PCR machine and run the PCR programme as one cycle of denaturation at 92°C for 2 min, 30 cycles of PCR amplification (denaturation at 92°C for 30 s, annealing at 64°C for 30 s, and extension at 68°C for 90 s), and one cycle of extension at 68°C for 10 min. Remove the samples for the PCR machine, allow to cool to RT, and either directly load onto 1.5% agarose gel or freeze samples at –20°C.

7. Examine by agarose gel electrophoresis and densitometry. Use a small (50 mL) horizontal 3% agarose gel for verification of annealing of leash to PMO, and a large (150 mL) horizontal 1.5% agarose gel for analysis of nested RT-PCR products.

   Dilute 5× TBE stock in H_2O to give a working 1× TBE stock (200 mL 5× TBE stock + 800 mL H_2O). For a large 1.5% agarose gel, weigh out 2.25 g of agarose into sealable glass bottle and add 150 mL of 1× TBE. For a small 3% agarose gel, weigh out 1.5 g of agarose into a sealable glass bottle and add 50 mL of 1× TBE. Heat in a microwave for 30 s at a time on full power until the agarose has completely melted into the TBA and formed a homogenous solution.

   Allow to cool to below 60°C, and add ethidium bromide to allow visualization of DNA in the agarose gel. For the small 3% agarose gel, add 2.5 μL of 10 mg/mL ethidium bromide, cap bottle, and swirl gently to mix. For the large 1.5% agarose gel, add 7.5 μL of 10 mg/mL ethidium bromide, cap bottle, and swirl gently to mix. Carefully pour the gel into casting trays with the correct combs inserted to provide enough wells for loading samples and markers. Leave gel to set for at least 20 min. Once the gel has set, it will turn a cloudy colour and will be firm to the touch.

Pour over 1× TBE to completely cover the gel to a depth of around 1 cm. Remove any tape or buffer dams used to cast the gel, and carefully pull out the combs in a vertical motion. Load samples into the wells of the gel, using coloured tape under the gel tank to allow visualization of the wells if necessary. For the leashed PMO samples, load the 5 μL samples prepared as described above. For the nested RT-PCR samples, directly load 10 μL of second round PCR product; the PCR master mix used in the second round contains glycerol and cresol red obviating the need for addition of loading buffer. Run samples against 7.5 μL of Hyperladder V for the verification of annealing of leash to PMO, and against 7.5 μL of Hyperladder IV for the analysis of the nested RT-PCR products. It is important to load ladders at either end and also in the middle of the gel. Run the small agarose gel at 50 V and the large agarose gel at 150 V until bromophenol blue or cresol red marker dye has run about 5 cm. Visualize the gel under UV light and document findings.

Perform semi-quantitative analysis of the levels of exon skipping by densitometric analysis of the full-length and skipped amplicons using GeneTools software. Calculate the percentage skip using the following formula.

$$\%\text{ Skip} - \frac{(\text{Skipped band pixel count-background pixel count})\times 100\%}{\begin{array}{c}(\text{Skipped pixel count-background pixel})\\ +(\text{full length pixel count-backgroung pixel count})\end{array}}$$

Average the percentage skip obtained over the number of repeats performed, and then use statistical analysis to express the strength of the significance of skipping induced.

8. Sequence the skipping products. Separate the full-length and skipped products electrophoretically on an agarose gel. Extract the products from the gel using Qiaquick gel extraction kit. Sequence the extracted PCR product to confirm the correct skip. Carry out all centrifugations at 17,900 × *g* in a conventional table-top microcentrifuge.

   Excise the DNA fragment from the agarose gel with a clean, sharp scalpel. This is made easier by viewing the gel on UV light, using face protection, lab coat, and gloves to prevent exposure of skin to UV light. It is important to minimize the size of the gel slice by cutting away the extra agarose. Weigh the gel slice in a colourless tube. Add 3 volumes of Buffer QG to 1 volume of gel (100 mg–100 μL). For example, add 300 μL of Buffer QG to each 100 mg of gel. The maximum amount of gel slice per column is 400 mg; for gel slices >400 mg, use more than one QIAquick column. Incubate at 50°C for 10 min, or until the gel slice has completely dissolved. To help dissolve the gel, vortex every 2–3 min during the incubation.

After the gel slice has dissolved completely, check if the colour of the mixture is yellow (similar to Buffer QG without dissolved agarose). DNA absorption to the QIAquick silica-gel membrane is pH-dependent. Buffer QG contains a pH indicator, and to be at the correct DNA absorption pH of ≤7.5, the mixture should be yellow in colour. If the pH of the QG buffer is >7.5 due to frequently used or incorrectly prepared agarose gel electrophoresis buffer, the gel slice/Buffer QG mixture turns orange or violet. The pH can be easily corrected by addition of small volume (10 μL) of 3 M sodium acetate, pH 5.0, before proceeding with the protocol.

Add 1 gel volume of isopropanol to the sample and mix. For example, if the agarose gel slice is 100 mg, add 100 μL isopropanol. This step increases the yield of DNA fragments <500 bp and >4 kb. Place a QIAquick spin column in a 2 mL collection tube provided in the kit. To bind the DNA to the column, apply the sample to the QIAquick column and centrifuge for 1 min. The maximum volume of the column reservoir is 800 μL; for sample volumes of more than 800 μL, discard flowthrough and simply load and spin again.

Discard the flowthrough and place the QIAquick column back in the same collection tube. Add 0.5 mL of Buffer QC to the Qiaquick column and centrifuge for 1 min. This will remove all traces of agarose, which is vital when the DNA is to be used for direct sequencing. To wash, add 0.75 mL of Buffer PE to the Qiaquick column. Allow the column to stand for 2–5 min after the addition of the Buffer PE, and then centrifuge for 1 min. Discard the flowthrough and centrifuge the QIAquick column for an additional 1 min.

Place the QIAquick column into a clean 1.5-mL microcentrifuge tube. To elute the DNA, add 30 μL of Buffer EB to the centre of the QIAquick membrane, let the column stand for 1 min for increased DNA concentration, and then centrifuge for 1 min. For higher yield of DNA, use the same 30 μL of Buffer EB to elute from a number of QIAquick columns bound with the same PCR product. Store DNA at -20°C. Measure concentration and estimate quality of DNA spectrophotometrically on a Nanodrop, using a 1 μL sample. Outsource sequencing from external contractor, e.g. Eurofins MWG Operon.

Studies should be extended as described in Notes 8–10 and 13–15, where deemed appropriate.

4. Notes

1. When designing PMOs for the targeted skipping of exons on the *DMD* gene, the potential for the PMOs to have their target sites over sequences that may contain single nucleotide polymorphisms (SNPs) should be considered. The target sites containing the SNPs should be avoided. Use the following websites for investigating SNPs in the *DMD* gene including http://www.genecards.org/cgi-bin/carddisp.pl?gene=DMD, http://www.ncbi.nlm.nih.gov/SNP, http://www.ensembl.org/, http://pupasuite.bioinfo.cipf.es, and http://www.uniprot.org/.
2. The potential for a PMO to have off-target effects across the human genome should be considered by performing BLAST analysis, using the website http://www.ensembl.org/Multi/blastview.
3. This technique was provided as a service by Nyrion Ltd., of Edinburgh, UK, which is no longer trading. The technique is under patent protection.
4. There are other primary human skeletal muscle cells available from other suppliers. We would not like to imply superiority of any particular supplier over another. However, our work has concentrated on the techniques optimized for cells supplied by TCS Cellworks. Standard cell culture optimization is required whatever cells are used.
5. As for the cells, there are other skeletal muscle cell growth and differentiation media suppliers. Our techniques have been optimized using PromoCell media in conjunction with TCS Cellwork cells. We would not like to recommend one supplier over another. Standard cell culture optimization is required whichever supplier is used. For the PromoCell medium, it is important not to freeze the growth or differentiation medium as this can result in high salt concentrations and irreversible precipitation of medium components. To preserve fidelity of the added growth factors and antibiotics, only warm the amount of medium required for the experiment. Store the supplemented medium at 4°C.
6. Primary cells have a limited number of population doubling. It is therefore important to plan your experiments well in advance, to ensure the cells are viable for the course of the experiment. It is good practice to maintain your stock of cryopreserved cells by freezing a vial each time you passage your cells.
7. The methods we describe are for using lipofectin as transfection reagent. There are other transfection reagents available (e.g. lipofectamine 2000, oligofectamine, ExGen 500, PEI).

Although we would not recommend one reagent/technique over another, the cost, reproducibility, and ease of use are important factors in your choice of reagent. For our work, lipofectin was the transfection reagent of choice. For any reagent/technique used, it is important to optimize the transfection conditions such as the DNA to transfection reagent ratio, cell density at time of transfection, and incubation time. If serum-free medium is required for transfection, you need to examine compatibility of medium with transfection reagent. Once optimized, it is vital to maintain the same seeding conditions between experiments.

8. To establish that PMOs are entering the cell efficiently, run parallel transfections using 3′-carboxyfluorescein fluorescently labelled inverse control PMOs from Gene Tools.

9. For a PMO to have the therapeutic potential, it is desirable for it to be bioactive at low concentrations, and for its action to be persistent. This would have implications on the required dose and the frequency of re-administration. To pick up the optimal PMO, one should include dose responses and time course during functional analysis (*see* Figs. 6 and 7).

10. The potential of continued exposure to PMO having an effect on cell viability should be investigated using trypan blue staining. Briefly, pellet trypsinized cells by centrifugation, resuspend in 100 μL of PBS, add 100 μL of 0.4% trypan blue, and microscopically count the percentage of blue (i.e. dead) cells.

11. It should be noted that there are other methods including commercial kits available for harvesting RNA, RT-PCR, PCR, and recovery of PCR products from agarose gels for sequencing. We would not like to comment on the merits of using one technique over another, but optimization is imperative.

12. If it is not possible to design primers over the exon/exon junctions, there is the potential that contaminating genomic DNA within the RNA may be amplified. Although the products derived from genomic DNA will be much larger and so distinguishable from the products derived from mRNA, it is possible that the presence of genomic DNA will interfere with the amplification of the full-length and skipped transcripts and lead to skewed results. To avoid this, treat the harvested RNA with RQ1 RNase-free DNase (Promega UK, Southampton, Hampshire, UK) prior to nested RT-PCR.

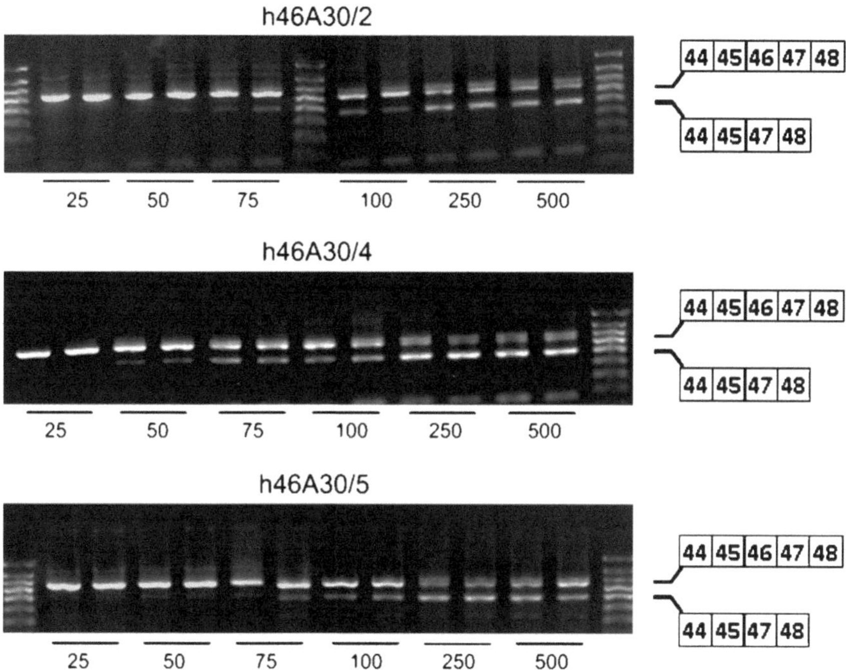

Fig. 6. Low-dose efficacy of exon 46 skipping of the most bioactive PMOs in normal hSkMCs. Normal hSkMC myoblasts were transfected with the PMOs indicated over a concentration range of 25–500 nM, at the concentrations indicated, using lipofectin. RNA was harvested after 24 h and subjected to nested RT-PCR, and products visualized by agarose gel electrophoresis. Samples were loaded in duplicate. The position of full-length product (exons 44, 45, 46, 47, and 48) and skipped product (exons 44, 45, 47, and 48) are indicated, differing in size by 156 bp (i.e. size of exon 46). The higher bioactivity of h46A30/4 at the lower concentrations used, relative to the other PMOs tested, is clearly evident.

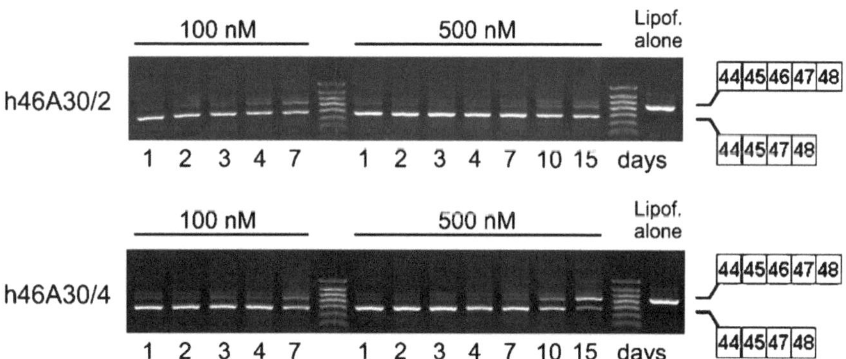

Fig. 7. Time course of exon 46 skipping of the most bioactive PMOs in normal hSkMCs. Normal hSkMC myoblasts were transfected with 100 and 500 nM concentrations of h46A30/2 and h46A30/4, respectively, using lipofectin. RNA was harvested at the time points indicated following transfection and subjected to nested RT-PCR, and products visualized by agarose gel electrophoresis. The position of full-length product (exons 44, 45, 46, 47, and 48) and skipped product (exons 44, 45, 47, and 48) are indicated, differing in size by 156 bp (i.e. size of exon 46). The persistence of exon 46 skipping by both PMOs for up to 15 days after transfection (i.e. for the lifetime of the cultures) is clearly evident.

13. Exon skipping in normal skeletal myotubes will in all likelihood disrupt the reading frame of the *DMD* gene, resulting in no dystrophin protein expression, so that the only means available to analyse skipping is just at the level of the transcript, i.e. by RT-PCR. Once designed PMO sequences have been screened in normal skeletal myotubes, functional analysis should be extended in suitable *DMD* patient cells. Availability of patient cells carrying appropriate deletion should be investigated at the following websites: http://www.eurobiobank.org and http://www.cnmd.ac.uk/coreactivities/biobanking. As for normal skeletal myotubes, the optimization of cell culture, transfection protocol, and downstream analysis will be required. The functional analysis of transfected patient cells will involve nested RT-PCR analysis of the transcript, and also Western blot analysis of protein, since skipping of targeted exons should restore the reading frame of the *DMD* gene, leading to expression of shortened dystrophin protein.
14. The functional analyses of PMOs in vitro, whether in normal or patient myotubes, will not reveal their in vivo skipping efficacy. Work should therefore be extended into an animal model of DMD. However, there are a number of differences in the human and mouse *DMD* gene sequences. PMOs designed to target human *DMD* exons are unlikely to be as effective in a mouse model due to the number of mismatches between different species. A transgenic mouse (hDMD mouse), expressing a complete copy of the human *DMD* gene, has been generated (10). This model has been used in previous studies to test the action of AOs in vivo (6). In this case, 20 μg PMOs are injected into the gastrocnemius muscle of hDMD mice. RNA is extracted from the muscles and analysed for exon skipping relative to a contralateral saline control. It should be noted that the levels of exon skipping by each particular PMO may be variable, which is likely due to the poor uptake into the non-dystrophic muscle of the hDMD mouse. However, this does not compromise the importance that studies in the hDMD mouse have in establishing induction of correct exon skip in vivo.
15. Different chemical modifications of PMOs are continually being developed. For example, vivo-morpholino is comprised of a PMO with a unique covalently linked delivery moiety called Vivo-Porter, which is comprised of an octaguanidine dendrimer. It uses the active component of arginine-rich delivery peptides (the guanidinium group) with improved stability, low toxicity, and reduced cost. It is important to keep abreast of the rapid advances in the field and apply them to your work.

References

1. Monaco AP, Bertelson CJ, Liechti-Gallati S, Moser H, Kunkel LM. (1988) An explanation for the phenotypic differences between patients bearing partial deletions of the DMD locus. *Genomics* 2, 90–95.
2. Bertoni C. (2008) Clinical approaches in the treatment of Duchenne muscular dystrophy (DMD) using oligonucleotides. *Front Biosci* 13, 517–527.
3. Trollet C, Athanasopoulos T, Popplewell L, Malerba A, Dickson G. (2009) Gene therapy for muscular dystrophy: current progress and future prospects. *Expert Opin Biol Ther* 9, 849–866.
4. Aartsma-Rus A, Janson AA, Kaman WE, Bremmer-Bout M, Den Dennen JT, Baas F, et al. (2003) Therapeutic antisense-induced exon skipping in cultured muscle cells from six different DMD patients. *Hum Mol Genet* 12, 907–914.
5. van Deutekom JC, Bremmer-Bout M, Janson AA, Ginjaar IB, Baas F, den Dunnen JT, et al. (2001) Antisense-induced exon skipping restores dystrophin expression in DMD patient derived muscle cells. *Hum Mol Genet* 10, 1547–1554.
6. Arechavala-Gomeza V, Graham IR, Popplewell LJ, Adams AM, Aartsma-Rus A, Kinali M, et al. (2007) Comparative analysis of antisense oligonucleotide sequences for targeted skipping of exon 51 during dystrophin pre-mRNA splicing in human muscle. *Hum Gene Ther* 18, 798–810.
7. Aartsma-Rus A, Janson AA, Kaman WE, Bremmer-Bout M, den Dunnen JT, Baas F, et al. (2003) Therapeutic antisense-induced exon skipping in cultured muscle cells from six different DMD patients. *Hum Mol Genet* 12, 907–914.
8. Aartsma-Rus A, Janson AA, Kaman WE, Bremmer-Bout M, van Ommen GJ, den Dunnen JT et al. (2004) Antisense-induced multiexon skipping for Duchenne muscular dystrophy makes more sense. *Am J Hum Genet* 74, 83–92.
9. Aartsma-Rus A, Janson AA, van Ommen GJ, van Deutekom JCT. (2007) Antisense-induced exon skipping for duplications in Duchenne muscular dystrophy. *BMC Med Genet* 8, 43–51.
10. Bremmer-Bout M, Aartsma-Rus A, de Meijer EJ, Kaman WE, Janson AA, Vossen RH, et al. (2004) Targeted exon skipping in transgenic hDMD mice: a model for direct preclinical screening of human-specific antisense oligonucleotides. *Mol Ther* 10, 232–240.
11. Graham IR, Hill VJ, Manoharan M, Inamati GB, Dickson G. (2004) Towards a therapeutic inhibition of dystrophin exon 23 splicing in mdx mouse muscle induced by antisense oligoribonucleotides (splicomers): target sequence optimisation using oligonucleotide arrays. *J Gene Med* 6, 1149–1158.
12. Lu QL, Rabinowitz A, Chen YC, Yokota T, Yin H, Alter J, et al. (2005) Systemic delivery of antisense oligoribonucleotide restores dystrophin expression in body-wide skeletal muscles. *Proc Natl Acad Sci U S A* 102, 198–203.
13. Mann CJ, Honeyman K, Cheng AJ, Ly T, Lloyd F, Fletcher S, et al. (2001). Antisense-induced exon skipping and synthesis of dystrophin in the mdx mouse. *Proc Natl Acad Sci U S A* 98, 42–47.
14. Jearawiriyapaisarn N, Moulton HM, Buckley B, Roberts J, Sazani P, Fucharoen S, et al. (2008) Sustained dystrophin expression induced by peptide-conjugated morpholino oligomers in the muscles of mdx mice. *Mol Ther* 16, 1624–1629.
15. van Deutekom JC, Janson AA, Ginjaar IB, Franzhuzen WS, Aartsma-Rus A, Bremmer-Bout M, et al. (2007) Local antisense dystrophin restoration with antisense oligonucleotide PRO051. *N Engl J Med* 357, 2677–2687.
16. Kinali M, Arechavala-Gomeza V, Feng L, Cirak S, Hunt D, Adkin C, et al. (2009) Local restoration of dystrophin expression with the morpholino oligomer AVI-4658 in Duchenne muscular dystrophy: a single-blind, placebo-controlled, dose-escalation, proof-of-concept study. *Lancet Neurol* 8, 918–928.
17. Arora V, Devi GR, Iversen PL. (2004) Neutrally charged phosphorodiamidate morpholino antisense oligomers: uptake, efficacy and pharmacokinetics. *Curr Pharm Biotechnol* 5, 431–439.
18. Heemskerk HA, de Winter CL, de Kimpe SJ, van Kuik-Romeijn P, Heuvelmans N, Platenburg GJ et al. (2009) In vivo comparison of 2′O-methyl-PS and morpholino antisense oligonucleotides for DMD exon skipping. *J Gene Med* 11, 257–266.
19. Gebski BL, Mann CJ, Fletcher S, Wilton SD. (2003) Morpholino antisense oligonucleotide induced dystrophin exon 23 skipping in mdx mouse muscle. *Hum Mol Genet* 12, 1801–1811.
20. Alter J, Lou F, Rabinowitz A, Yin H, Rosenfeld J, Wilton SD, et al. (2006) Systemic delivery of morpholino oligonucleotide restores dystrophin expression bodywide and improves dystrophic pathology. *Nat Med* 12, 175–177.

21. Malerba A, Thorogood FC, Dickson G, Graham IR. (2009) Dosing regimen has a significant impact on the efficiency of morpholino oligomer-induced exon skipping in mdx mice. *Hum Gene Ther* 20, 955–965.
22. McClorey G, Fall AM, Moulton HM, Iversen PL, Rasko JE, Ryan M, et al. (2006) Induced dystrophin exon skipping in human muscle explants. *Neuromuscul Disord* 16, 583–590.
23. McClorey G, Moulton HM, Iversen PL, Fletcher S, Wilton SD. (2006) Antisense oligonucleotide-induced exon skipping restores dystrophin expression *in vitro* in a canine model of DMD. *Gene Ther* 13, 1373–1381.
24. Aartsma-Rus A, Fokkema I, Verschuuren J, Ginjaar I, van Deutekom J, van Ommen GJ, et al. (2009) Theoretic applicability of antisense-mediated exon skipping for Duchenne muscular dystrophy mutations. *Hum Mutat* 30, 292–299.
25. Aartsma-Rus A, De Winter CL, Janson AAM, Kaman WE, van Ommen G-JB, Den Dunnen JT, et al. (2005) Functional analysis of 114 exon-internal AONs for targeted DMD exon skipping: indication for steric hindrance of SR protein binding sites. *Oligonucleotides* 15, 284–297.
26. Wilton SD, Fall AM, Harding PL, McClorey G, Coleman C, Fletcher S. (2007) Antisense oligonucleotide-induced exon skipping across the human dystrophin gene transcript. *Mol Ther* 15, 1288–1296.
27. Aartsma-Rus A, van Vliet L, Hirschi M, Janson AA, Heemskerk H, de Winter CL, et al. (2009) Guidelines for antisense oligonucleotide design and insight into splice-modulating mechanisms. *Mol Ther* 17, 548–553.
28. Popplewell LJ, Trollet C, Dickson G, Graham IR. (2009) Design of phosphorodiamidate morpholino oligomers (PMOs) for the induction of exon skipping of the human DMD gene. *Mol Ther* 17, 554–561.

Chapter 11

Engineering Exon-Skipping Vectors Expressing U7 snRNA Constructs for Duchenne Muscular Dystrophy Gene Therapy

Aurélie Goyenvalle and Kay E. Davies

Abstract

Duchenne muscular dystrophy (DMD) is a fatal muscle wasting disorder caused by mutations in the dystrophin gene. In most cases, the open-reading frame is disrupted which results in the absence of a functional protein. Antisense-mediated exon skipping is one of the most promising approaches for the treatment of DMD and has recently been shown to correct the reading frame and restore dystrophin expression in vitro and in vivo. Specific exon skipping can be achieved using synthetic oligonucleotides or viral vectors encoding modified snRNAs, by masking important splicing sites.

We have recently demonstrated that enhanced exon skipping can be induced by a U7 snRNA carrying binding sites for the heterogeneous ribonucleoprotein A1. In DMD patient cells, bifunctional U7 snRNAs harboring silencer motifs induce complete skipping of exon 51 and thus restore dystrophin expression to near wild-type levels. Furthermore, we have confirmed the efficacy of these constructs in vivo in transgenic mice carrying the entire human DMD locus after intramuscular injection of AAV vectors encoding the bifunctional U7 snRNA. These new constructs are very promising for the optimization of therapeutic exon skipping for DMD, but also offer powerful and versatile tools to modulate pre-mRNA splicing in a wide range of applications. Here, we outline the design of these U7 snRNA constructs to achieve efficient exon skipping of the dystrophin gene. We also describe methods to evaluate the efficiency of such U7 snRNA constructs in vitro in DMD patient cells and in vivo in the transgenic hDMD mouse model, using lentiviral and recombinant adeno-associated viral vectors, respectively.

Key words: AAV vector, Duchenne muscular dystrophy, Lentiviral vector, Exon skipping, U7 snRNA, Gene therapy, Antisense, Exonic splicing silencer

1. Introduction

Duchenne muscular dystrophy (DMD) is a severe neuromuscular disorder caused by mutations in the dystrophin gene that result in the absence of functional protein. The majority of mutations causing DMD disrupt the open-reading frame and give rise to

Dongsheng Duan (ed.), *Muscle Gene Therapy: Methods and Protocols*, Methods in Molecular Biology, vol. 709, DOI 10.1007/978-1-61737-982-6_11,

prematurely truncated proteins, thus leading to progressive muscle wasting. There is currently no effective treatment for DMD. One of the most promising strategies aims to convert an out-of-frame mutation into an in-frame mutation, which would give rise to internally deleted but still functional dystrophin (1, 2). This can be achieved by using antisense oligonucleotides (AON) that interfere with splice sites or regulatory elements within the exon and thus induce the skipping of specific exons at the pre-mRNA level (3, 4). Alternatively, the antisense sequence can be delivered to cells using viral vectors carrying a gene from which the antisense sequence can be transcribed, such as modified small nuclear RNA (snRNA) genes (5, 6). U7 snRNA is normally involved in histone pre-mRNA 3′-end processing, but can be converted into a versatile tool for splicing modulation by a small change in the binding site for Sm/Lsm proteins (7). The antisense sequence embedded into a snRNP particle is therefore protected from degradation and accumulates in the nucleus where splicing occurs. We previously demonstrated that a single treatment with an adeno-associated virus (AAV) 2/1 vector containing appropriately modified U7 snRNA induces widespread exon skipping of the dystrophin pre-mRNA and results in the sustained correction of muscular dystrophy in both the *mdx* mouse (8) and the golden retriever muscular dystrophy dog (*GRMD* dog) (unpublished work). This long-term restoration of dystrophin represents a strong advantage of this approach over synthetic AON by eliminating the need for repeated injections.

Considering the diversity of mutations among DMD patients, the translation of this strategy to human requires specific tools adapted to different human dystrophin exons. As with the synthetic AON approach, choice of the antisense sequence is one of the most important parameters to ensure efficient exon skipping. The first targets usually considered to induce exon skipping are the donor and acceptor splice sites, together with the branch point, and these sites have indeed been successfully targeted on the DMD gene (9, 10). Exon skipping can also be achieved by targeting exon internal sites and, in particular, exonic splicing enhancers (ESE) (11), but these ESE motifs are only loosely defined and their identification is still complex (12–14). Therefore, even after several years of research in the exon skipping field, there are still no clear rules to guide investigators in the design of their antisense sequences. The choice of the antisense sequence to insert into the U7 snRNA constructs to achieve an efficient skipping of the dystrophin exons can be based on previously identified sequences as many dystrophin exons have already been targeted by AON (15–17). However, the most effective antisense sequence identified using AON will not necessarily be the best in the U7 snRNA context, which implies steps of optimization in each case.

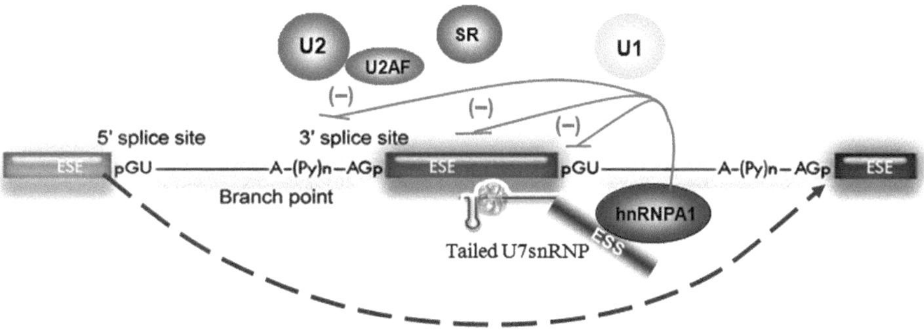

Fig. 1. Schematic representation of the silencing tail action. When the U7 snRNA construct binds to the exon of interest through a complementary sequence, the silencing tail is thought to induce exon skipping via recruitment of hnRNPA1 proteins in its near vicinity. hnRNPA1 represses the normal splicing process by inhibiting the binding of splicing factors such as U1snRNP, U2AF, and U2snRNP either by direct steric encumbrance or by competing with the SR proteins normally recruited by exonic splicing enhancer (ESE) motifs.

In order to avoid this lengthy optimization necessary to target each individual exon, we have developed a more "universal" U7 snRNA carrying a complementary sequence to the exon and a free tail harboring canonical binding sites for the heterogeneous nuclear ribonucleoproteins (hnRNP) A1/A2 that are powerful splicing repressors (18). Therefore, this tail is thought to induce skipping of any targeted exon via recruitment of these proteins in its near vicinity (Fig. 1). We have demonstrated that such tailed U7 snRNA constructs can achieve efficient exon 51 skipping in human myoblasts and restore dystrophin expression in cells from DMD patients (Fig. 2). We have also confirmed the efficacy of these U7 snRNA constructs in vivo after inserting them into AAV2/1 vectors which were subsequently injected into the tibialis anterior muscle of a mouse model transgenic for the entire human dystrophin locus (19, 20) (Fig. 3).

In this protocol, we outline the procedures to design U7 snRNA constructs to achieve efficient exon skipping of the dystrophin gene and describe methods to evaluate their efficacy in vitro in DMD patient cells and in vivo in the transgenic hDMD mouse model, using lentiviral and recombinant AAV vectors, respectively.

2. Materials

2.1. Generating the Modified U7 snRNA Vectors

2.1.1. Engineering U7smOpt from wt U7 snRNA Gene

1. The *Mus musculus* U7 snRNA gene sequence is required to design primers for PCR and can be found in GenBank (X54165).
2. pBluescript SK+ (Stratagene, La Jolla, CA, USA).
3. The QuikChange™ Site-Directed Mutagenesis Kit (Stratagene) allows a quick and easy modification of the U7 SmWT into the consensus U7 SmOpt.

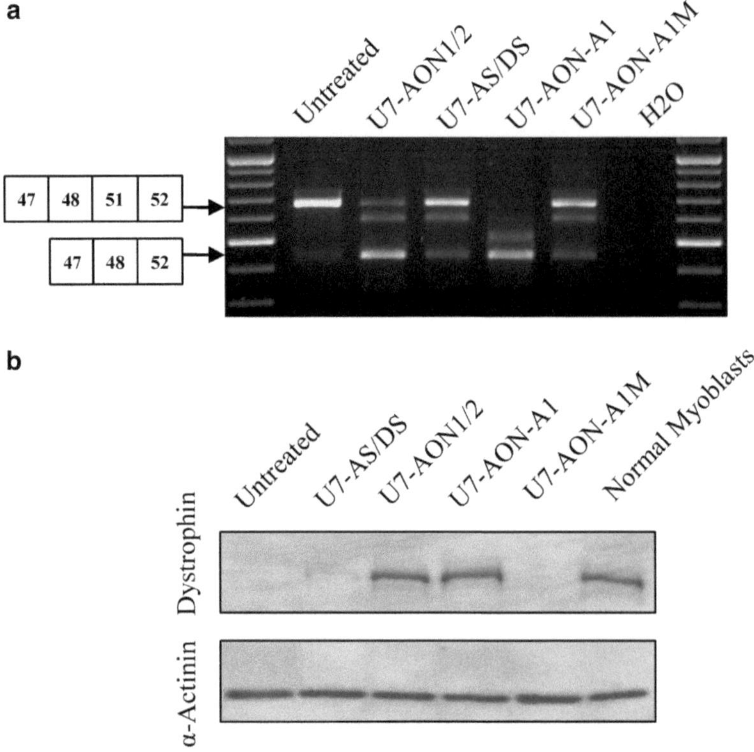

Fig. 2. Comparison of exon 51 skipping in DMD myoblasts and subsequent dystrophin rescue. (**a**) DMD myoblasts carrying the Δ49–50 deletion were transduced with lentiviral vectors encoding the different U7 snRNA constructs. Exon 51 skipping was assessed by nested RT-PCR and a fragment of the expected size was detected in each case. Additional bands due to heteroduplex formation and activation of a cryptic splice site within the exon 51 (previously described (25) are visible in same analyses. (**b**) Western blot probed with dystrophin antibody (top gel) and α-actinin antibody (bottom gel). Note that the dystrophin detected corresponds to an internally deleted protein missing exons 49, 50, and 51. Adapted by permission from Macmillan Publishers Ltd: Molecular Therapy (18), copyright (2009).

2.1.2. Engineering Modified U7 snRNA Specific to Dystrophin Exons

To select the targeted exonic sequences, there is no special material needed besides the dystrophin gene sequence itself.

1. *Pfu* DNA polymerase (Stratagene) is a high fidelity DNA polymerase (see Note 1).
2. Cis-plasmid for recombinant AAV packaging – pSub201 can be obtained from University of North Carolina Chapel Hill (for more details see the website http://genetherapy.unc.edu/mta.htm).
3. SURE competent cells (Stratagene). AAV ITRs can be re-arranged or deleted by endogenous DNA repair when they are propagated in bacterial cells. SURE cells are defective in DNA repair and recombination pathways (see Note 2). Store at –80°C.

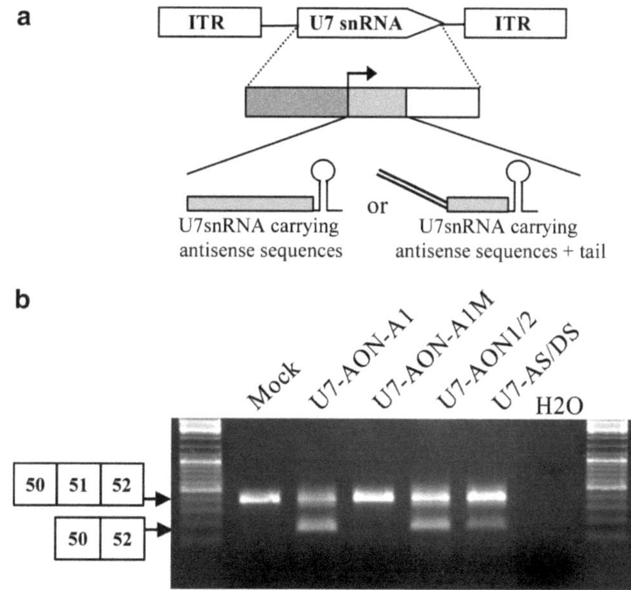

Fig. 3. Comparison of exon 51 skipping in vivo in hDMD mice induced by AAV vectors encoding different engineered U7 snRNA. (**a**) *Top*: Structure of the AAV vectors encoding the different U7 snRNA cassettes. The U7 snRNA cassette is inserted between two AAV2 inverted terminal repeats (ITR) and is constituted of the engineered U7 snRNA sequence (*gray box*) carrying different antisense sequences with or without tail as previously described (bottom), placed under the control of its natural U7 promoter (*hatched box*) and 3′ downstream elements (*open box*). (**b**) AAV2/1 vectors encoding different engineered U7 snRNA were injected in the tibialis anterior of adult hDMD mice. Four weeks after the injection, muscles were harvested and analyzed by nested RT-PCR for exon 51 skipping. Reprinted by permission from Macmillan Publishers Ltd: Molecular Therapy (18), copyright (2009).

4. SOC medium (Gibco-BRL, Grand Island, NY, USA). Store at room temperature.
5. Amp selection LB agar plates (100 μg/mL, ampicillin). Store at 4°C.
6. EndoFree Plasmid maxi kit (Qiagen, Valencia, CA, USA).

2.1.3. Generating AAV Vectors Encoding the Modified U7 snRNA

1. 293 cells (ATCC). This is an adenovirus E1 gene transformed human kidney cell line. They are used as a packaging cell line for recombinant AAV production. These cells are propagated in 4.5 g/L glucose Dulbecco's modified Eagle's medium (DMEM, Gibco) containing 10% fetal bovine serum (FBS, Hyclone) and 1% penicillin G/streptomycin (PS, Gibco).
2. 1× Trypsin-EDTA: 0.25% Trypsin, 1 mM EDTA/4Na (Gibco).
3. 10 mM Polyethylenime (PEI MM=25, Sigma, St Louis, MO, USA).

4. 150 mM NaCl.
5. Transfection medium: DMEM 4.5 g/L glucose 1% FBS+ 1% PS.
6. Posttransfection medium: DMEM 1 g/L glucose 10% FBS+ 1% PS.
7. Helper plasmids: pXX6-80 (adenoviral helper plasmid) and pXR1 (containing the AAV rep2 and cap1 coding regions) can be obtained from University of North Carolina Chapel Hill (for more details see the website http://genetherapy.unc.edu/mta.htm).
8. Cell scraper (Greiner Bio One, Monroe, NC, USA).
9. Lysis buffer: Dulbecco's Phosphate-Buffered Saline with Ca^{++} 0.901 mM and Mg^{++} 0.493 mM (Gibco), Hepes 50 mM pH 7.6, NaCl 150 mM. Sterilize by filtration and store at 4°C for 1 month.
10. Polyallomer tubes with screw-on cap, 50 mL (#357003) (Beckman Coulter, Brea, CA, USA).
11. Benzonase (Sigma).
12. Cesium Chloride (Sigma). CsCl 1.35 g/mL and CsCl 1.5 g/mL solutions in Dulbecco's Phosphate-Buffered Saline with 0.901 mM Ca^{++} and 0.493 mM Mg^{++} (Gibco).
13. 12 mL ultra clear tubes (#344059) (Beckman Coulter).
14. 9 mL Beckman tube (#361623) (Beckman Coulter).
15. Whatman minifield 1 dot-blot array system (Sigma).
16. Dot-blot buffer: 0.4 M NaOH, 10 mM EDTA, 2× SSC.
17. Hybond N+ nylon membrane (GE Healthcare, Waukesha, WI, USA).
18. AlkPhos direct labeling and detection system with CDP-Star (GE Healthcare). This system contains the labeling kit to make dot-blot probe from the digested cis-plasmid, the blocking reagent to prepare the hybridization buffer and wash buffer 1, and the CDP-Star reagent for detection.
19. Wash buffer 1: Urea 2 M, SDS 0.1 %, sodium phosphate 50 mM, NaCl 150 mM, $MgCl_2$ 1 mM, and blocking reagent 0.2%. Store at 4°C for up to 1 week.
20. Wash buffer 2 (20×stock): Tris base 1 M, NaCl 2 M, pH 10.0. Store at 4°C for up to 4 months.
21. Wash buffer 2 working dilution: dilute stock 1:20 and add 2 mL/L of 1 M $MgCl_2$ (final concentration 2 mM). This buffer cannot be stored.
22. Amersham Hyperfilm ECL (GE Healthcare).
23. Dialysis cassettes Slide-a-lyser (Pierce, Rockford, IL, USA).

24. Dnase I and stop solution (Promega, Madison, WI, USA).
25. Proteinase K, PCR grade (Roche, Pleasanton, CA, USA).
26. 2× Proteinase K buffer: 0.1 M EDTA, 0.5% SDS.
27. Precision Master Mixture with SYBR green for titration by qPCR (Primer design, Southampton, UK).

2.1.4. Generating the Lentiviral Vectors Encoding the Modified U7 snRNA

1. The transfer vector for lentiviral production pRRLSIN.cPPT. PGK-GFP.WPRE (plasmid 12252 from Addgene) and the packaging plasmids (pLP1, pLP2, and pLP/VSVG from Invitrogen) (see Note 3).
2. 293FT HEK cells (Invitrogen). The 293FT cell line is a fast-growing, highly transfectable clonal isolate derived from human embryonal kidney cells. These cells are propagated in 4.5 g/L glucose DMEM (Gibco) containing 10% FBS (Hyclone) and 1% PS (Gibco).
3. 2× HBS buffer: 0.3 M NaCl, 1.5 mM Na_2HPO_4, 40 mM Hepes, pH 7.05. Sterilize by filtration and store at −20°C.
4. 0.5 M $CaCl_2$. Sterilize by filtration and store at 4°C.
5. Thickwall polycarbonate 25×89 mm tubes (ref. 355631) (Beckman coulter).
6. NIH/3 T3 cells (ATCC). This mouse embryonic fibroblast cell line is used to titrate lentiviral vectors.
7. Genomic DNA purification kit (Qiagen).
8. Precision Master Mixture with SYBR green for titration by qPCR (Primer design).

2.2. In Vitro Evaluation of Engineered U7 snRNA Construct in Human Myoblasts

2.2.1. Lentiviral Transduction of Human Myoblasts

1. Immortalized human myoblasts (Vincent Mouly, Institut de Myologie, Paris, France).
2. Human myoblast culture medium: Four parts of DMEM (4.5 mg/mL glucose) to one part medium 199. Supplement with 20% FBS, 100 U/mL penicillin, and 100 U/mL streptomycin.

2.2.2. Skipping Analysis by RT-PCR

1. RNeasy kit (Qiagen).
2. Access RT-PCR kit and PCR master mixture (Promega).

2.2.3. Quantifying Dystrophin Expression by Western Blot

1. Newcastle buffer for homogenization protein samples. 3.8% SDS, 75 mM Tris-HCl pH 6.7, 4 M urea, 10% β-mercaptoethanol, 10% glycerol, 0.001% bromophenol blue.
2. Bicinchoninic acid (BCA) protein assay kit (Perbio Science, UK).
3. 10% nonfat milk in phosphate-buffered saline-Tween (sigma) buffer.

4. NCL-DYS1 primary antibody (monoclonal antibody to dystrophin R8 repeat; Novocastra, Newcastle upon Tyne, UK).
5. Alpha-actinin primary antibody (Santa Cruz Biotechnology, Santa Cruz, CA, USA). This antibody is used as a loading control.
6. Mouse horseradish peroxidase-conjugated secondary antibody and goat horseradish peroxidase-conjugated secondary antibody to detect dystrophin and alpha-actinin, respectively (GE Healthcare).
7. ECL analysis system (GE Healthcare).

2.3. In Vivo Evaluation of AAV-U7 snRNA Vectors

2.3.1. Intramuscular Injection

1. The transgenic human DMD (hDMD) mice carrying the entire human DMD locus. This strain is obtained from Annemieke Aarstma-Rus and Gert-Jan Van Ommen (Leiden University Medical Center, the Netherlands).
2. Isoflurane anesthetic machine (Parkland Scientific, Coral Springs, Fl, USA).
3. 29G insulin syringe.

2.3.2. Skipping Analysis by RT-PCR

1. TRIzol reagent (Invitrogen)
2. Access RT-PCR kit and PCR master mixture (Promega).

3. Methods

3.1. Generating the Modified U7 snRNA Vectors

3.1.1. Engineering U7smOpt from wt U7 snRNA Gene

1. Design the primers using the GenBank sequence (X54165) as the template. Flank the primers with restriction sites. The forward primer is U7-F-XbaI 5′-GGGTCTAGATAA CAACATAGGAGCTGTGA-3′. The reverse primer is U7-R-NheI 5′-AAAGCTAGCCACAACGCGTTTCCTAGGA.
2. Amplify a 450 bp product containing the complete U7 snRNA gene from mouse genomic DNA by PCR.
3. Clone the wild-type (wt) U7 snRNA gene into pBluescript SK+.
4. Convert the wt U7 snRNA into a versatile tool for splicing modulation by changing the binding site for Sm/Lsm proteins (7). Exchange the wt Sm binding site of U7 snRNA (AAUUUGUCUAG = U7SmWT) for the consensus SmOpt binding site (AAUUUUUGGAG = U7SmOPT) using the QuikChange™ Site-Directed Mutagenesis kit from Stratagene.

3.1.2. Engineering Modified U7 snRNA Specific to Dystrophin Exons

1. Replace the 18 nt natural sequence complementary to histone pre-mRNAs in U7smOpt with the selected antisense sequences to promote dystrophin exon skipping by PCR-mediated mutagenesis (see Note 4) (Fig. 4).

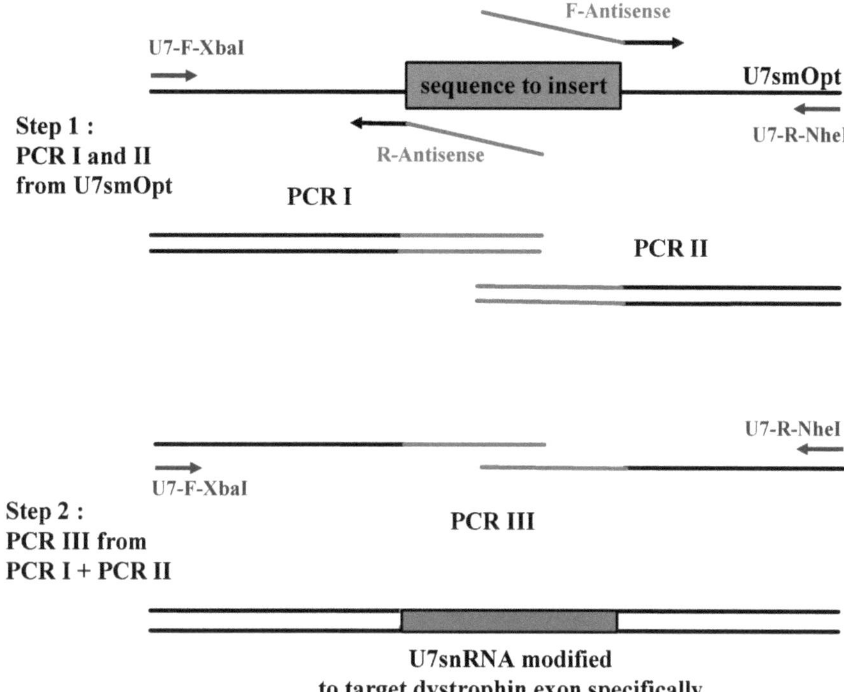

Fig. 4. Schematic representation of the mutagenesis directed by PCR procedure. The first step consists of two PCR reactions (PCR I and PCR II) performed using U7smOpt as a matrix with primers containing the antisense sequences to the insert (F-Antisense and R-Antisense). In a second step, PCR III is carried out using purified PCR I and II products with the external primers U7-F-XbaI and U7-R-NheI. PCR III product is then cloned in a cis-plasmid for recombinant AAV packaging.

2. Design the primers F-Antisense and R-Antisense for PCR I and PCR II anneal on approximately 20 nt of the U7 gene and carry 2/3 of the sequence to insert, therefore overlapping on 1/3 of it for the PCR III (Fig. 4).
3. Perform PCR I and II using U7smOpt as the substrate. Purify PCR I and PCR II products from agarose gel to eliminate any trace of the original U7smOpt.
4. Perform PCR III with external primers using one tenth of purified PCR I and II products as the substrate.
5. Clone PCR III product in pSub201 for recombinant AAV packaging. Select positive clones after sequencing.

3.1.3. Generating AAV Vectors Encoding the Modified U7 snRNA

1. The day before transfection, seed 3×10^7 293 cells in 15-cm dishes to achieve approximately 90% confluency the next morning. Seed a total of 15 plates for one production. Maintain 293 cells in complete DMEM/10% FBS and split cells before they reach confluency by trypsinizing with Trypsin/EDTA. Cells should be transfected within 24 h after seeding.

2. For transfection of five 15-cm dishes, combine the following in a 50 mL falcon tube: 125 μg pXX6-80 helper plasmid, 62.5 μg rAAV vector plasmid containing the U7 snRNA transgene (pSub201 modified with U7 snRNA), 62.5 μg pXR1. Bring up the final volume to 5 mL with 150 mM NaCl. In another 50 mL falcon tube, introduce 562.5 μL of 10 mM PEI and then bring the volume to 5 mL with 150 mM NaCl. Carefully add (drop by drop) the 5 mL PEI solution to the 5 mL plasmid mixture. Rest for 1 min between each milliliter. Do not vortex. Repeat these steps 3 times for 15 dishes. Incubate for 15–20 min at room temperature.
3. Mix the PEI/plasmid cocktail with the transfection medium (12 mL per dish). For 15 dishes, this means 30 mL of the PEI/plasmid cocktail and 180 mL of transfection medium.
4. Remove the culture medium from the cells and replace with 14 mL/dish of the mixture generated in step 3. Incubate for 6 h at 37°C. Then, add 12 mL of posttransfection medium to the cells.
5. 48–72 h after transfection, harvest cells with a Cell scraper. Scrap the cells (and the medium) directly into a 50 mL Falcon tube (2 dishes per one 50 mL Falcon tube). Centrifuge for 15 min at 500 g. Resuspend pellets in lysis buffer (1 mL per one 50 mL Falcon tube) and then pool the tubes. Freeze/thaw cell lysate 4 times using dry ice/ethanol bath and a 37°C water bath. Vortex vigorously between each freeze/thaw cycle (after the 4th freezing cycle, tube can be kept at -80°C).
6. Centrifuge for 20 min at 1,000 × *g* at 4°C. Transfer the supernatant (around 7 mL) into a 50-mL polyallomer Beckman tube with screw-on cap. Add Benzonase (50 unit per 1 mL of lysate). Incubate for 45 min in a 37°C water bath. Shake the tube every 5 min. Centrifuge cell lysate for 20 min at 12,000 × *g* at 4°C using a JA-20 rotor in a Beckman centrifuge (type Avanti® J series). Transfer the supernatant into a 15-mL falcon tube before loading the gradient (final volume should be around 7 mL per 15 plates of original cell culture).
7. Prepare the first CsCl gradient in ultra clear Beckmann tubes. In each tube, introduce 2.5 mL of CsCl (1.3 g/mL). Then carefully add 2.5 mL of CsCl (1.5 g/mL) under the first layer with a plastic Pasteur pipette. Carefully add the crude viral lysate (~7 mL) on the top. Add two drops of mineral oil on the top and carefully load the tubes onto a SW41 swinging bucket rotor. Ultracentrifuge for 2 h at 151,000 × *g* at 20°C (Make sure to use the slowest acceleration and deceleration speed).

8. After the first round of ultracentrifugation, recombinant AAV particles should be located between the two CsCl layers, approximately 2 cm above the bottom of the tube. Very carefully harvest the cloudy viral fraction with a plastic Pasteur pipette. You should expect to get approximately 1–1.5 mL.
9. Introduce the fraction in a 9 mL-Beckmann tube and complete with CsCl (1.41 g/mL) until the top. Make sure there are no bubbles in the tube. Close the tubes. Ultracentrifuge for 24 h and 30 min at 220,000 g at 8°C with slow acceleration and no brake. Collect fractions from the bottom of the tube with a 20G needle (approximately 500 μL/fraction).
10. Identify the viral containing fractions by dot blot. Use 20 μL of each fraction for the dot blot. Add 380 μL of the dot-blot buffer to each aliquot. Denaturate the sample for 10 min at 95°C. Then immediately chill on ice. Prepare the dot-blot manifold apparatus with one sheet of whatman paper and the Hybond N+ membrane presoaked in 2× SSC. Add 400 μL H_2O in each well and aspirate. Load samples in each well and aspirate (400 μL). Wash with 400 μL of the dot-blot buffer and aspirate. Take off the membrane and mark it (on top right for example). Rinse the membrane in 2× SSC and place it in a hybridization tube (Make sure the DNA side is facing the inside). Replace 2× SSC with 8 mL of hybridization buffer (prewarmed to 55°C). Incubate for 20 min at 55°C (with slow shaking). Add the probe into the tube and incubate overnight at 55°C under slow shaking. The next morning, prewarm the wash buffer 1 at 55°C. Eliminate the hybridization buffer. Wash 2× 10 min with 100 mL of wash buffer 1 at 55°C. Then wash 2× 5 min at room temperature with 2× 200 mL of wash buffer 2 on a belly dancer shaker. Take off the membrane from the wash bath and put it on a saran warp (make sure the DNA side is facing up). Add 4 mL of CDP-Star detection reagent and incubate for 3 min in the dark. Expose with a Hyperfilm ECL for approximately 30 min.
11. Select the fractions that are the most positive and pool them. The final volume for a 15 plates production should be 1–2 mL. Introduce these fractions in a dialysis cassette with an 8G needle syringe. Put the cassette in 1 L of phosphate-buffered saline 1× and incubate at least 1 h under shaking at 4°C. Repeat this step 5 times (5 washes of 1 h in 1 L) and finish with an overnight wash in 1 L. Aliquot the recombinant AAV vector and store at −80°C. Keep approximately 5 μL for the titration.
12. Determine the viral particle titer by qPCR. Digest 5 μL of virus with Dnase I for 30 min at 37°C. This step digests any DNA that may be present, but has not been packaged into virions. Inactivate Dnase I with 2 μL of stop solution and

incubate for 10 min at 65°C. Place samples on ice. Digest virus particle with proteinase K for a minimum of 1 h at 50°C. Place the reactions on ice. Prepare dilutions of the digested samples (final dilutions 1/200, 1/500, 1/1,000). Run the qPCR program using a standard curve of the cis-plasmid (10^8–10^2 copies). Analyze the Ct results and calculate how many molecules of the plasmid standards correspond to a given dilution of the rAAV stock. (Remember to take into consideration that the plasmid standards are double-stranded, whereas the rAAV virions harbor only a single strand).

3.1.4. Generating the Lentiviral Vectors Encoding the Modified U7 snRNA

1. Clone the modified U7 snRNA construct into the transfer vector, pRRLSIN.cPPT.PGK-GFP-WPRE.
2. Two days before viral production, seed 293FT cells into 10-cm culture plates. Seed a total of 10 plates for one production. Maintain 293FT cells in complete DMEM/10% FBS and split cells before they reach confluency by trypsinizing with Trypsin/EDTA. Change to fresh culture medium about 2–3 h before transfection.
3. Prepare DNA-calcium-phosphate precipitate in a 50 mL tube. For a 10×10 cm dish preparation, use 130 μg of transfer vector, 30 μg of pLP1, 130 μg of pLP2, and 37.5 μg of pVSVG. Mix all plasmid thoroughly in 2.5 mL of H_2O. Add 2.5 mL of 0.5 M $CaCl_2$. Generate the DNA-calcium-phosphate precipitate by slowly dropping the DNA/$CaCl_2$ mixture into 5 mL of 2× HBS.
4. Gently apply the DNA-calcium-phosphate precipitate to 293FT cells drop by drop. Swirl the culture plate during the process. Incubate at 37°C for 16–20 h. Then change to fresh culture medium.
5. At 48–60 h after transfection, harvest the viral supernatant and spin for 7 min at 3,000×*g* at 4°C in 50 mL Falcon tubes. Filter the supernatant through a 45 μm filter and transfer into four thick wall polycarbonate ultracentrifuge tubes. Concentrate the viral preparation by ultracentrifugation for 130 min at 64,300×*g* in a 70 Ti rotor at 4°C. Pour off the supernatant and allow the remaining liquid to drain by resting the inverted tubes on paper towels. The pellet should be barely visible as a small translucent spot. Add 50 μL of 1× PBS on each viral pellets and let them resuspend overnight at 4°C. Next morning, homogenize and pool all pellets. Avoid frothing. Rinse the tubes with 50 μL of 1× PBS. Pool all volumes to obtain 250 μL final volume (from four thickwall tubes). Aliquot lentiviral vector and store at –80°C.
6. Determine the infectious viral titer (infectious particles per mL, IP/mL) by transduction of 10^5 NIH3T3 cells with serial

dilutions of the vector preparation in a 12-well plate. 72 h later, extract genomic DNA from transduced cells using a genomic DNA purification kit. Quantify vector copy number using real-time PCR analysis of genomic DNA with WPRE forward primer 5′-GGCACTGACAATTCCGTGGT-3′ and WPRE reverse primer 5′-AGGGACGTAGCAGAAGGACG -3′ normalized to mDAG1F 5′-CCA AGG AGC AGA TCA TAG GGC-3′ and mDAG1R primers 5′-AGA GCA TTG GAG AAG GCA GG-3′ using SYBR Green master mix on an ABI Prism 7700 machine with dissociation curve enabled and the $-\Delta\Delta$Ct method. Calculate titers according to the following formula: $IP/mL = (C \times N \times D \times 1{,}000)/V$, where C=proviral copies per genome, N= number of cells at time of transduction (corresponding to about 1×10^5 NIH3T3 cells per well), D=dilution of vector preparation, V=volume of diluted vector added in each well for transduction. Average values are obtained from at least two of the vector dilutions to obtain an accurate titer.

3.2. In Vitro Evaluation of Engineered U7 snRNA Construct in Human Myoblasts

Evaluation of the U7 snRNA constructs efficiency to induce exon skipping can be achieved in either normal human myoblasts or DMD patient myoblasts. However, skipping efficiencies are likely to appear higher in DMD patient cells (see Note 5). Moreover, dystrophin restoration assay can only be performed from DMD patient cells since normal myoblasts already express normal level of dystrophin.

3.2.1. Lentiviral Transduction of Human Myoblasts

1. Grow immortalized normal myoblasts in human myoblast culture medium.
2. For lentiviral transduction, seed 1×10^5 myoblasts/well in a 12-well plate. After 24 h, incubate cells with 10^6–10^7 ip of lentiviral vector in a total volume of 500 μL of media. 4–6 h after transduction, add 1 mL of medium. Replace with fresh medium next morning. Cells can be subsequently amplified for further analysis (RNA and protein analysis) (see Note 6).

3.2.2. Skipping Analysis by RT-PCR

1. Extract total RNA from the transduced human myoblasts using the RNeasy kit.
2. Perform RT-PCR using the Access RT-PCR kit and the external primers surrounding the exon of interest. Use aliquots of 200 ng of total RNA in a 50 μL reaction volume.
3. After 30 cycles of the primary PCR of 94°C (30 s), 58°C (1 min), and 72°C (2 min), perform cDNA synthesis at 45°C for 45 min.
4. Perform 28 cycles of a nested PCR of 94°C (30 s), 58°C (1 min), and 72°C (2 min) using the internal primers and 2 μL of the reaction product from step 3. Analyze the final

product on 2% agarose gels. Figure 2 shows typical results from exon 51 skipping analysis.

3.2.3. Quantifying Dystrophin Expression by Western Blot

1. Treat myoblast with the Newcastle buffer and extract protein. Determine the total amount of protein with the BCA protein assay kit.
2. Denature the protein sample at 95°C for 5 min and load 50 μg of protein in a 5 % polyacrylamide gel with a 4 % stacking gel. Electrophorese for 4–5 h at 100 V. Transfer protein to a PVDF membrane overnight at 40 V.
3. Block the membrane for 1 h with 10% nonfat milk in PBS–Tween buffer. Detect dystrophin and α-actinin by probing the membrane with 1:100 dilution of NCL-DYS1 antibody and 1:200 dilution of α-actinin antibody, respectively. Incubate with a mouse horseradish peroxidase-conjugated secondary antibody (1:2,000) and goat horseradish peroxidase-conjugated secondary antibody (1:160,000) to visualize dystrophin and α-actinin, respectively, using the ECL analysis system.

   Scan membranes into digital pictures and quantify the band intensity using the ImageJ 1.33a software (http://rsb.info.nih.gov/ij/).

3.3. In Vivo Evaluation of AAV-U7 snRNA Vectors

The U7 snRNA designed specifically to target human dystrophin exons can be tested in vivo in a transgenic mouse model carrying the entire human DMD locus. These transgenic hDMD mice were generated at the Leiden University Medical Center (19, 20). For all animal experiments, one needs to get approval from the Home Office (for the UK) or similar institution for other countries and follow their guidelines and protocols. For in vivo evaluation of the U7 constructs, the tibialis anterior (TA) muscle is chosen for intramuscular injection.

3.3.1. Intramuscular Injection in the TA Muscle

1. Anesthetize 6 to 8-week-old hDMD mice under isoflurane and shave their legs to visualize the TA muscle.
2. Dilute 10^{11} vg of AAV2/1 vector encoding the different U7 snRNA constructs into a final volume of 25 μL of saline. Slowly inject AAV into the middle belly of the TA muscle using a 29G insulin syringe. Inject AAV into muscle while slowly backing out the injection needle.
3. Remove the animal from the anesthetic machine and monitor until it recovers.

3.3.2. Skipping Analysis by RT-PCR

1. Sacrifice mice at 4 weeks after the injections. Snap freeze the TA muscles in liquid nitrogen-cooled isopentane. Crush muscle samples in liquid nitrogen by hand and extract total RNA from the resulting powder using TRIzol reagent.

2. Use 200 ng of total RNA for RT-PCR analysis using the Access RT-PCR kit in a 50 µL reaction using external primers surrounding the exon of interest.
3. After 20 cycles of the primary PCR of 94°C (40 s), 60°C (40 s), and 72°C (40 s), carry out cDNA synthesis at 45°C for 45 min. Reamplify 2 µL of the reaction product in a nested PCR by 30 cycles of 94°C (40 s), 60°C (40 s), and 72°C (40 s) using internal primers. Analyze PCR product on a 2% agarose gel.

4. Notes

1. As mutations in the U7 snRNA gene and, in particular, in the antisense sequence may affect the skipping efficiency, we strongly recommend using high fidelity Taq polymerase for all PCR-mediated cloning, such as the *pfu* DNA polymerase from Stratagene. It is also essential to confirm the cloned PCR product by sequencing.
2. When growing the rAAV vector plasmid, do not allow the culture to remain in stationary phase too long, as ITR deletions occur more rapidly after the culture has proceeded past log phase. The pXX6-80 plasmid is also unstable in bacteria. After receiving the plasmid, transform pSub201 and pXX6-80 into SURE cell (Stratagene) and check the integrity of the miniprepped plasmid by restriction digest. Check the integrity of AAV ITRs by SmaI restriction digest. After selection of an intact clone, individual glycerol stocks should be prepared. When growing the plasmid, grow the bacteria at 30°C for no longer than 12 h. This comment also applies to the transfer vectors used for lentiviral production (plasmid 12252: pRRLSIN.cPPT.PGK-GFP.WPRE from Addgene). These plasmids contain the long terminal repeats.
3. The packaging plasmids (pLP1, pLP2, and pLP/VSVG from Invitrogen) provide the required *trans*-acting factors, namely Gag-Pol, Rev, and the envelope protein VSVG, respectively. Plasmid pLP1 encodes a Gag-Pol precursor protein that is eventually processed into an integrase, reverse transcriptase and structural proteins. While the structural proteins are an absolute requirement of particle production, integrase and reverse transcriptase (packaged into the viral particle) are involved in events subsequent to infection. Rev interacts with a cis-acting element (RRE) in the transfer vector enhancing export of unspliced, full-length genomic transcripts. The presence of VSVG in the viral envelope membrane confers the viral particle with the ability to transduce a broad range of cell types, including primary cells.

4. There are several possibilities to consider when deciding to skip an exon. If the exon of interest has already been efficiently skipped using AONs and sequences are available in the literature, then these can be inserted into U7smOpt. When choosing that option, the optimal result will be obtained by targeting two sequences (i.e., inserting two antisense sequences into U7) (21). The most favorable length of antisense sequence inserted into U7smOpt to replace its natural histone annealing sequence is approximately 40 nt (i.e., 2 antisense sequences of 20 nt each).

5. If the exon of interest has not been skipped yet using AON, the tailed U7 construct option may be chosen. In that case, a 16-nt long nonhybridizing tail carrying two high-affinity binding sites (underlined in the tail sequence) for the hnRNP A1 protein (TATGATAGGGACTTAGGGTG) (22) will be added in 5′ of a 20–25 nt long complementary sequence to the targeted exon. These sequences can be chosen either randomly on the exon or to target any particular splice site or enhancers of splicing to increase the chance of efficient skipping. The splicing silencer effect of the tail and masking effect of the antisense sequence will be cumulative, making their combination likely to be effective (18).

   For each tailed U7 snRNA engineered, a corresponding control targeting the same sequence and carrying a mutated version of the tail A1 should be designed (GGG–CGC mutations) to evaluate the contribution of the tail compared to the antisense sequence.

5. U7 snRNA constructs may appear more effective when tested in DMD patient cells when compared to normal cells if they target an out-of-frame exon. This may occur even though viral infections are performed under exactly the same conditions. This observation has previously been reported by others (15, 23) and could be explained by the nonsense-mediated mRNA decay (NMD). NMD protects the cell from potentially detrimental dominant negative truncated proteins (24). Thus, in human control cells, the skipping of an out-of-frame exon like exon 51 results in frame-shifted transcript. This resulting transcript containing a stop codon will be targeted for NMD. In patient myoblasts with an appropriate genotype, however, the original transcripts are subjected to degradation. In this case, in-frame mRNA resulting from the skipping of exon 51 would escape NMD and would therefore benefit from a form of enrichment.

6. Transduction conditions described in Subheading 3.2.1 generally lead to 100% of transduction efficiency. Since lentiviral vectors induce stable gene transfer through genome integration, transduced myoblasts can divide without losing the U7

snRNA transgene. Therefore, cells can be expanded to increase starting material for further analysis like RNA extraction or Western blot.

Acknowledgments

We would like to thank Vincent Mouly (Institut de Myologie, Paris) for providing the immortalized myoblasts used in this study. We also thank Annemieke Aarstma-Rus and Johan T. den Dunnen (Leiden University Medical Center, the Netherlands) for providing the transgenic human DMD mice. This work was supported by Action Duchenne, the Association Monegasque contre les myopathies, and Duchenne Parent Project de France. A.G. was supported by an EMBO long-term postdoctoral fellowship.

References

1. Koenig, M., Beggs, A. H., Moyer, M., Scherpf, S., Heindrich, K., Bettecken, T., Meng, G., Muller, C. R., Lindlof, M., Kaariainen, H., and et al. (1989) The molecular basis for Duchenne versus Becker muscular dystrophy: correlation of severity with type of deletion. *Am J Hum Genet* 45, 498–506.
2. Monaco, A. P., Bertelson, C. J., Liechti-Gallati, S., Moser, H., and Kunkel, L. M. (1988) An explanation for the phenotypic differences between patients bearing partial deletions of the DMD locus. *Genomics* 2, 90–95.
3. Aartsma-Rus, A., Janson, A. A., Heemskerk, J. A., De Winter, C. L., Van Ommen, G. J., and Van Deutekom, J. C. (2006) Therapeutic modulation of DMD splicing by blocking exonic splicing enhancer sites with antisense oligonucleotides. *Ann NY Acad Sci* 1082, 74–76.
4. Mann, C. J., Honeyman, K., Cheng, A. J., Ly, T., Lloyd, F., Fletcher, S., Morgan, J. E., Partridge, T. A., and Wilton, S. D. (2001) Antisense-induced exon skipping and synthesis of dystrophin in the mdx mouse. *Proc Natl Acad Sci USA* 98, 42–47.
5. De Angelis, F. G., Sthandier, O., Berarducci, B., Toso, S., Galluzzi, G., Ricci, E., Cossu, G., and Bozzoni, I. (2002) Chimeric snRNA molecules carrying antisense sequences against the splice junctions of exon 51 of the dystrophin pre-mRNA induce exon skipping and restoration of a dystrophin synthesis in Delta 48–50 DMD cells. *Proc Natl Acad Sci USA* 99, 9456–9461.
6. Brun, C., Suter, D., Pauli, C., Dunant, P., Lochmuller, H., Burgunder, J. M., Schumperli, D., and Weis, J. (2003) U7 snRNAs induce correction of mutated dystrophin pre-mRNA by exon skipping. *Cell Mol Life Sci* 60, 557–566.
7. Schumperli, D., and Pillai, R. S. (2004) The special Sm core structure of the U7 snRNP: far-reaching significance of a small nuclear ribonucleoprotein. *Cell Mol Life Sci* 61, 2560–2570.
8. Goyenvalle, A., Vulin, A., Fougerousse, F., Leturcq, F., Kaplan, J. C., Garcia, L., and Danos, O. (2004) Rescue of dystrophic muscle through U7 snRNA-mediated exon skipping. *Science* 306, 1796–1799.
9. Dunckley, M. G., Manoharan, M., Villiet, P., Eperon, I. C., and Dickson, G. (1998) Modification of splicing in the dystrophin gene in cultured Mdx muscle cells by antisense oligoribonucleotides. *Hum Mol Genet* 7, 1083–1090.
10. Mann, C. J., Honeyman, K., McClorey, G., Fletcher, S., and Wilton, S. D. (2002) Improved antisense oligonucleotide induced exon skipping in the mdx mouse model of muscular dystrophy. *J Gene Med* 4, 644–654.
11. Aartsma-Rus, A., De Winter, C. L., Janson, A. A., Kaman, W. E., Van Ommen, G. J., Den Dunnen, J. T., and Van Deutekom, J. C. (2005) Functional analysis of 114 exon-internal AONs for targeted DMD exon skipping: indication for steric hindrance of SR protein binding sites. *Oligonucleotides* 15, 284–297.

12. Fairbrother, W. G., Yeh, R. F., Sharp, P. A., and Burge, C. B. (2002) Predictive identification of exonic splicing enhancers in human genes. *Science* 297, 1007–1013.
13. Cartegni, L., and Krainer, A. R. (2003) Correction of disease-associated exon skipping by synthetic exon-specific activators. *Nat Struct Biol* 10, 120–125.
14. Zhang, X. H., and Chasin, L. A. (2004) Computational definition of sequence motifs governing constitutive exon splicing. *Genes Dev* 18, 1241–1250.
15. Aartsma-Rus, A., Bremmer-Bout, M., Janson, A. A., den Dunnen, J. T., van Ommen, G. J., and van Deutekom, J. C. (2002) Targeted exon skipping as a potential gene correction therapy for Duchenne muscular dystrophy. *Neuromuscul Disord* 12 Suppl 1, S71–S7.
16. Wilton, S. D., Fall, A. M., Harding, P. L., McClorey, G., Coleman, C., and Fletcher, S. (2007) Antisense oligonucleotide-induced exon skipping across the human dystrophin gene transcript. *Mol Ther* 15, 1288–1296.
17. Popplewell, L. J., Trollet, C., Dickson, G., and Graham, I. R. (2009) Design of phosphorodiamidate morpholino oligomers (PMOs) for the induction of exon skipping of the human DMD gene. *Mol Ther* 17, 554–561.
18. Goyenvalle, A., Babbs, A., van Ommen, G. J., Garcia, L., and Davies, K. E. (2009) Enhanced exon-skipping induced by U7 snRNA carrying a splicing silencer sequence: Promising tool for DMD therapy. *Mol Ther* 17, 1234–1240.
19. Bremmer-Bout, M., Aartsma-Rus, A., de Meijer, E. J., Kaman, W. E., Janson, A. A., Vossen, R. H., van Ommen, G. J., den Dunnen, J. T., and van Deutekom, J. C. (2004) Targeted exon skipping in transgenic hDMD mice: A model for direct preclinical screening of human-specific antisense oligonucleotides. *Mol Ther* 10, 232–240.
20. t Hoen, P. A., de Meijer, E. J., Boer, J. M., Vossen, R. H., Turk, R., Maatman, R. G., Davies, K. E., van Ommen, G. J., van Deutekom, J. C., and den Dunnen, J. T. (2008) Generation and characterization of transgenic mice with the full-length human DMD gene. *J Biol Chem* 283, 5899–5907.
21. Suter, D., Tomasini, R., Reber, U., Gorman, L., Kole, R., and Schumperli, D. (1999) Double-target antisense U7 snRNAs promote efficient skipping of an aberrant exon in three human beta-thalassemic mutations. *Hum Mol Genet* 8, 2415–2423.
22. Burd, C. G., and Dreyfuss, G. (1994) RNA binding specificity of hnRNP A1: significance of hnRNP A1 high-affinity binding sites in pre-mRNA splicing. *Embo J* 13, 1197–1204.
23. van Deutekom, J. C., Bremmer-Bout, M., Janson, A. A., Ginjaar, I. B., Baas, F., den Dunnen, J. T., and van Ommen, G. J. (2001) Antisense-induced exon skipping restores dystrophin expression in DMD patient derived muscle cells. *Hum Mol Genet* 10, 1547–1554.
24. Maquat, L. E. (1995) When cells stop making sense: effects of nonsense codons on RNA metabolism in vertebrate cells. *RNA* 1, 453–465.
25. Aartsma-Rus, A., Janson, A. A., Kaman, W. E., Bremmer-Bout, M., den Dunnen, J. T., Baas, F., van Ommen, G. J., and van Deutekom, J. C. (2003) Therapeutic antisense-induced exon skipping in cultured muscle cells from six different DMD patients. *Hum Mol Genet* 12, 907–914.

Chapter 12

Application of MicroRNA in Cardiac and Skeletal Muscle Disease Gene Therapy

Zhan-Peng Huang, Ronald L. Neppl Jr., and Da-Zhi Wang

Abstract

MicroRNAs (miRNAs) are a class of small ~22 nt noncoding RNAs. miRNAs regulate gene expression at the posttranscriptional levels by destabilization and degradation of the target mRNA or by translational repression. Numerous studies have demonstrated that miRNAs are essential for normal mammalian development and organ function. Deleterious changes in miRNA expression play an important role in human diseases. We and others have previously reported several muscle-specific miRNAs, including miR-1/206, miR-133, and miR-208. These muscle-specific miRNAs are essential for normal myoblast differentiation and proliferation, and they have also been implicated in various cardiac and skeletal muscular diseases. miRNA-based gene therapies hold great potential for the treatment of cardiac and skeletal muscle disease(s). Herein, we introduce the methods commonly applied to study the biological role of miRNAs, as well as the techniques utilized to manipulate miRNA expression.

Key words: miRNA, Muscle, Cardiac, Gene expression, Posttranscriptional regulation, Gene therapy

1. Introduction

Cardiac and skeletal muscle disorders are a group of diseases caused by different mechanisms, including defective structural proteins, disorganized sarcomeres, and perturbed regulation of growth/maturation signaling pathways (1). In cardiac muscles, these diseases can be classified into congenital heart defects such as hypoplastic left/right heart syndrome and cardiac septum defects, as well as cardiomyopathies such as dilated cardiomyopathy and ischemic cardiomyopathy. In skeletal muscle, these diseases can be classified as neuromuscular such as multiple sclerosis or musculoskeletal such as Duchenne muscular dystrophy and myotubular myopathy. In the adult heart, pathological cardiac hypertrophy is the end-term consequence of stresses such as

Dongsheng Duan (ed.), *Muscle Gene Therapy: Methods and Protocols*, Methods in Molecular Biology, vol. 709,
DOI 10.1007/978-1-61737-982-6_12, © Springer Science+Business Media, LLC 2011

hypertension, ischemia, and neurohormone disorders (2) and is often accompanied by myocardial alterations such as fibrosis and necrosis. As a result, pathological cardiac hypertrophy diminishes cardiac output and pumping efficiency, which in many cases eventually leads to heart failure. Despite the fact that cardiovascular disease is the leading cause of death worldwide, the underlying molecular mechanisms for these and other muscle-related diseases are not completely understood.

Recently, a large class of ~22 nt noncoding RNAs has been discovered and is collectively referred to as microRNAs (miRNAs). To date, thousands of miRNA genes have been identified in multiple organisms from nematodes and plants, to fish and mammals (3). Similar to protein encoding genes, expression of a miRNA begins with the transcription of the miRNA gene by RNA polymerase II. Though many miRNAs are under the control of their own promoter, some miRNA genes are found in clusters sharing a single promoter (4), while others are encoded within an intron and coexpressed with their host gene (5). After transcription, the primary miRNA transcript is processed in the nucleus by the microprocessor complex (Drosha/DGCR8) into a hairpin intermediate commonly referred to as a pre-miRNA. However, a small subgroup of miRNAs found within short introns is known to bypass this step (6). Pre-miRNAs are then exported from the nucleus by exportin-5 (7), where they are further processed into miRNA duplexes by the cytosolic RNase III enzyme Dicer (8). Finally, the functional strand of the miRNA duplex is loaded into the RNA-induced silencing complex (RISC) to facilitate target mRNA degradation and/or translational repression (9).

Evolutionarily conserved miRNAs have been identified in multiple eukaryotes from *Caenorhabditis elegans*, to *Drosophila melanogaster*, *Mus musculus*, and *Homo sapiens*. The *C. elegans* genome contains a single mammalian miR-1 ortholog (10), whereas in higher eukaryotes there are multiple genes encoding miR-1 (identical coding sequences of *miR-1-1* and *miR-1-2*). miR-1 differs from miR-206 by only four nucleotides (11). Several mammalian miRNAs, including the miR-1/206 and miR-133 families, and miR-208a/b, are specifically expressed in cardiac and skeletal muscle (12–14). miR-1 is known to regulate the *C. elegans* neuromuscular junction (15) as well as skeletal muscle differentiation and proliferation in C2C12 myoblasts (*M. musculus*) (13). In addition, miR-1 expression is dependent upon serum response factor in both the *D. melanogaster* (16) and *M. musculus* (12) model systems. Together, these data strongly suggest that both the function and regulation of miR-1 are conserved throughout eukaryotic evolution.

miRNAs play important roles in cell proliferation, differentiation, migration, and apoptosis during both normal development and disease progression (17, 18). miR-1 and miR-133 are clustered on the same chromosomal locus and transcribed together as a single transcript. However, they represent two distinct miRNAs, each with its own biological function (13). miR-1 overexpression has been shown to negatively regulate skeletal muscle differentiation in the *Xenopus laevis* model system (13). miR-1 also significantly impairs normal cardiac development in both the *X. laevis* and *M. musculus* model systems (12, 13). Further, miR-1 is overexpressed in individuals with coronary artery disease and miR-1 overexpression induces arrhythmogenesis through negative regulation of *Kcnj2* and *Gja1* in mouse model (19). Cardiac-specific miR-208a is known to potentiate stress-dependent cardiac hypertrophy and remodeling by downregulating the expression of thyroid hormone receptor-associated protein 1 (14). In addition, the expression of many miRNAs is altered under pathological conditions. Subsets of miRNAs are found to be both positively and negatively regulated in clinical human samples and animal models of cardiac and skeletal myopathies (20–23). In the *M. musculus* model system, overexpression of miR-195 in cardiomyocytes in vivo is sufficient to drive dilated cardiomyopathy and heart failure (21). In addition, dystrophin-deficient mice are found to have significantly decreased expression of miR-133a and miR-206 (24). Together, these results indicate that the proper expression of miRNAs is necessary for both normal development and function of skeletal and cardiac muscles.

As proper miRNA expression is critical for normal development and homeostasis, it is necessary to better understand the biological function of miRNAs. Strategies commonly used to investigate the function of a particular miRNA are the gain-of-function and loss-of-function studies. Gain-of-function studies are commonly performed in vitro, where cells can be transiently transfected with an expression construct encoding the pre-miRNA. Alternatively, synthetic miRNA duplexes and virus-based miRNA expression systems may also be employed. In vitro loss-of-function studies can be accomplished with either 2 -*O*-methyl miRNA antisense oligonucleotides or locked nucleic acid (LNA)-miRNA antisense oligonucleotides which will block the function of an endogenous miRNA. The in vivo determination of a miRNA's function is best examined utilizing conventional transgenic and gene knockout strategies. Recently, a lentivirus targeting strategy that overexpresses short RNA fragments containing multiple miRNA target sequences has been shown to phenocopy a genetic miRNA knockout mouse (25). In addition, intravenous delivery of cholesterol-modified miRNA antisense oligonucleotides (antagomiRs) can inhibit miRNA function in vivo (26). These molecular approaches are invaluable in elucidating the biological function of miRNAs and may potentially lend themselves to future gene-based therapies.

In this chapter, we are going to discuss how to determine the expression of miRNAs by Northern blotting analysis. Then, we will discuss the method of studying the regulation of miRNAs on their targets in vitro by luciferase reporter assays. Finally, we will talk about how to manipulate the function level of muscle miRNAs in C2C12 myoblast cell line and determine their biological function on muscle differentiation.

2. Materials

2.1. Detecting the Expression of Muscle miRNAs by Northern Blotting Analysis

1. Hoefer SE 400 vertical slab gel electrophoresis unit (Amersham Biosciences, Piscataway, NJ, USA).
2. Heofer TE77 Semidry transfer unit (Amersham Biosciences).
3. UV stratalinker 1800 (Stratagene, Cedar Creek, TX, USA).
4. Trizol reagent (Invitrogen, Carlsbad, CA, USA).
5. 40% acrylamide: AccuGel 29:1 (National Diagnostics, Atlanta, GA, USA).
6. 10× TBE buffer: 0.9 M Tris base, 0.9 M Boric acid, 0.02 M EDTA (pH 8.0), autoclave for 20 min.
7. Urea (molecular biology grade, Sigma-Aldrich, St. Louis, MO, USA).
8. 10% (w/v) Ammonium persulfate solution (APS). Aliquot and store at −20°C.
9. *N,N,N,N'*-Tetramethyl-ethylenediamine (TEMED, Fisher Bioreagent, Fair Lawn, NJ, USA).
10. Formamide (Fisher Bioreagent).
11. Bromophenol blue solution: 10% (w/v) bromophenol blue.
12. Filter paper, sheet, grade 3, 460 × 570 mm (Whatman, Clifton, NJ, USA).
13. Zeta-Probe GT genomic-tested blotting membranes (Bio-Rad, Hercules, CA, USA).
14. T4 polynucleotide kinase (Promega, Madison, WI, USA).
15. Mini Quick Spin Oligo Columns (Roche, Indianapolis, IN, USA).
16. Adenosine 5′-triphosphate (γ-^{32}P), 3,000 Ci/mmol (PerkinElmer, Waltham, MA, USA).
17. Anti-miRNA probe: the synthetic antisense oligonucleotide of the target miRNA.
18. Diethylpyrocarbonate (DEPC)-treated water: 1 mL DEPC in 1 L double distilled H_2O. Stir at room temperature for 1 h and autoclave.
19. 1 M phosphate buffer: 71 g of anhydrous Na_2HPO_4, 4 mL of 85% H_3PO4. Add DEPC-treated water to 1 L.

20. Hybridization buffer: 0.5 M phosphate buffer; 1 mM EDTA at pH 8.0; 7% (w/v) of sodium dodecyl sulfate (SDS); 1% (w/v) of bovine serum albumen (BSA); in DEPC-treated water.
21. 20× SSC: 3 M sodium chloride and 300 mM Trisodium citrate dihydrate, pH 7.0.
22. Wash buffer: 1× SSC supplemented with 0.1% SDS.
23. Stripping buffer: 0.1× SSC supplemented with 0.1% SDS.
24. Storage phosphor screen (Amersham Biosciences).

2.2. Studying the Regulation of Muscle miRNAs on Their Targets by Luciferase Reporter Assays

1. Mouse genomic DNA.
2. Primers for reporter construction: HDAC4-UTR-F, 5′-ATCGGAGCTCCAGCACTGGTGATAGACTTGG-3′; HDAC4-UTR-R, 5′-GTCTTATTGAACTTATTCTTAAGCTCGAGATCG-3′; HDAC4-Mut-F, 5′-GTTTCTTTCCTCAGA**TTTAAAA**TTCTTCACTGGTCACAGCCACG-3′; HDAC4-Mut-R, 5′-GTGACCAGTGAAGAA**TTTTAAA**TCTGAGGAAAGAAACACAACC-3′.
3. PfuTurbo DNA Polymerase (Stratagene).
4. pGL3cM vector (modified by Chen JF and Wang DZ). The backbone is the pGL3-Control vector (Promega).
5. Sac I restriction endonuclease (New England Biolabs, Ipswich, MA, USA).
6. Xho I restriction endonuclease (New England Biolabs).
7. T4 DNA ligase (New England Biolabs).
8. pShuttle-CMV-lacZ Vector (Stratagene).
9. NucleoBond plasmid Maxi kit (Macherey-Nagel, Bethlehem, PA, USA).
10. HEK293T cells (ATCC, Manassas, VA, USA).
11. CELLSTAR 12- and 24-well tissue culture plate (Greiner Bio-One, Monroe, NC, USA).
12. Cell culture medium: DMEM, high glucose with L-glutamine (Gibco-BRL, Langley, OK, USA).
13. Fetal bovine serum (FBS) (Hyclone, Logan, UT, USA).
14. Penicillin G-Streptomycin (PS): Penicillin 100 U/mL DMEM culture medium and Streptomycin 100 μg/mL DMEM culture medium (Gibco-BRL).
15. 1× Trypsin-EDTA: 0.25% Trypsin, 1 mM EDTA/4Na (Gibco-BRL).
16. Lipofectamine LTX and Plus reagent (Invitrogen, Carlsbad, CA, USA).
17. Opti-MEM I reduced-serum medium (Gibco-BRL, Langley).
18. miR-1 miRIDIAN miRNA mimic (Dharmacon, Lafayette, CO, USA).

19. 10× Phosphate-buffered saline (PBS) solution: 80.6 mM sodium phosphate, 19.4 mM potassium phosphate, 27 mM KCl and 1.37 M NaCl at pH 7.4.
20. Luciferase assay system (Promega).
21. LacZ buffer: 0.06 M of Na_2HPO_4, 0.045 M of NaH_2PO_4, 0.01 M of KCl, 2 mM of $MgSO_4$.
22. LacZ substrate: to 5 mL of LacZ buffer add 1 mL of o-Nitrophenyl-β-galactopyranoside (ONPG) stock solution (4 mg/mL), and 13 μL of β-Mercaptoethanol.
23. 1 M Na_2CO_3.

2.3. Overexpression and Knockdown of Muscle miRNAs In Vitro

1. Primers for miR-22 overexpression vector construction: miR22-F 5′-TAGCAGGTACCTTATTCAAGAACCCCTCATTAG-3′, miR22-R 5′-GTATCTCTAGATTTCCCTCCCATAAAGCCAT-3′.
2. pcDNA3.1(+) vector (Invitrogen).
3. Anti-miR-22 probe: antisense oligonucleotide to miR-22.
4. C2C12 cells (ATCC).
5. Kpn I restriction endonuclease (New England Biolabs).
6. Xba I restriction endonuclease (New England Biolabs).
7. 2′-*O*-methyl miR-133 antisense oligonucleotide (Dharmacon).
8. Growth medium for C2C12: DMEM medium with 10% FBS and 1% PS.
9. Differentiation medium for C2C12: DMEM medium with 2% horse serum and 1% PS.
10. Nonidet P-40 (NP-40) (Fisher Bioreagent).
11. 4% Paraformaldehyde (PFA) solution: 4% (w/v) PFA in 1× PBS.
12. Antibody buffer: 0.1% NP40 and 3% BSA in 1× PBS.
13. Anti-phospho-histone H3 antibody (Upstate, Lake Placid, NY, USA).
14. Alexa-488 or Alexa-495 conjugated goat anti-rabbit IgG antibody (Molecular Probes, Eugene, OR, USA).
15. 4′,6-Diamidino-2-phenylindole dihydrochloride (DAPI) (Molecular Probes).

3. Methods

3.1. Detecting the Expression of Muscle miRNAs by Northern Blotting Analysis

1. Prepare total RNA from tissue or cultured cells with Trizol reagent according to manufacturer's protocol (see Note 1).
2. Prepare 15% denaturing gel for electrophoresis separation of miRNAs. Carefully wash, dry, and assemble the Hoefer SE 400 vertical slab gel electrophoresis unit. Prepare denaturing

gel containing 18.75 mL of 40% acrylamide, 2.5 mL of 10× TBE buffer, 12.5 mL of DEPC-treated water, and 20 g of urea. Mixture may need to be gently heated in 37°C water bath in order for urea to completely dissolve. To polymerize, add 400 μL of 10% APS; 40 μL of TEMED, mix well, and quickly pour. Allow the gel polymerize for 1 h.

3. Prerun denaturing gel for 30 min at 200 V. Use 0.5× TBE for running buffer.
4. Prepare RNA samples for electrophoresis. Mix the RNA sample (40 μg) 1:1 (v/v) with formamide, and incubate at 65°C for 10 min. Chill RNA on ice for 3 min and add 2 μL of bromophenol blue solution. Mix well.
5. Load the sample(s) into the well(s) and run the gel at 250 V (see Note 2). Use 0.5× TBE for running buffer. Voltage can be stopped when the loading dye reaches the bottom of the plate.
6. Transfer the RNA from the gel to the membrane with Heofer TE77 Semidry transfer unit. Soak the membrane and six pieces of filter papers in 0.5× TBE. Set up the transfer in the order from top (−) to bottom (+) as: three pieces of filter paper, gel, membrane, three pieces of filter paper (see Note 3). Transfer with constant current (0.8 mA/cm^2 of gel area) for 1 h.
7. After transfer, wash the membrane with 0.5× TBE and perform UV crosslink using the auto crosslink option.
8. Mix 5 μL of Adenosine 5'-triphosphate [γ-^{32}P], 5 μL of 1 μM anti-miRNA probe, 2 μL of 10× PNK buffer, 1 μL of T4 polynucleotide kinase, and 7 μL of double distilled water and incubate at 37°C for 1 h.
9. Purify the [γ-^{32}P]-labeled probe using a mini Quick Spin Oligo Column according to manufacturer's protocol (see Note 4).
10. Prehybridize the membrane for 1 h at 37°C with 5–10 mL of hybridization buffer.
11. Add the labeled anti-miRNA probe into the hybridization buffer and incubate overnight at 37°C.
12. Remove the hybridization buffer and wash the membrane 3 times with wash buffer (10 min per wash).
13. Expose the membrane to the storage phosphor screen for 4–24 h. The length of exposure depends upon strength of signal and will vary with different miRNA probes (see Note 5).
14. Scan the screen with Typhoon phosphor-imager.
15. If you want to probe the membrane with a different miRNA probe or if you want to store the membrane for long term, add the membrane to heated stripping buffer (>95°C). Incubate for approximately 10 min while rocking.

16. After stripping, rinse membrane with fresh stripping buffer, and then allow to dry. The membrane may be reprobed immediately following the steps outlined above. The membrane can also be stored at –20°C for future use.

3.2. Studying the Regulation of Muscle miRNAs on Their Targets by Luciferase Reporter Assays

Here we show an example using the luciferase reporter vectors which contain either the wild-type or the mutant HDAC4 3′ UTR.

1. Generate the ~400 bp HDAC4 gene 3′ UTR DNA fragment containing the seed sequence for miR-1 by PCR reaction using mouse genomic DNA as the template and the HDAC4-UTR-F and HDAC4-UTR-R primers. The Sac I and Xho I sites are introduced at the 5′ and 3′-end, respectively, by the PCR primers. The UTR PCR products are cloned into the Sac I/Xho I sites of the pGL3cM vector (see Note 6). The resulting Luc-WT-UTR reporter contains the wild-type 3′ UTR of the HDAC4 gene.
2. Generate the Luc-Mut-UTR reporter using the same method described in step 1 except for the primers. Here, the HDAC4-Mut-F and HDAC4-Mut-R primers are used in the PCR reaction (Fig 1).
3. Prepare high-quality Luc-WT-UTR reporter, Luc-Mut-UTR reporter, and pShuttle-CMV-lacZ plasmid for reporter assays with NucleoBond Plasmid Maxi Kit. These plasmids will be used for HEK293T cell transfection.
4. At one day before transfection, plate HEK293T cells in a 24-well plate at 5×10^4 cells per well in 500 μL of growth medium (see Note 7). This will yield 50–80% confluence at the day of transfection.
5. To generate the transfection complex for one well, add 25 ng reporter (either Luc-WT-UTR or Luc-Mut-UTR), 25 ng pShuttle-CMV-lacZ plasmid, and 0.5 μL of 10 μM miRIDIAN miRNA mimic to 100 μL of Opti-MEM I reduced-serum medium and mix gently. To this mixture, add 0.5 μL of PLUS reagent, mix gently, and incubate for 5–10 min at room temperature. Finally, add 1.25 μL of Lipofectamine LTX Reagent, mix gently, and incubate for 30 min at room temperature.
6. Add the transfection complex (~100 μL) to the well. Mix gently by rocking the plate back and forth.
7. At 24 h after transfection, remove cell culture medium and wash cells twice with 1× PBS.
8. Lyse cells with 100 μL of cell lysis buffer and perform one freeze–thaw cycle (see Note 8).
9. Mix 40 μL of cell lysate with 40 μL of luciferase substrate and measure the signal with a scintillation counter (see Note 9).

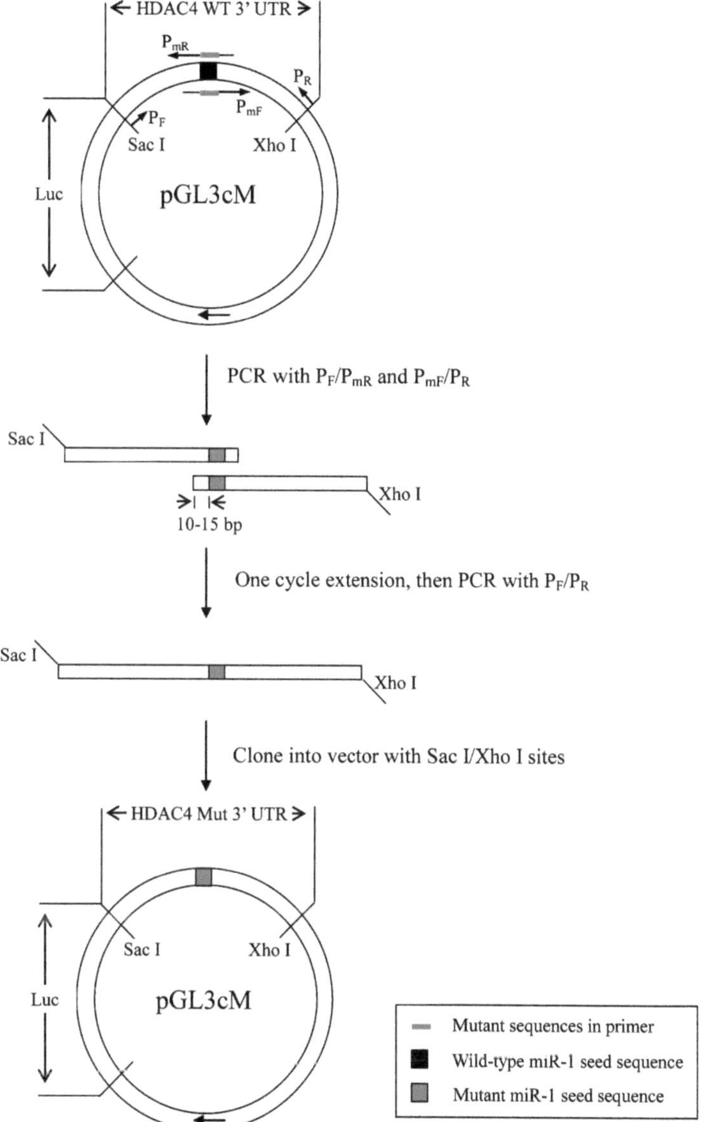

Fig. 1. The strategy for the construction of the Luc-Mut-UTR reporter. The primer pair (P_{mF}/P_{mR}) for mutation introduction contains complementary mutated nucleotides (see Subheading 2.2, sequences in bold) for miR-1 seed sequence mutant. Besides the mutated sequence, these primers contain 20–25 nt 3′ end sequence and 10–15 nt 5′ end sequence. PCR with P_F/P_{mR} and P_{mF}/P_R produces two DNA fragments with 30–40 bp overhang, containing the miR-1 seed sequence mutant. The Mut-UTR template is generated by one cycle of RCR reaction.

10. Mix 20 μL of cell lysate with 150 μL of the LacZ substrate and incubate at 37°C with gentle rocking until the mixture turns yellow. Add 50 μL of 1 M Na_2CO_3 to stop the reaction. The signal is measured by scintillation counter.
11. Normalize luciferase activity to LacZ activity.

3.3. Overexpression or Knockdown of Muscle miRNAs In Vitro

In this section, we describe steps to overexpress miR-22 in HEK293T cells and to knockdown miR-133 in C2C12 myoblasts.

1. For the overexpression study, use PCR to generate a ~350 bp DNA fragment containing the intact hairpin for the miR-22 precursor plus the flanking sequences on both ends (see Note 10). Use mouse genomic DNA as the template. The Kpn I and Xba I sites are introduced at the 5′ and 3′-end, respectively, by the PCR primers. Clone the PCR product into the Kpn I/Xba I sites in the pcDNA3.1(+) vector (see Note 11). The resulting construct is termed the miR-22 overexpression vector (Fig. 2a).
2. Prepare high-quality plasmid for transfection with NucleoBond Plasmid Maxi Kit.
3. Transfect the miR-22 overexpression vector into HEK293T cells following the steps outlined in Subheading 3.2.
4. Extract total RNA from cells 48 h after transfection using the Trizol reagent according to manufacturer's protocol.
5. Evaluate the overexpression of miR-22 by Northern blot according to the protocol described in Subheading 3.1.

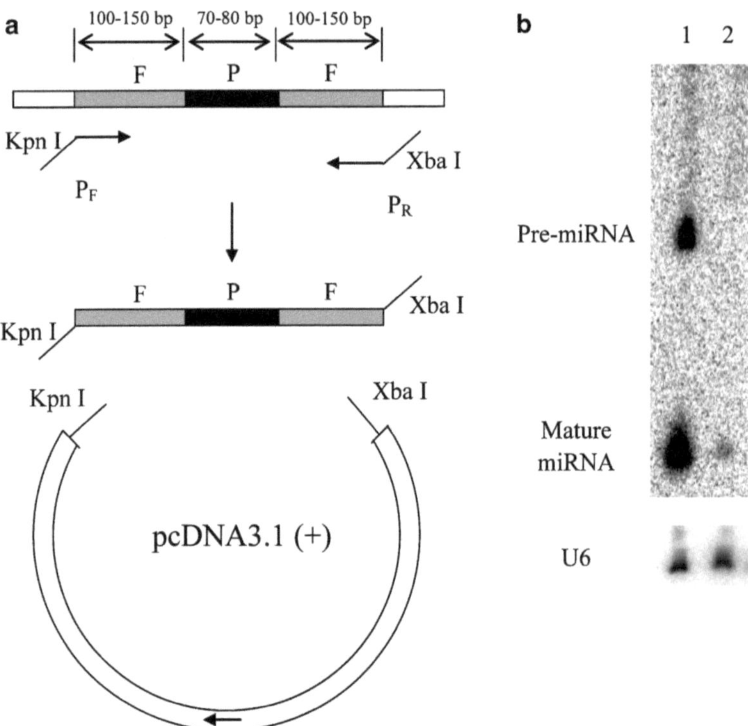

Fig. 2. (**a**) The strategy for the construction of miR-22 overexpression vector. *P*, miRNA hairpin precursor; *F*, flanking sequences. (**b**) Evaluate the overexpression of miR-22 by Northern blotting analysis. Lane 1, cells transfected with the miR-22 overexpression vector; Lane 2, cells transfected with a control vector. U6 snRNA serves as loading control.

Similarly, the miR-22 overexpression vector can also be evaluated in other cells such as C2C12 myoblasts (Fig. 2b).

6. For miR-133 knockdown study, plate C2C12 myoblasts in a 12-well plate at 6×10^4 cells per well in 1 mL of growth medium one day before transfection. This will yield 50–80% confluence at the day of transfection.
7. Transfect C2C12 myoblasts with 200 nM 2′-*O*-methyl miR-133 antisense oligonucleotides (see Note 12). Use the same protocol described in Subheading 3.2, except to double all volumes. The final volume should be 200 μL per well.
8. Change growth medium 4–6 h after transfection and continue to culture the cells for additional 24 h.
9. 24 h after transfection, replace growth medium with differentiation medium and culture the cells for additional 12 h.
10. Confirm miR-133 knockdown by Northern blotting analysis according to the protocol described in Subheading 3.1.
11. Examine cell proliferation in C2C12 cells treated with 2′-*O*-methyl miR-133 antisense oligonucleotides. Wash the cells twice with 1× PBS, and then fix with 4% PFA for 5 min at room temperature. Wash the cells 3 times with 1× PBS containing 0.1% NP-40 (10 min each wash). Dilute the primary antibody, anti-phospho-histone H3, to 1:100 with the antibody buffer. Phospho-histone H3 signal indicates mitotic cells. Incubate overnight at 4°C. Wash the cells 3 times with 1× PBS containing 0.1% NP-40 (10 min each wash). Dilute the secondary antibody, Alexa Fluor 488 or Alexa Fluor 495-conjugated goat anti-rabbit IgG antibody, to 1:1,000 with the antibody buffer. Incubate at room temperature for 1 h. Wash the cells 3 times with 1× PBS containing 0.1% NP-40 (10 min each wash). Rinse the cells with 1× PBS. Dilute DAPI to 1:50,000 dilution in 1×PBS. Incubate at room temperature for 10 min. Rinse with 1× PBS 3 times. Observe immunofluorescence signal using an inverted fluorescence microscope (see Note 13) (Fig. 3).

4. Notes

1. RNase(s) rapidly degrade RNA and are abundant in the environment. When extracting total RNA from samples, RNase-free tubes, DEPC-treated water, and solutions made with DEPC-treated water are highly recommended. RNA samples can be preserved in pellet for more than 1 year if stored in 100% ethanol at −80°C.
2. Prior to loading the RNA sample into the denaturing gel, wash the well by flushing with 0.5× TBE running buffer.

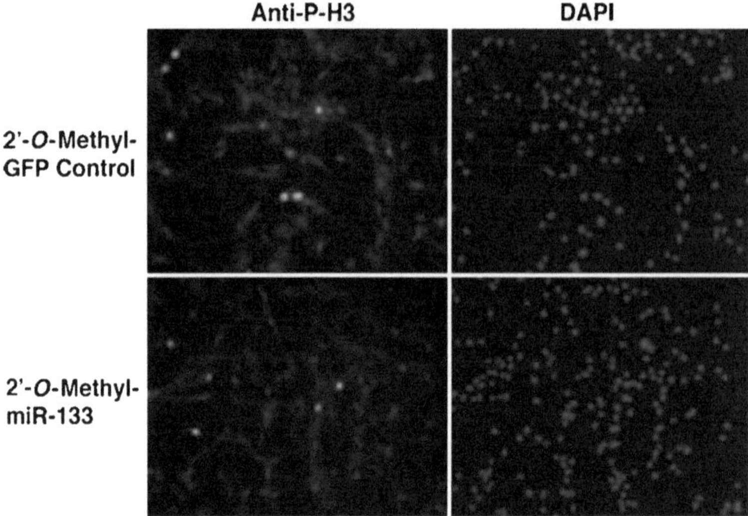

Fig. 3. Detecting phospho-histone H3 by immunostaining after miR-133 knockdown using 2′-*O*-methyl miR-133 antisense oligonucleotide in C2C12 cells. 2′-*O*-methyl GFP antisense oligonucleotide serves as control. DAPI counterstained nuclei.

Excess urea in the well will prevent the RNA sample from sinking to the bottom of the well.

3. Exclude air bubbles when assembling the "sandwich" for RNA transfer.
4. It is important to have enough protection when conducting the isotope-related experiments. Always wear personal protective equipment when handling radioisotopes.
5. Besides the phosphor-imager system, Northern blot can also be imaged with X-ray autography. In general, the membrane needs to be exposed to film for 1 day to 1 week.
6. To generate the pGL3cM vector, the multiple cloning sites (MCS) are removed from pGL3-control vector by Kpn I+Bgl II and filled in by Klenow. The 53 bp oligonucleotides containing the MCS are then introduced into the Xba I site.
7. At least 12 wells are needed for one experiment to examine four combinations of transfection reagents including Luc-WT-UTR reporter and miR-1 miRIDIAN mimic, Luc-WT-UTR reporter and control miRIDIAN mimic, Luc-Mut-UTR reporter and miR-1 miRIDIAN mimic, and Luc-Mut-UTR reporter and control miRIDIAN mimic. Each combination of transfection reagents is performed in triplicate.
8. Cells are frozen at –80°C for 1 h and then thawed at room temperature by rocking for 30 min. Once thawed, incubate the lysate on ice to prevent proteolysis.

9. The luciferase substrate should be preserved in –80°C and protected from light.
10. Different cloning strategies can be applied to generate a miRNA overexpression vector. In this protocol, our strategy is to clone the fragment containing the whole hairpin (miRNA precursor) and a 100–150 bp flanking sequence on both the 5′ and 3′ ends of the miRNA precursor sequence. Alternatively, the full-length noncoding transcript can be cloned into the expression vector. However, this is only applicable for miRNAs generated from a nonprotein coding gene.
11. Besides pcDNA3.1(+), other expression vectors can be used to construct a miRNA overexpression plasmid. Virus-based expression vectors have already been reported for miRNA overexpression (27).
12. In this protocol, a 2′-*O*-methyl miRNA antisense oligonucleotide is used to knockdown the endogenous miRNA. Alternatively, LNA antisense oligonucleotides can be used to obtain similar effects.
13. The immunofluorescence signal can be preserved in 1× PBS at 4°C for up to 1 month.

Acknowledgments

We thank members of the Wang laboratory for discussion and support. Research in the Wang lab was supported by the March of Dimes Birth Defect Foundation, National Institutes of Health and Muscular Dystrophy Association. DZ Wang is an established investigator of the American Heart Association.

References

1. Wagner, K.R. (2002) Genetic diseases of muscle. *Neurol Clin.* 20, 645–678.
2. Wakatsuki, T., Schlessinger, J., and Elson, E.L. (2004) The biochemical response of the heart to hypertension and exercise. *Trends Biochem Sci.* 29, 609–617.
3. Bartel, D.P. (2004) MicroRNAs: genomics, biogenesis, mechanism, and function. *Cell.* 116, 281–297.
4. He, L., Thomson, J.M., Hemann, M.T., Hernando-Monge, E., Mu, D., Goodson, S., Powers, S., Cordon-Cardo, C., Lowe, S.W., Hannon, G.J., and Hammond, S.M. (2005) A microRNA polycistron as a potential human oncogene. *Nature.* 435, 828–833.
5. Rodriguez, A., Griffiths-Jones, S., Ashurst, J.L., and Bradley, A. (2004) Identification of mammalian microRNA host genes and transcription units. *Genome Res.* 14, 1902–1910.
6. Ruby, J.G., Jan, C.H., and Bartel, D.P. (2007) Intronic microRNA precursors that bypass Drosha processing. *Nature.* 448, 83–86.
7. Yi, R., Qin, Y., Macara, I.G., and Cullen, B.R. (2003) Exportin-5 mediates the nuclear export of pre-microRNAs and short hairpin RNAs. *Genes Dev.* 17, 3011–3016.
8. Hutvágner, G., McLachlan, J., Pasquinelli, A.E., Bálint, E., Tuschl, T., and Zamore, P.D. (2001) A cellular function for the RNA-interference enzyme Dicer in the maturation

of the let-7 small temporal RNA. *Science.* 293, 834–838.

9. Schwarz, D.S., Hutvágner, G., Du, T., Xu, Z., Aronin, N., and Zamore, P.D. (2003) Asymmetry in the assembly of the RNAi enzyme complex. *Cell.* 115, 199–208.
10. Lee, R.C., and Ambros, V. (2001) An extensive class of small RNAs in *Caenorhabditis elegans. Science.* 294, 862–864.
11. Williams, A.H., Liu, N., van Rooij, E., and Olson, E.N. (2009) MicroRNA control of muscle development and disease. *Curr Opin Cell Biol.* 21(3):461–469.
12. Zhao, Y., Samal, E., and Srivastava, D. (2005) Serum response factor regulates a muscle-specific microRNA that targets Hand2 during cardiogenesis. *Nature.* 436, 214–220.
13. Chen, J.F., Mandel, E.M., Thomson, J.M., Wu, Q., Callis, T.E., Hammond, S.M., Conlon, F.L., and Wang, D.Z. (2006) The role of microRNA-1 and microRNA-133 in skeletal muscle proliferation and differentiation. *Nat Genet.* 38, 228–233.
14. van Rooij, E., Sutherland, L.B., Qi, X., Richardson, J.A., Hill, J., and Olson, E.N. (2007) Control of stress-dependent cardiac growth and gene expression by a microRNA. *Science.* 316, 575–579.
15. Simon, D.J., Madison, J.M., Conery, A.L., Thompson-Peer, K.L., Soskis, M., Ruvkun, G.B., Kaplan, J.M., and Kim, J.K. (2008) The microRNA miR-1 regulates a MEF-2-dependent retrograde signal at neuromuscular junctions. *Cell.* 133, 903–915.
16. Kwon, C., Han, Z., Olson, E.N., and Srivastava, D. (2005) MicroRNA1 influences cardiac differentiation in Drosophila and regulates Notch signaling. *Proc Natl Acad Sci USA.* 102, 18986–18991.
17. Callis, T.E., and Wang, D.Z. (2008) Taking microRNAs to heart. *Trends Mol Med.* 14, 254–260.
18. Chen, J.F., Callis, T.E., and Wang, D.Z. (2009) MicroRNAs and muscle disorders. *J Cell Sci.* 122, 13–20.
19. Yang,B., Lin, H., Xiao, J., Lu, Y., Luo, X., Li, B., Zhang, Y., Xu, C., Bai, Y., Wang, H., Chen, G., and Wang, Z. (2007) The muscle-specific microRNA miR-1 regulates cardiac arrhythmogenic potential by targeting GJA1 and KCNJ2. *Nat Med.* 13,486–491.
20. Eisenberg, I., Eran, A., Nishino, I., Moggio, M., Lamperti, C., Amato, A.A., Lidov, H.G,, Kang, P.B., North, K.N., Mitrani-Rosenbaum, S., Flanigan, K.M., Neely, L.A., Whitney, D., Beggs, A.H., Kohane, I.S., and Kunkel, L.M. (2007) Distinctive patterns of microRNA expression in primary muscular disorders. *Proc Natl Acad Sci USA.* 104, 17016–17021.
21. van Rooij, E., Sutherland, L.B., Liu, N., Williams, A.H., McAnally, J., Gerard, R.D., Richardson, J.A., and Olson, E.N. (2006) A signature pattern of stress-responsive microRNAs that can evoke cardiac hypertrophy and heart failure. *Proc Natl Acad Sci USA.* 103, 18255–18260.
22. van Rooij, E., Sutherland, L.B., Thatcher, J.E., DiMaio, J.M., Naseem, R.H., Marshall, W.S., Hill, J.A., and Olson, E.N. (2008) Dysregulation of microRNAs after myocardial infarction reveals a role of miR-29 in cardiac fibrosis. *Proc Natl Acad Sci USA.* 105, 13027–13032.
23. Tatsuguchi, M., Seok, H.Y., Callis, T.E., Thomson, J.M., Chen, J.F., Newman, M., Rojas, M., Hammond, S.M., and Wang, D.Z. (2007) Expression of microRNAs is dynamically regulated during cardiomyocyte hypertrophy. *J Mol Cell Cardiol.* 42, 1137–1141.
24. McCarthy, J.J., Esser, K.A., and Andrade, F.H. (2007) MicroRNA-206 is overexpressed in the diaphragm but not the hindlimb muscle of mdx mouse. *Am J Physiol Cell Physiol.* 293, C451–C457.
25. Gentner, B., Schira, G., Giustacchini, A., Amendola, M., Brown, B.D., Ponzoni, M., and Naldini, L. (2009) Stable knockdown of microRNA in vivo by lentiviral vectors. *Nat Methods.* 6, 63–66.
26. Krützfeldt, J., Rajewsky, N., Braich, R., Rajeev, K.G., Tuschl, T., Manoharan, M., and Stoffel, M. (2005) Silencing of microRNAs in vivo with 'antagomirs'. *Nature.* 438, 685–689.
27. Stegmeier, F., Hu, G., Rickles, R.J., Hannon, G.J., Elledge, S.J. (2005) A lentiviral microRNA-based system for single-copy polymerase II-regulated RNA interference in mammalian cells. *Proc Natl Acad Sci USA.* 102, 13212–13217.

Chapter 13

Molecular Imaging of RNA Interference Therapy Targeting PHD2 for Treatment of Myocardial Ischemia

Mei Huang and Joseph C. Wu

Abstract

Coronary artery disease is the number one cause of morbidity and mortality in the Western world. It typically occurs when heart muscle receives inadequate blood supply due to rupture of atherosclerotic plaques. During ischemia, up-regulation of hypoxia inducible factor-1 alpha (HIF-1α) transcriptional factor can activate several downstream angiogenic genes. However, HIF-1α is naturally degraded by prolyl hydroxylase-2 (PHD2) protein. Recently, we cloned the mouse PHD2 gene by comparing the homolog gene in human and rat. The best candidate shRNA sequence for inhibiting PHD2 was inserted behind H1 promoter, followed by a separate hypoxia response element (HRE)-incorporated promoter driving a firefly luciferase (Fluc) reporter gene. This construct allowed us to monitor gene expression noninvasively and was used to test the hypothesis that inhibition of PHD2 by short hairpin RNA interference (shRNA) can lead to significant improvement in angiogenesis and contractility as revealed by in vitro and in vivo experiments.

Key words: RNA interference, Molecular imaging, Hypoxia inducible factor (HIF), Prolyl hydroxylases (PHD), Ischemic heart disease

1. Introduction

Coronary artery disease (CAD) is the leading cause of morbidity and mortality in the Western world (1). Conventional treatment for CAD consists of medical therapy as the first-line strategy, followed by percutaneous coronary intervention (PCI) or coronary artery bypass graft (CABG). However, a significant number of patients will still have refractory angina despite these treatments (2). Over the past decades, a better understanding of the molecular and genetic bases of different diseases has made gene therapy an increasingly viable treatment option (3). With the use of gene transfer techniques, it is now possible to modify somatic cells in ischemic myocardium, to overexpress beneficial or inhibitory

Dongsheng Duan (ed.), *Muscle Gene Therapy: Methods and Protocols*, Methods in Molecular Biology, vol. 709,
DOI 10.1007/978-1-61737-982-6_13,

pathologic proteins, and to achieve positive therapeutic effects (4). Indeed, several clinical trials evaluating both viral and nonviral delivery of vascular endothelial growth factor (VEGF) and fibroblast growth factor (FGF) have been completed to date (5).

A growing body of evidence suggests that the expression of a single angiogenic factor alone is *not* sufficient for the functional revascularization of ischemic tissues (6). Newer approaches based on the transcriptional factor HIF-1α may be a more natural choice. HIF-1α is known to control the expression of over 60 genes that affect cell survival and metabolism in adverse conditions, including VEGF (7), insulin-like growth factor (8), erythropoietin (9), nitric oxide synthase (10), among others. However, during normoxia, HIF-1α subunits have an exceptionally short half-life (~3–5 min) and low steady-state levels (11). This is due to hydroxylation of two prolyl residues (Pro402 and Pro564) by a family of prolyl-4-hydroxylases (PHDs) (12). Hydroxylation of HIF-1α allows recognition by the von Hippel–Lindau (VHL) tumor suppressor, which targets HIF-1α for proteosomal destruction (13). In contrast, increasing the severity of hypoxia retards degradation of HIF-1α subunits, allowing nuclear localization, dimerization with HIF-1β subunits, and formation of a stable DNA-binding HIF complex (14). Thus, the short hairpin RNA (shRNA) plasmid for knockdown of PHD2 (shPHD2) can potentially be used as a novel gene therapy for treatment of ischemic heart disease.

To date, the majority of cardiac gene therapy studies have relied on ex vivo quantification of gene expression (e.g., GFP or lacZ) in small animals or indirect markers (e.g., changes in perfusion or contractility) in clinical trials (15, 16). In order to *characterize*, *visualize*, and *quantify* biological processes at the molecular and cellular levels within intact living organisms, in vivo imaging techniques are needed. Over the past 10 years, molecular imaging has been widely used for oncology studies, but applications in cardiology have been a recent development (17). One such example is the use of reporter genes that can be transferred into cells via a delivery vector and regulated by constitutive, inducible, or tissue-specific promoters.

In this protocol, we outline the procedures used to address the two issues mentioned above – better therapeutic gene and more sophisticated tracking method. We show that the inhibition of HIF-1α degradation via shRNA knockdown of PHD2 in the ischemic heart represents a novel angiogenic therapy approach. At the same time, we track the shRNA vector in vivo through novel molecular imaging technology (18).

2. Material

2.1. Generating the shPHD2 Construct

1. Mouse ES cell (ATCC). Primary mouse ES cell line at passage 10 from SV129 mouse strain.
2. Cell culture medium: DMEM, high glucose with L-glutamine (Gibco-BRL, Grand Island, NY, USA). Store at 4°C.
3. Fetal bovine serum (FBS) (Hyclone, Logan, Utah, USA). Store at –20°C.
4. Penicillin G-Streptomycin: Penicillin 100 U/mL DMEM culture medium and Streptomycin 100 μg/mL DMEM culture medium (Gibco-BRL). Store at –20°C.
5. Leukemia inhibitory factor (LIF) (Sigma, St Louis, MO, USA). Store at 4°C.
6. Feeder cell. Primary fibroblast cells from a 12.5-day-old mouse embryo.
7. 1× Trypsin–EDTA. 0.05% Trypsin, 1 mM EDTA/4Na (Gibco-BRL). Store at 4°C.
8. RNA easy kit (Qiagen, Valencia, CA, USA). Store at room temperature.
9. iScript cDNA synthesis kit (Biorad, Hercules, CA, USA). Store at –20°C.
10. pGEM-T (Promega, Madison, WI, USA). Store at –20°C.
11. pSuper shRNA vector (Oligoengine, Seattle, WA, USA). Store at –20°C.
12. Five tandem repeats of hypoxia responsive element (5×HRE) (synthesized by Stanford Protein and Nucleic Acid Facility).
13. Plasmid pGL3 including SV40-Firefly luciferase (Promega). Store at –20°C.
14. Restriction endonucleases and T4 ligase. (New England Biolabs, Ipswich, MA, USA). Store at –20°C.
15. S.O.C. medium (Gibco-BRL). Store at room temperature.
16. Amp selection LB agar plates (100 μg/mL, ampicillin). Store at 4°C.
17. Chemical-competent *Escherichia coli* Top 10 cells (Invitrogen, Carlsbad, CA, USA). Store at –80°C.

2.2. Testing the shPHD2 Construct In Vitro

1. Mouse myoblast cell line C2C12 (ATCC). Culturing under 37°C at 5% CO_2.
2. Lipofectamine 2000 (Invitrogen, Carlsbad, CA, USA). Store at 4°C.
3. 0.25% Trypsin (Gibco-BRL). Store at 4°C.

4. D-Luciferin (Synchem, Elk Grove Village, IL, USA). Store at −20°C.
5. We inserted firefly luciferase under the CMV promoter in pcDNA3.1 (Invitrogen) at the SalI restriction enzyme site. Store at −20°C.
6. RIPA buffer (Sigma). Store at 4°C.
7. Trizol reagent (Invitrogen). Store at 4°C.
8. 10% SDS-PAGE gel (Biorad). Store at 4°C.
9. Hybond-P memberane (GE, Fairfield, CT, USA).
10. Nonfat dry milk (Biorad).
11. Mouse HIF-1 α antibody (Novus, Littleton, CO, USA). Store at 4°C.

2.3. Evaluation of the shPHD2 Construct In Vivo

1. Two-month-old FVB female mice (Charles River Laboratories, Wilmington, MA).
2. Anesthetic reagents and instruments for mouse heart injection: 3% isoflurane + 97% oxygen, sterile forceps and scissors (World Precision Instruments, Inc., Sarasota, FL, USA), needle holders (Accurate Surgical & Scientific Instruments Corp., Westbury, NY, USA), Guthrie double hook retractor (Fine Science Tools, Inc., Foster City, CA, USA), 5-0 Sofsilk suture (Auto Suture Company, Norwalk, CT, USA), 0.5 mL gas-tight Hamilton syringe, and 30 gauge needle (Hamilton Company, Reno, NV, USA).
3. 70% ethanol. Rossville Gold Shield Alcohol, proof 200 grade.
4. Standard materials for large-scale plasmid preparation.
5. D-Luciferin (Synchem, Felsberg, Germany). 150 mg/kg body weight for intraperitoneal injection.
6. Xenogen In Vivo Imaging System 200 (IVIS 200) small animal imaging system (Calipfer Lise Sciences, Hopkinton, MA).
7. 0.5 mL gas-tight Hamilton syringe and 30-gauge needle (Hamilton Company, Reno, NV).
8. Siemens-Acuson Sequioa C512 Small Animal Echocardiogram System. This system is used to compare heart function changes between shPHD2 treated group and shScramble control group.
9. Muscle homogenization solution: 20 mM $Na_4P_2O_4$, 20 mM $NaHPO_4$, 1 mM $MgCl_2$, 0.5 mM EDTA, and 303 mM sucrose.
10. Muscle microsome preparation wash buffer: 20 mM Tris–HCl (pH 7.0), 60 mM KCl, 303 mM sucrose.
11. Muscle microsome storage solution: 20 mM Tris–HCl (pH 7.0) and 303 mM sucrose.

12. Tissue-Tek OCT compound (Sakura Finetek Inc., Torrance, CA, USA).
13. Rat anti-CD31 (BD Biosciences, San Jose, CA) monoclonal antibody. This antibody binds to endothelial CD31 markers expressed on small blood vessels.
14. 2-methylbutane (Sigma).
15. Rabbit anti-rat IgG (Sigma, final concentration 0.46 mg/mL).
16. Papain (Sigma, final concentration 45 μg/mL).
17. L-cystein (Sigma, final concentration 22 mM).
18. Iodoacetic acid (Sigma, final concentration 10 mM).
19. 20% rabbit serum (Jackson ImmunoResearch Laboratories, Inc.).
20. Alexa 488 conjugated rabbit anti-rat antibodies (Molecular Probe).
21. Alexa 594 conjugated rabbit anti-rat antibodies (Molecular Probe).
22. KPBS: 356 μM KH_2PO_4, 1.64 mM K_2HPO_4, 160 mM NaCl.
23. Gelatin (Sigma).
24. Vector Mouse on Mouse (M.O.M) Kits (Vector Laboratories, Inc., Burlingame, CA).

3. Methods

3.1. Generating the shPHD2 Construct

1. Compare the human and rat PHD2 gene sequences. Because the mouse PHD2 cDNA cannot be found in gene bank, we compared the exact sequence that has been isolated from both. By comparing human and rat PHD2 genes in the Sanger center (http://www.sanger.ac.uk), we selected the 350 bp fragment sequence in both human and rat PHD2 gene, which is located at the 3′-end of both sequence. We hypothesized this sequence is the same as in mouse genome.
2. Clone the mouse PHD2 gene from mouse ES cells using PCR method. Approximately 350 bp fragment in 3′ end mouse PHD2 gene.
3. Design RNA interference targeting sequences using a commercially available web-based software (http://www.ambion.com/techlib/misc/siRNA_finder.html). Four sequences of sites were designed as the shRNA target. The targeting sequences are shown in Fig. 1a. Sequence for the control short hairpin scramble (shScramble) antisense is TGTGAGGAACTTGAGATCT.

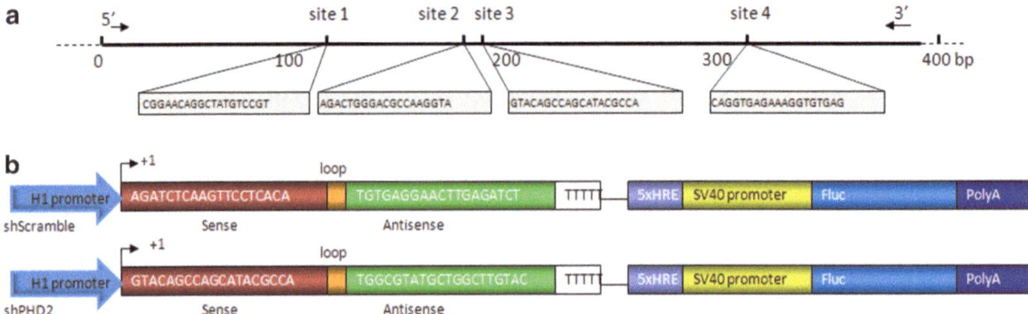

Fig. 1. Schema of the shPHD2 knockdown sites and the reporter constructs. (**a**) Individual sequences of four small interfering RNA target sites against the PHD2 gene. (**b**) Schema of a classic hairpin carrying the site-2 sequence (shPHD2) and a control hairpin carrying the scramble sequence (shScramble). The H1 promoter drives the expression of a hairpin structure in which the sense and antisense strands of the small interfering RNA are connected by a 9-bp long loop sequence. In addition, a separate 5× HRE-SV40 promoter driving firefly luciferase (Fluc) is used to track shRNA activity in vitro and in vivo. *5× HRE* 5 repeat of hypoxia response elements.

3.2. Testing the shPHD2 Construct In Vitro

1. Clone the four shRNA fragments under H1 promoter. To determine the site that possesses the optimal knockdown efficiency, we cloned the sense and antisense downstream of the H1 promoter by the PCR method (Fig. 1b).
2. Test the best knocking down efficiency fragment. These four shRNA constructs were used to transfect C2C12 myoblasts in 6-well plates in conjunction with the pCMV-luciferase plasmid, which confirmed equal transfection efficiency. After 48 h of cell culture, mRNA levels of PHD2 within C2C12 cells were measured by RT-PCR. Construction of the H1 promoter driving sense and antisense was performed as described (19). The No. 2 fragment demonstrated the best efficiency.
3. Clone the best fragment into pSuper shRNA vector. The fragment No. 2 knocking down site was inserted after H1 promoter in the vector pSuper as described in the Oligoengine™ manual.
4. Insert 5×HRE fragment as hypoxia indication. A hypoxia sensing 5×HRE-SV40 promoter driving Fluc cassette was also inserted into the backbone of pSuper vector. The five copies of hypoxia response element (5×HRE) derived from the erythropoietin gene are activated through binding of the HIF-1 complex (20), enabling us to monitor the efficacy of the upstream shPHD2 knockdown compared to the upstream shScramble control (Fig. 2a). SpeI site on the pSuper was cut in order to insert the hypoxia response element HRE-SV40-firefly luciferase cassette.

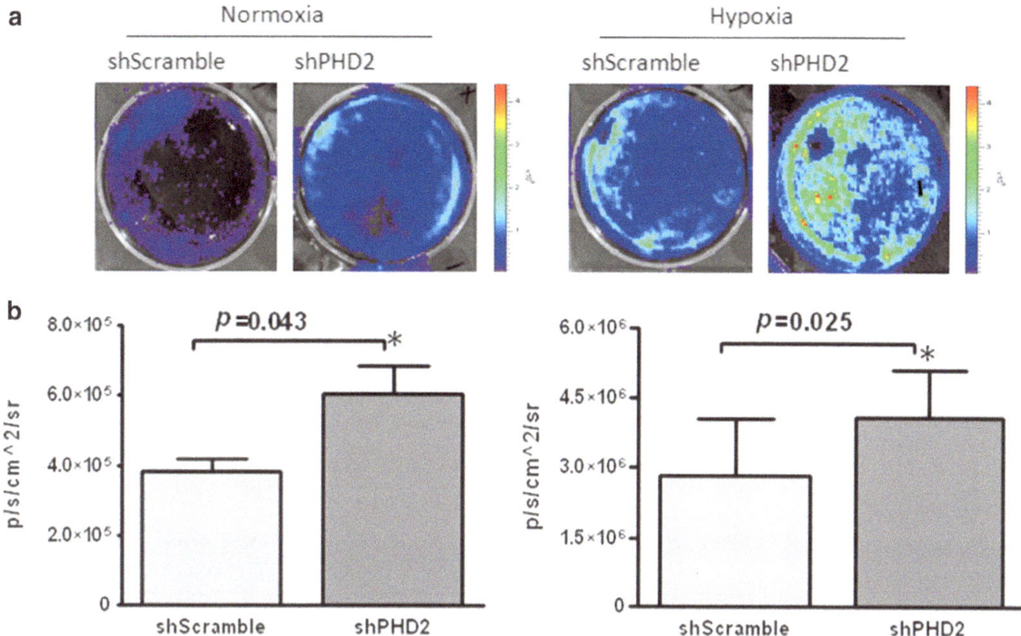

Fig. 2. In vitro characterization of mouse shPHD2. (**a**) In vitro imaging results indicate that Fluc signals increased significantly in response to shPHD2 therapy during both normoxia and hypoxia conditions by binding of HIF-1α protein on the 5×HRE binding site. (**b**) Quantitative analysis of Fluc bioluminescence signals from cell plates (reproduced from ref. (24) with permission).

3.3. Evaluation of the shPHD2 Construct In Vitro

1. Culture mouse C2C12 myoblasts in DMEM medium (high glucose) supplemented with 10% FBS as described in the ATCC protocol. Use Lipofectamine 2000 for the transfection. Culture cells for 1 day after shRNA fragment transfection before placing cells in a hypoxia chamber filled with 5% CO_2, 1% O_2, and 94% N_2 at 37°C. Keep cells under hypoxic conditions for 48 h. At the end of the hypoxic treatment, harvest cells immediately to extract RNA and protein.
2. Use RT-PCR to compare the expression of antigenic genes (bFGF, tranferin, FLT, KDR, TGF, PAI-1) in transfected cells under nomoxia vs. hypoxia. Prepare total RNA from C2C12 cells with Trizol reagent. The primer sets used in the amplification reaction are shown in Table 1. Separate PCR products in 1% agarose gel and quantify the intensity of the bands with Labworks 4.6 image acquisition and analysis software.
3. To confirm the shPHD2 knocking down efficiency, isolate nuclear extracts and detect HIF-1α protein by western blot. After 48 h of hypoxia culture, wash the C2C12 cells with phosphate buffered saline (PBS) and homogenize cells with 200 μL of homogenization buffer (RIPA). Fractionate the supernatants on SDS-PAGE and blot to a Hybond-P membrane. Block the membranes with 5% nonfat dry milk in 1× TBS containing 0.05% Tween 20. Incubate with the antibodies

Table 1
Primers used in PCR

bFGF	Forward	5′-CTTCAAGGACCCCAAGCGGGCTCTA-3′
	Reverse	5′-CGAGTTTATACTGCCCAGTT-3′
Transferrin	Forward	5′-CTTCAAGGACCCCAAGCGGCTCTAC-3′
	Reverse	5′-GTTCGTTTCAGTGCCACATACCAAC-3′
FLT	Forward	5′-TGAAGTCTGCTCGCTATTTGGTA-3′
	Reverse	5′-CTATGGTGCATGGTTCTGTTGTT-3′
KDR	Forward	5′-GAAGCTACTGCCGTCCGATTGAG-3′
	Reverse	5′-TGCTGGCTTTGGTGAGGTTTGAT-3′
TGF	Forward	5′-AAATTCGACATGATCCAGGGACT-3′
	Reverse	5′-TGCACTTACACGACTTCACCACC-3′
PAI	Forward	5′-ATGGCTCAGAGCAACAAGTTCAA-3′
	Reverse	5′-GACAAAGGCTGTGGAGGAAGACG-3′

against HIF-1α. Detect with secondary HRP-conjugated antibodies and the ECL detection system. For the in vivo western blot, take heart samples at days 1, 4, 7, 14, and 28 after plasmid injection (see Note 1).

3.4. Testing the shPHD2 Efficacy In Vivo

1. Get approval from the Animal Research Committee of the institute.
2. Ligate the mid left anterior descending (LAD) artery of adult female FVB mice. Anesthetize mice using isoflurane inhalation. Place a tube through the mouth into the trachea and connect the tube to a Harvard rodent ventilator. After stabilizing the respiratory status of the animal, make a thoracotomy incision in the fourth intercostal space and use a surgical retractor to expose the heart. Ligate the LAD permanently with a 7-0 nonabsorbable surgical suture. Confirm myocardial infarction by myocardial blanching and EKG changes. Wait for 10 min. Inject 25 μg (1 μg/μL) of shRNA plasmid at the peri-infarct zone or 25 μg (1 μg/μL) of shScramble plasmid as control using a 30-gauge Hamilton needle (see Notes 2 and 3).
3. Perform cardiac bioluminescence imaging with the Xenogen IVIS. After intraperitoneal injection of the reporter probe D-luciferin (150 mg/kg body weight), image animals for 1–10 min. Scan the same mice repetitively for 4 weeks according to the specific study design (Fig. 3). Quantify bioluminescence signals in units of maximum photons per second per

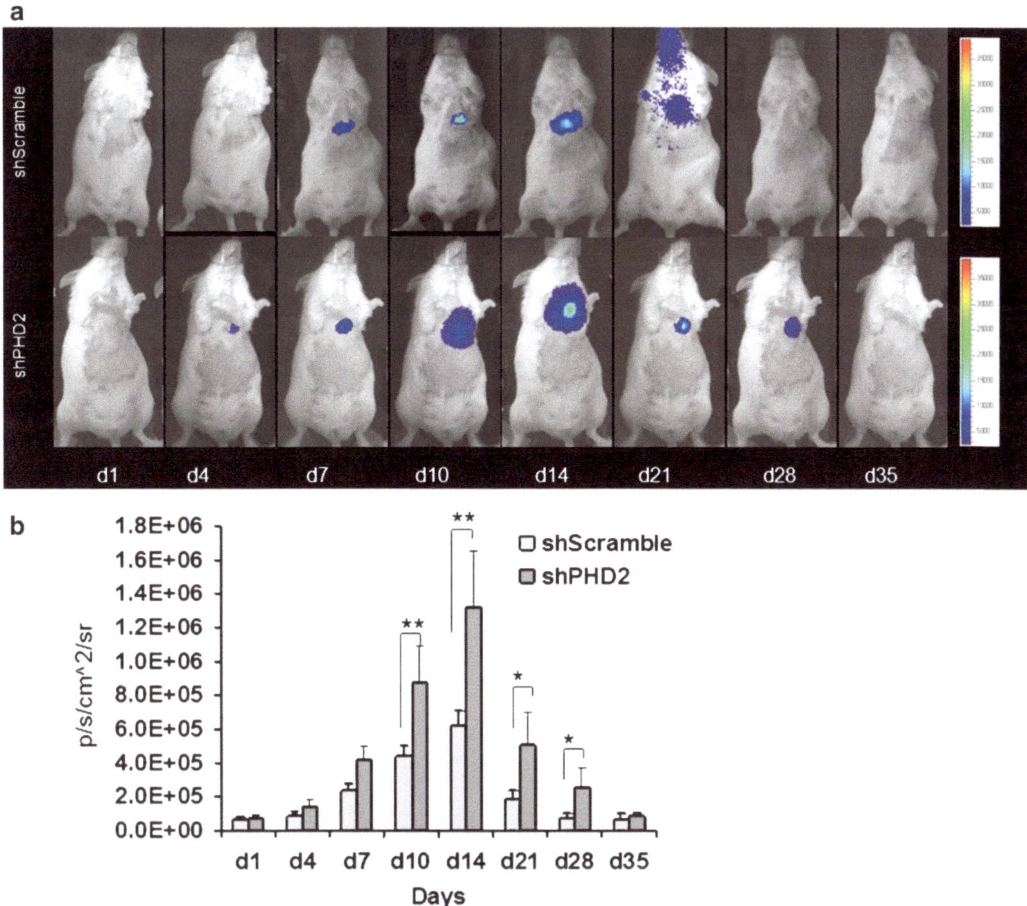

Fig. 3. Molecular imaging of shRNA plasmid fate after intramyocardial delivery. (**a**) Following myocardial infarction, activation of HIF-1α protein binds to 5×HRE site to activate Fluc expression. Infarcted mice injected with shPHD2 (*bottom row*) had more robust Fluc signals compared to infarcted mice injected with shScramble (*top row*) due to knocking down of PHD2, which result in more HIF-1α protein binding to 5× HRE site. Peak expression occurred within week 1–2 as reflected by the Fluc imaging signals. (**b**) Detailed quantitative analysis of Fluc signals from all animals injected with shPHD2 or shScramble plasmid with LAD ligation. Signal activity is expressed as p/s/cm^2/sr (reproduced from ref. (24) with permission).

centimeter squared per steradian (p/s/cm^2/sr) using the built-in software on the IVIS system (18, 21).

4. Perform echocardiography before (day 7) and after (week 2, 4, 8) the LAD ligation with the Siemens-Acuson Sequioa C512 system equipped with a multifrequency (8–14 MHz) 15L8 transducer. Analyze M-mode images using the Siemens built-in software. Measure left ventricular end diastolic diameter (EDD) and end-systolic diameter (ESD) and calculate left ventricular fractional shortening by the formula: LVFS = (EDD–ESD)/EDD.

5. Embed the explanted hearts into OCT compound. Process 5 μm frozen sections for immunostaining. Use a rat anti-CD31 to detect microvascular density in the peri-infarct area. Count the number of capillary vessels by a blinded investigator in ten randomly selected areas using a light microscope (×200 magnification). Use additional samples to examine the infarction size by Masson's trichrome staining.

4. Notes

1. We harvest the heart tissue under microscopy to isolate the peri-infarct and infarct area tissues.
2. To obtain the optimal quality of plasmids for in vivo injection, we recommend preparing the plasmids by $CsCl_2$-ethidium bromide equilibrium centrifugation.
3. To obtain the large yield of shPHD2 plasmid, we select DH5 alpha and fresh *E. coli* solution to do the isolation.

Acknowledgments

This work was supported in part by grants from the NIH HL095571 (JCW), NIH HL093172 (JCW), and AHA Western Postdoctoral Fellowship (MH).

References

1. Rosamond, W., Flegal, K., Friday, G., Furie, K., Go, A., Greenlund, K., Haase, N., Ho, M., Howard, V., Kissela, B., Kittner, S., Lloyd-Jones, D., McDermott, M., Meigs, J., Moy, C., Nichol, G., O'Donnell, C. J., Roger, V., Rumsfeld, J., Sorlie, P., Steinberger, J., Thom, T., Wasserthiel-Smoller, S. and Hong, Y. (2007) Heart disease and stroke statistics--2007 update: a report from the American Heart Association Statistics Committee and Stroke Statistics Subcommittee. *Circulation* 115, e69–e171.
2. Kim, M. C., Kini, A. and Sharma, S. K. (2002) Refractory angina pectoris: mechanism and therapeutic options. *J Am Coll Cardiol* 39, 923–934.
3. Nabel, G. J. (2004) Genetic, cellular and immune approaches to disease therapy: past and future. *Nat Med* 10, 135–141.
4. Yla-Herttuala, S. and Alitalo, K. (2003) Gene transfer as a tool to induce therapeutic vascular growth. *Nat Med* 9, 694–701.
5. Isner, J. M. (2002) Myocardial gene therapy. *Nature* 415, 234–239.
6. Luttun, A., Tjwa, M., Moons, L., Wu, Y., Angelillo-Scherrer, A., Liao, F., Nagy, J. A., Hooper, A., Priller, J., De Klerck, B., Compernolle, V., Daci, E., Bohlen, P., Dewerchin, M., Herbert, J. M., Fava, R., Matthys, P., Carmeliet, G., Collen, D., Dvorak, H. F., Hicklin, D. J. and Carmeliet, P. (2002) Revascularization of ischemic tissues by PlGF treatment, and inhibition of tumor angiogenesis, arthritis and atherosclerosis by anti-Flt1. *Nat Med* 8, 831–840.
7. Vincent, K. A., Feron, O. and Kelly, R. A. (2002) Harnessing the response to tissue hypoxia: HIF-1 alpha and therapeutic angiogenesis. *Trends Cardiovasc Med* 12, 362–367.
8. Fukuda, R., Hirota, K., Fan, F., Jung, Y. D., Ellis, L. M. and Semenza, G. L. (2002) Insulin-like growth factor 1 induces hypoxia-inducible factor 1-mediated vascular

endothelial growth factor expression, which is dependent on MAP kinase and phosphatidylinositol 3-kinase signaling in colon cancer cells. *J Biol Chem* 277, 38205–38211.

9. Stolze, I., Berchner-Pfannschmidt, U., Freitag, P., Wotzlaw, C., Rossler, J., Frede, S., Acker, H. and Fandrey, J. (2002) Hypoxia-inducible erythropoietin gene expression in human neuroblastoma cells. *Blood* 100, 2623–2628.
10. Sandau, K. B., Fandrey, J. and Brune, B. (2001) Accumulation of HIF-1alpha under the influence of nitric oxide. *Blood* 97, 1009–1015.
11. Jewell, U. R., Kvietikova, I., Scheid, A., Bauer, C., Wenger, R. H. and Gassmann, M. (2001) Induction of HIF-1alpha in response to hypoxia is instantaneous. *FASEB J* 15, 1312–1314.
12. Chan, D. A., Sutphin, P. D., Denko, N. C. and Giaccia, A. J. (2002) Role of prolyl hydroxylation in oncogenically stabilized hypoxia-inducible factor-1alpha. *J Biol Chem* 277, 40112–40117.
13. Ivan, M., Kondo, K., Yang, H., Kim, W., Valiando, J., Ohh, M., Salic, A., Asara, J. M., Lane, W. S. and Kaelin, W. G., Jr. (2001) HIFalpha targeted for VHL-mediated destruction by proline hydroxylation: implications for O_2 sensing. *Science* 292, 464–468.
14. Jiang, B. H., Semenza, G. L., Bauer, C. and Marti, H. H. (1996) Hypoxia-inducible factor 1 levels vary exponentially over a physiologically relevant range of O_2 tension. *Am J Physiol* 271, C1172–C1180.
15. Pislaru, S., Janssens, S. P., Gersh, B. J. and Simari, R. D. (2002) Defining gene transfer before expecting gene therapy: putting the horse before the cart. *Circulation* 106, 631–636.
16. Yla-Herttuala, S., Markkanen, J. E. and Rissanen, T. T. (2004) Gene therapy for ischemic cardiovascular diseases: some lessons learned from the first clinical trials. *Trends Cardiovasc Med* 14, 295–300.
17. Wu, J. C. and Yla-Herttuala, S. (2005) Human gene therapy and imaging: cardiology. *Eur J Nucl Med Mol Imaging* 32 Suppl 2, S346–S357.
18. Huang, M., Chan, D. A., Jia, F., Xie, X., Li, Z., Hoyt, G., Robbins, R. C., Chen, X., Giaccia, A. J. and Wu, J. C. (2008) Short hairpin RNA interference therapy for ischemic heart disease. *Circulation* 118, S226–S233.
19. Miyagishi, M. and Taira, K. (2002) U6 promoter-driven siRNAs with four uridine 3 overhangs efficiently suppress targeted gene expression in mammalian cells. *Nat Biotechnol* 20, 497–500.
20. Ruan, H., Su, H., Hu, L., Lamborn, K. R., Kan, Y. W. and Deen, D. F. (2001) A hypoxia-regulated adeno-associated virus vector for cancer-specific gene therapy. *Neoplasia* 3, 255–263.
21. Cao, F., Lin, S., Xie, X., Ray, P., Patel, M., Zhang, X., Drukker, M., Dylla, S. J., Connolly, A. J., Chen, X., Weissman, I. L., Gambhir, S. S. and Wu, J. C. (2006) In vivo visualization of embryonic stem cell survival, proliferation, and migration after cardiac delivery. *Circulation* 113, 1005–1014.

Chapter 14

Lentiviral Vector Delivery of shRNA into Cultured Primary Myogenic Cells: A Tool for Therapeutic Target Validation

Emmanuel Richard, Gaelle Douillard-Guilloux, and Catherine Caillaud

Abstract

RNA interference has emerged as a powerful technique to down-regulate gene expression. The lentiviral vector-mediated expression of small hairpin RNAs (shRNAs) from polymerase III promoters allows permanent down-regulation of a specific gene in a wide range of cell types both in vitro and in vivo. In this chapter, we describe a method permitting the expression of shRNA from lentiviral vectors in primary murine myogenic cells. We designed shRNAs targeted to the muscular glycogen synthase isoform (shGYS1), a highly regulated enzyme responsible for glycogen synthesis, in order to modulate the muscle glycogen biosynthetic pathway and to improve the phenotype in primary myogenic cells from a murine model of glycogen storage disease type II (GSDII). This method based on shRNA-mediated down-regulation could be applied to other muscular disorders to evaluate new therapeutic options.

Key words: Lentiviral vector, RNA interference, Muscular disorder, Glycogen synthesis, Glycogen synthase, Glycogen storage disease, Pompe disease, Lysosomal storage disorder

1. Introduction

RNA interference has emerged as a pathway constituting a new regulation level of gene expression and it has introduced a new tool for studying gene function by inducing posttranscriptional gene silencing (1). Briefly, short 19–23 nucleotides double-stranded small interfering RNAs (siRNAs) are incorporated into a multicomponent nuclease complex (RISC, RNA induced silencing complex) that selects and degrades mRNAs containing a target complementary to the siRNA antisense strand. In mammals, siRNA can be delivered either by transient transfection of synthetic siRNAs or by stable viral-mediated transduction of constructs

Dongsheng Duan (ed.), *Muscle Gene Therapy: Methods and Protocols*, Methods in Molecular Biology, vol. 709,
DOI 10.1007/978-1-61737-982-6_14,

expressing shRNAs from polymerase III (pol III) promoters. Expressed shRNAs are converted by the ribonuclease III enzyme Dicer into small siRNAs that are incorporated into the RISC complex mediating the subsequent degradation of the target mRNAs (2). Lentiviral vectors derived from HIV-1 are able to infect a wide variety of dividing and nondividing cells and to stably integrate into the host genome, resulting in long-term expression of the transgene (3, 4). The presence of vesicular stomatitis virus glycoprotein (VSV-G) in the viral envelop membrane confers to the viral particle, the ability to transduce a broad range of cell types, including primary myogenic cells (5–7). Lentiviral vectors can be designed to express constitutively and conditionally both transgenes and shRNAs either in single units or in multiple combinations (8, 9).

Glycogen storage disease type II (GSDII), also termed Pompe disease (MIM 232300), is an autosomal recessive disorder caused by a deficiency of the lysosomal enzyme, acid maltase, or alpha-glucosidase (GAA; EC 3.2.1.20) (10). This enzyme catalyzes the complete hydrolysis of its natural substrate glycogen by cleaving both α-1,4 and α-1,6 glucosidic linkages at an acidic pH, allowing glucose to be liberated into the cytoplasm. GAA deficiency results in massive glycogen accumulation in various tissues, among which skeletal and cardiac muscle are the most affected. Clinically, GSDII manifests as a continuum of phenotypes as the disease severity largely correlates with the level of residual enzyme activity. Near complete or total loss of GAA activity results in severe myopathy and cardiomyopathy in infants leading to death before 2 years of life due to cardiorespiratory failure. Partial enzyme deficiency results in late-onset forms of GSDII with slowly progressive muscle weakness without cardiac involvement and death occurring by respiratory failure (11). Animal models mimicking the human disease were created by targeted disruption of the GAA gene (12, 13). We have recently demonstrated that substrate reduction therapy (SRT), already used in other lysosomal storage disorders (14) and based on the inhibition of storage material biosynthesis, could be efficient to rescue Pompe disease by modulating glycogen synthesis (15). In our approach, lentiviral vector-mediated delivery of shRNA targeted to muscular glycogen synthase *GYS1* was used, resulting in a specific knockdown gene expression and an impaired glycogen synthesis.

In this protocol, we outline the procedures to generate lentiviral vectors mediating a permanent shRNA delivery into primary myogenic cells. As an example, we describe the efficient inhibition of *GYS1* gene expression in primary myoblasts from GSDII mice. This substrate reduction strategy could constitute a novel approach for GSDII treatment, which may complement or be used as an alternative to ERT.

2. Materials

2.1. Generation of shRNA-Expressing Lentiviral Vectors

2.1.1. Selection and Cloning of shRNA Cassettes

1. S.O.C. medium. Store at –20°C.
2. Electroporation-competent *Escherichia coli* Stbl2 cells (Invitrogen, Carlsbad, CA, USA). Stbl2 cells show superior stability of direct repeat sequences like LTR-containing lentiviral vector backbones. Store at –80°C.
3. Gene pulser apparatus.
4. Amp selection LB agar plates (100 μg/ml Ampicillin). Store at 4°C.
5. Restriction enzymes, modifying enzymes, and T4 DNA ligase (New England Biolabs, Ipswich, MA, USA).
6. Complementary oligonucleotides purified by HPLC or PAGE and phosphorylated at the 5′ end (Sigma-Aldrich, Saint Louis, MO, USA) (see Note 1).
7. The H1 promoter-shRNA cassette. The integrity of the cassette is sequenced using the primer 5′-GGGAAATCA CCATAAACGTG-3′.
8. Annealing buffer (AB) for shRNA: 100 mM potassium acetate, 30 mM HEPES pH 7.4, 2 mM magnesium acetate.
9. Plasmids for lentiviral vector production (15).

2.1.2. Production of Self-Inactivated Recombinant Lentiviral Vectors

1. HEK293T cells (ATCC, Manassas, VA, USA). This is an adenovirus E1 gene transformed human kidney cell line. These cells are propagated in high-glucose Dulbecco's modified Eagle's medium (DMEM) containing 10% fetal bovine serum (FBS) and 1% penicillin G/streptomycin. C2-C12 murine myogenic cells are propagated in DMEM containing 20% FBS and 1% penicillin G/streptomycin.
2. Cell culture medium: DMEM, high glucose (4,500 mg/L) with L-glutamine and OPTI-MEM (Invitrogen). Store at 4°C.
3. FBS (Invitrogen). Store at –20°C.
4. Penicillin G-Streptomycin: penicillin 100 IU/ml and streptomycin 100 μg/ml. Store at –20°C.
5. 1× Trypsin-EDTA: 0.25% Trypsin, 1 mM EDTA/4Na (PAA, Pasching, Austria). Store at 4°C.
6. 2.5 mM $CaCl_2$ (Sigma-Aldrich). Prepared in Milli-Q H_2O and sterilized by filtration at 0.22 μM.
7. HEPES buffer saline, HBS 2× (Sigma-Aldrich).
8. Poly-l lysine, cell culture tested (Sigma-Aldrich).
9. Plasmids. The helper plasmids encoding packaging proteins (Pax2) are from Dr. Didier Trono (Geneva, Switzerland)

and the VSV-G encoding plasmid is from Dr. Inder Verma (Salk Institute, San Diego, CA). Plasmids are amplified in the Stbl2 bacteria strain. Endotoxin-free plasmids are obtained from bacteria pellets using NucleoBond PC 500 Plasmid DNA extraction kit (Macherey-Nagel, Duren, Germany).

2.2. Isolation of Primary Satellite Cells

1. Murine primary satellite cells and myoblasts are propagated in myoblast culturing medium (MCM): low-glucose (1,000 mg/L) DMEM/HamF12 (Invitrogen) containing 20% fetal calf serum (FCS), Ultroser G 2% and 1% penicillin G/streptomycin.
2. Initiating myoblast culturing medium (IMCM): For growing cells in the first 5 days. DMEM/HamF12 containing 10% FBS, Ultroser G 2%, 1% penicillin G/streptomycin, and 1% fungizone (Invitrogen).
3. Protease digestion medium: DMEM/HamF12 medium supplemented with 0.14% protease, 1% penicillin G/Streptomycin, and 1% fungizone (Invitrogen).
4. Differentiation medium: DMEM/HamF12 containing 2% Horse Serum (HS), 1% penicillin G/streptomycin.
5. FCS (Invitrogen). Store at –20°C.
6. HS (Invitrogen). Store at –20°C.
7. Penicillin G-Streptomycin: penicillin 100 IU/ml and streptomycin 100 μg/ml. Store at –20°C.
8. Protease from *Streptomyces griseus* (Sigma-Aldrich). Store at –20°C.
9. Gelatin from bovine skin (Sigma-Aldrich). Keep powder at room temperature and make dilution in sterile water. Prepare 0.1% stock solution and store at –20°C.
10. Ultroser G (Pall Corporation, East Hills, NY). Store aliquot of 1 ml at –20°C. Use at 2% final concentration.
11. Matrigel (BD Biosciences, Bedford, MA): dilute at 1:15 in DMEM/Ham12 (v/v). Aliquot the solution and store at –20°C. Dispense 1 ml per well (12-well plate), incubate 30 min at 37°C, and remove the solution before use.

2.3. Evaluation of Silencing Efficiency by Quantitative RT-PCR Analysis

1. mRNA mini kit (Qiagen, Hilden, Germany).
2. Two-step RT-PCR-&G Kits (MP Biomedicals, Irvine, CA). Use master mixes dedicated to RT and PCR (standard and high fidelity).
3. Primers: *GYS1* (249 bp amplicon): forward 5′-TATCGCTG GCCGCTATGAGTT-3′, reverse 5′-CACTAAAAGGGATT CATAGAG-3′; actin (210 bp amplicon): forward 5′-GGCATAGAGGTCTTTACGG-3′, reverse 5′-GTGGCA TCCATGAAACTACAT-3′.

4. LightCycler Carousel-based PCR system (Roche, Mannheim, Germany).
5. SYBR green PCR master mix reagent and LightCycler RNA amplification Kit SYBR Green (Roche).

3. Methods

3.1. Generation of shRNA-Expressing Lentiviral Vectors

1. Design lentiviral vector. Lentiviral vectors are designed for coexpressing EGFP and shRNAs, respectively, under the control of the human phosphoglycerate kinase (PGK) promoter and the polymerase III (Pol III) H1 promoter. The two pol III promoters most commonly used for hairpin expression are H1 and U6 (human and mouse). Both promoters have efficient, ubiquitous expression and are essentially equivalent for in vitro experiments. However, differences in expression efficiency have been reported in vivo (16). Pol III promoters are ideally suited to express shRNAs consisting of approximately 60 nucleotides. It includes a 19–23 nucleotide sense sequence identical to the target mRNA sequence, followed by a 9-nucleotide loop and an antisense 19–23 nucleotide sequence. A stretch of five Ts provides a pol III transcriptional termination signal. Expression results in a short 19–23 bp hairpin loop. This loop is cleaved by the endonuclease Dicer. The resulting siRNA triggers degradation of the target gene mRNA. Development and validation of an efficient lentiviral silencing vector involves several steps. First, siRNA target sequences should be carefully determined. Second, shRNA needs to be cloned and validated in a lentiviral silencing vector. A number of algorithms have been developed to predict effective siRNA sequences and many of them are available online for free or commercial use (17, 18). Some siRNA design softwares allow the simultaneous use of several algorithms to predict effective siRNA sequences (ex. http://i.cs.hku.hk/~sirna/software/sirna.php). Generally, 3–4 shRNAs need to be generated and tested for each target gene. The target sequence should be 19–23 bases long, but lengths of up to 28 bases have been reported. The G/C content should be 40–55%. Several rules can be successfully applied to determine shRNA sequences capable of inducing highly effective gene silencing in mammalian cells: (a) A/T at the 5′end of the antisense strand; (b) At least five A/T residues in the 5′ one-third terminal of the antisense strand; and (c) Absence of any GC stretch of more than 9 nucleotides in length. The general strategy used to obtain shRNA-expressing lentiviral vectors is described in Fig. 1.
2. Clone the shRNA sequences into the pic20/H1 promoter plasmid (9). Get complementary shRNA oligonucleotides at

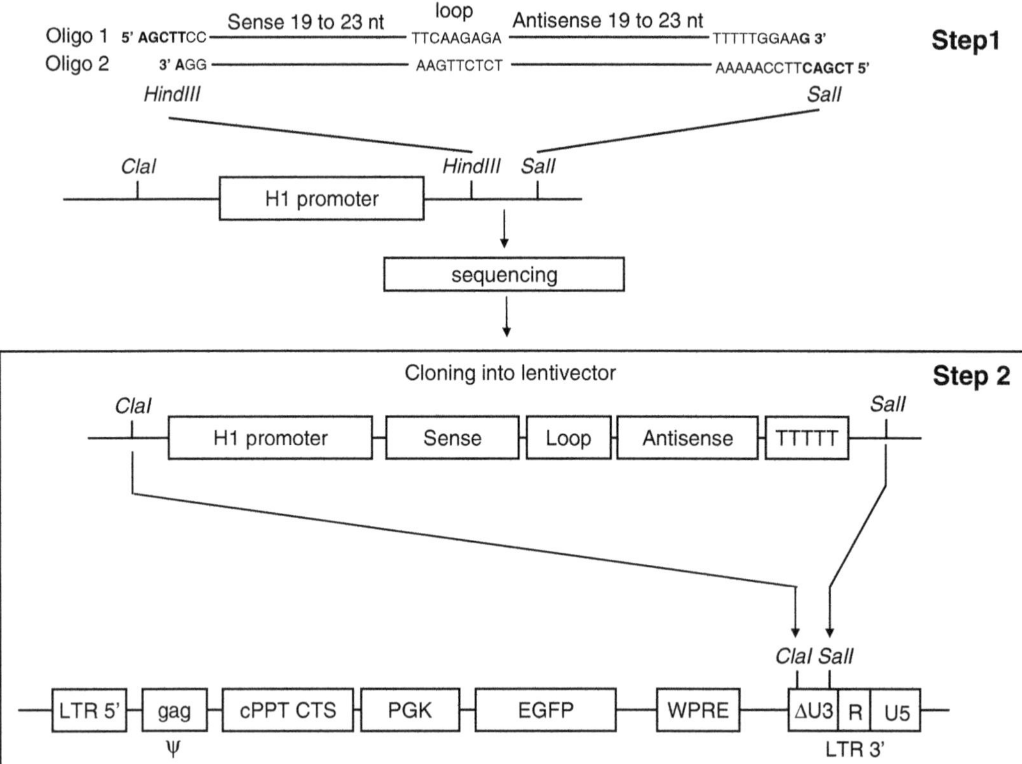

Fig. 1. Cloning of shRNA-expressing lentiviral vectors. *Step 1*: annealing and cloning of shRNA duplex into H1 promoter containing pic20/H1p plasmid using the HindIII and SalI enzyme restriction site. *Step 2*: inserting the H1p/shRNA cassette in the U3 deleted 3′ LTR region of the lentiviral backbone using the ClaI and SalI enzyme restriction sites. *LTR* long terminal repeat; *cPPT* central polypurine track; *CTS* central termination sequence; *PGK* phosphoglycerate kinase promoter; *EGFP* enhanced green fluorescent protein; *WPRE* woodchuck posttranscriptional regulatory element.

0.05 mM scale (Fig. 1). Dissolve both oligonucleotides in 50 μl of deionized H_2O. Take 2 μl of each oligonucleotide (forward and reverse) and add 48 μl of annealing buffer. Incubate 4 min at 95°C and 10 min at 70°C for denaturating oligonucleotides. Cool down oligonucleotides on the bench (can be stored at -20°C). For ligation, prepare pic20/H1p vector digested with Hind III and Sal I restriction enzyme. Mix 5 μl phosphorylated and annealed oligonucleotides, 1 μl pic20/H1p vector (20–100 ng) and 11 μl water, 2 μl 10× T4 ligase buffer and 1 μl T4 DNA ligase (1–5U). Incubate 1 h at room temperature and transform Stbl2 bacteria. Check minipreps by digestions and sequence the shRNA using the H1 primer sequence (see Note 2).

3. Clone the H1p/shRNA into lentiviral vector. Design three shRNA sequences targeting the muscular *GYS1* gene (shGYS) (NM 030678) and clone them downstream the H1 promoter using SalI–HindIII restriction enzymes (Fig. 1). Introduce a ClaI-SalI fragment containing the H1 promoter and shRNA

in the 3′ U3 LTR region of the SIN-U3 PGK-EGFP-WPRE vector (kindly provided by Dr P. Charneau, Institut Pasteur, Paris, France) in order to obtain a dual shRNA expression cassette from the integrated provirus.

4. Produce lentiviral vectors by triple plasmid transient transfection. Start with a nonconfluent, but actively proliferating HEK293T cell population (see Note 3). Precoat 10-cm culture dishes with 10 ml poly l-lysine. On day 1, seed HEK293T cells at a density of 5×10^6 cells per 10-cm tissue culture plates and incubate at 37°C for 24 h (see Note 3). On day 2, check cell confluence to confirm it is ~80–90% confluent. Carefully add 5 ml fresh culture medium to the plates. Transfect the plasmid mixture (pCMVΔ8.9 or Pax2 packaging plasmid, VSV-G pMD.G envelope plasmid, and lentiviral vector backbone expressing EGFP and shRNA) using $CaPO_4$ precipitation (see Note 4). Mix three plasmids into a 15-ml tube. For five 10-cm dish preparation, use 50 µg of the transfer vector, 50 µg of Pax2 or pCMVΔ8.9 (Gag/Pol), 20 µg of pVSVG (vesicular stomatitis virus glycoprotein), and dilute in deionized H_2O to the final volume of 2.25 ml. Add 250 µl of 2.5 M $CaCl_2$ to the filtered plasmid solution and mix gently. Add 2.5 ml 2×-HBS to an empty 15-ml tube and incubate 5 min at room temperature (22–26°C). The mixture must become slightly cloudy. Slowly add the filtered DNA/$CaCl_2$ solution, dropwise in a circular motion to distribute the DNA evenly onto the HEK293T cells. Swirl the dish gently to further distribute the DNA and incubate at 37°C overnight (16–20 h). On day 3, observe cells and change media. Remove the medium and carefully wash the cell layer with previously warmed OPTI-MEM to remove nontransfected plasmid precipitates. Add 10 ml fresh medium and incubate for 24–36 h at 37°C (see Note 4). On day 4, collect lentiviral vector-containing supernatant and concentrate viral preparation. Collect the viral vector supernatants from dishes. Eliminate cell debris by filtering through a 0.22 µM filter. Concentrate the virus by centrifugation of the supernatant at 60,000 × *g* for 2 h at 4°C with a Beckman SW28 rotor. Pour off the supernatant and carefully remove the remaining liquid from the pellet with a micropipette. The pellet should be barely visible as a small translucent spot. Resuspend the viral pellets in 100–200 µl of fresh OPTI-MEM medium and let the pellet dissolve at 4°C under slow agitation for 1 h. The resuspended viral preparation will look from clear to slightly milky. Make 20–50 µl aliquots of the supernatant, transfer to a screw-cap microfuge tube, and immediately cryopreserved at −80°C (see Note 4).

5. Titrate lentiviral vectors. For lentiviral vectors expressing both EGFP and shRNA, infectious titers were determined in transduced cells by measuring EGFP expression by flow cytometry analysis. Seed 10^5 C2-C12 cells in each well of a 24-well cluster plate in a final volume of 300 μl per well (see Note 5). Make a serial dilution of the lentiviral vector supernatant in the cultured medium (1/10 to 1/2,000) and add 10 μl of each viral dilution to the cells. Mix thoroughly but gently. 24 h after transduction, add 500 μl fresh culture medium. Grow the cells for at least 5 days in order to avoid pseudotransduction (see Note 5). Detach the cell layer using trypsinisation and wash with 10% FCS-containing medium to protect the cells from trypsin damages. Centrifuge the cells and resuspend the pellet in 500 μl phosphate buffer saline (PBS). Determine the percentage of labeled cells (GFP^+) by counting fluorescent vs. nonfluorescent cells in Malassez counting chamber using a fluorescence microscope. Ideally, the percentage of GFP^+ cells should be accurately determined by flow cytometry analysis. Calculate the biological titer (BT=TU/ml) according to the following formula: $TU/ml=((\%GFP^+ \text{ cells} \times N/V) \times 1000) \times 1/D$, where N=number of cells at time of transduction, V=volume of dilution added to each well=10 μl and D=dilution factor=10^{-1} (diluted 1/10), 10^{-2} (diluted 1/100). Appropriate dilutions give rise to similar titers while lower dilutions underestimate the titer. The range of dilution to test will depend on the concentration of the viral preparation and must be determined empirically.

3.2. Isolation of Primary Satellite Cells

1. For all animal experiments, get approval from your institute Animal Care and Use Committee and follow NIH guidelines.
2. Prepare gelatin-coated plates by dispensing 1 ml of a 0.02% gelatin solution per well (12-well plates). Let the plates for 1 h at room temperature and then remove the gelatin solution and let dry before use.
3. Sacrifice mice by cervical dislocation and remove the muscle. Eliminate the fibrous and translucide envelope that covers the muscle (see Note 6).
4. Place the muscle in a sterile plastic dish and rinse it with PBS. Under binocular microscope, gently clean muscle of any extraneous material.
5. Cut the muscle in small pieces in a 6-well plate and incubate 1 h in 3 ml Ham F12/DMEM medium (v/v) at 37°C in 5–7.5% CO_2.
6. Transfer muscle pieces in a 15 ml tube and incubate for 10 min at 37°C, in 10 ml of a protease digestion medium (0.14% protease solution).

7. Centrifuge at 500 × *g* for 30 s and resuspend the pellet in 10 ml of fresh protease digestion medium. Incubate for 10 min at 37°C.
8. Repeat steps 6 and 7 for 4 cycles of incubation/centrifugation and discard supernatant.
9. Following the fourth centrifugation, harvest the pellet and filter through 100 μm nylon gauze.
10. Wash the cells with 10 ml fresh Ham F12/DMEM medium (v/v) supplemented with 20% FBS.
11. Plate cells at a density of 200 by well in 12-well gelatin-coated (0.02%) plates, containing Ham F12/DMEM medium (v/v) supplemented with 10% FBS, 5 mM l-glutamine, 100 IU/ml penicillin, 100 μg streptomycin, and Ultroser G 2%. Incubate during 5 days at 37°C, 5% CO_2 (see Note 6).
12. Observe the cells under a phase contrast microscope and discard fibroblasts containing wells (see Note 7).
13. Trypsinize, pool, and seed satellite cells in 6-well gelatin-coated plates with 3 ml MCM.
14. Change MCM medium every 2 days. Cells should be trypsinized and seeded every 4 days at a density of $2–5 \times 10^4$ cells/flask (75 cm^2) (see Note 8). Primary myoblasts can divide about 20 times without any phenotypic modification (see Note 9).

3.3. Evaluation of Silencing Efficiency by Quantitative RT-PCR Analysis

1. Harvest primary myoblasts by trypsinization and seed cells in 6-well plates at a density of 5×10^4 cells/well in 3 ml MCM medium.
2. Add concentrated lentiviral vector at a multiplicity of infection between 10 and 100 IU (based on infectious titer calculated on C2-C12 cells).
3. Incubate at 37°C until you have enough biological material for silencing efficiency analysis (RT-PC, western blot or phenotypic analysis). Change the medium every 2 days and pass the cells every 4 days to maintain conditions that allow cellular expansion without differentiation.
4. Isolate lentivirally transduced cells (see Note 10). Five to seven days after lentiviral transduction, trypsinize the myoblasts. Transduce cells in 6-well plates with each viral supernatant with an estimated multiplicity of infection (m.o.i) of 2 and 50 IU (C2-C12 cells) or 10 and 100 IU (GSDII myoblasts), in order to select cells respectively expressing high or low copy number of shRNA. Isolate EGFP-positive cells using FACS (Fig. 2). Isolate and expand cell populations expressing either high or low level of shRNA products.

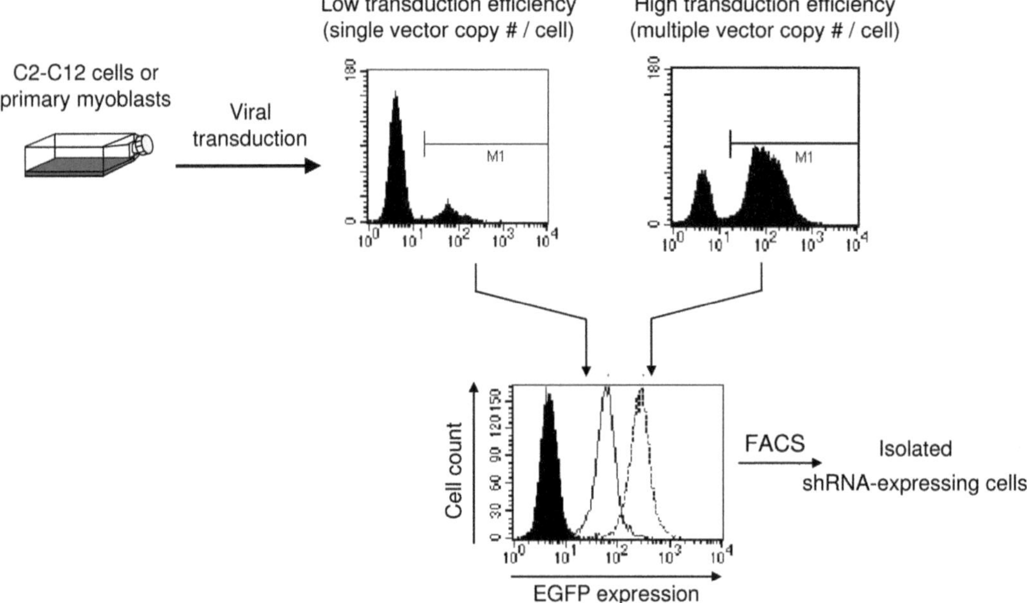

Fig. 2. Selection of shRNA-expressing cells by FACS. Transduced cells are analyzed by FACS for EGFP expression. The mean fluorescence intensity (*x* axis) is related to the proviral copy numbers integrated in the host DNA; low (LC) or high (HC) proviral copy number containing cells could be selected. These cells will present low or high shRNA expression that can influence the silencing efficiency.

5. Extract total RNAs from EGFP-positive C2-C12 cells and primary myoblasts using the mRNA mini kit. Perform quantitative real-time PCR assay to detect GYS mRNA level using specific primers and β-actin as an endogenous control. Carry out the PCR reaction using a LightCycler Carousel-Based System and SYBR green PCR master mix reagent. Express the results as a ratio vs. negative control (untransduced C2-C12 or GAA-KO primary myoblasts).

4. Notes

1. Because oligonucleotides are very long (up to 70 bp), they should be bought purified by HPLC or PAGE to ensure high cloning efficiency. Phosphorylation at 5′ end can be alternatively performed using polynucleotide kinase (New England Biolabs), but most oligonucleotide providers propose this modification.
2. Due to strong secondary structure of the shRNA, sequencing may be difficult to perform and requires optimization. To limit these difficulties, chose shRNA sequences with less 60% G/C content. For some shRNA with very high secondary structures and high G/C content, sequencing is impossible.

3. The transfection and lentiviral production efficiency is highly dependent on (a) Cellular density: HEK293T should undergo 2–3 cellular divisions after transfection to get plasmid DNA located in the nucleus and obtain the optimal efficiency. Transfection efficiency depends on the homogeneous distribution of the cells across the surface of the dish. (b) Number of producing cells: lentiviral titer will be dependent of the number of cells transfected with the three plasmids. (c) Viability of HEK293T cells: It is essential to plate highly dividing, but not confluent HEK293T cells at day 1.
4. Transfection can also be performed using ionic liposomes such as Lipofectamin (Invitrogen) or DOTAP (Roche). However, HEK293T cells are highly susceptible to transfection (up to 99% transfection efficiency) and the $CaPO_4$ method is shipper. On day 3, cells should reach confluence. As the lentivector expresses EGFP from the ubiquitous PGK promoter, transfection efficiency may be assessed visually. Ideally, transfection efficiency should be >90%. Use serum free OPTI-MEM medium in order to avoid protein precipitation during lentiviral vector concentration. Unconcentrated supernatant can be used to directly transduce cells. However, viral titer will be relatively low (10^6–10^7 IU/ml). During concentration, conical tubes (as opposed to round bottom tubes) and swinging bucket rotors (as opposed to fixed angle rotors) are recommended to make the pellet easier to localize. For concentrated virus, avoid repeated freeze-thaw cycles.
5. HEK293T cells could be used to determine the titer instead of murine myogenic C2-C12 cells which mimic the transduction of murine myogenic cells. Infectious titers are usually fivefold higher on HEK293T cells as compared with C2-C12 cells due to better transduction susceptibility. Pseudotransduction phenomenon is a well established bias of EGFP expression analysis using lentiviral vectors. It is associated with a transient viral mRNA expression of transduced cells as well as packaging of residual EGFP proteins into virions that gives a low but detectable fluorescence of infected cells. To overcome this time-limited events, grow the cells for at least 5–7 days before analysis. At this time, most of the lentiviral vectors will be retrotranscribed in DNA and integrated in the host genome (a minor fraction will be present as episomes).
6. The method is derived from Ohanna et al. (19). Primary satellite cells are extracted from tibialis anterior (TA) muscle, but other muscles could be used such as the gastrocnemius and soleus muscles. The connective tissues that ensure individual muscles (epimysium) contain numerous fibroblasts that will be copurified with myoblasts and actively proliferate and contaminate myoblast preparation. After filter through 100-μm nylon

gauze, your preparation will contain a highly enriched satellite cell preparation. Around 10,000 cells could be obtained per muscle and should be split in four 12-well plates.

7. During the first days, satellite cells are distinguishable from fibroblasts by microscopic analysis: satellite cells are small and round, while fibroblasts are fusiform with stellar cytoplasm.
8. Satellite cells or myoblasts should be cultured at a low density ($2–5 \times 10^4$ cells/flask 75 cm^2) to prevent spontaneous myotube differentiation. Cells should never exceed the density of 1,500 cells/cm^2. Myotube differentiation can be obtained by culturing myoblasts at a density of 3×10^5 cells/well in DMEM medium containing 2% HS (differentiation medium) for 5–10 days in 6-well plates precoated with Matrigel. The fusogenic ability of cells should be observed between 24 and 48 h after plating and myotubes could be kept during 6–10 days after the beginning of differentiation. The capacity of differentiation and the viability of myotubes need to be tested every ten population-doublings.
9. During their in vitro life span, the division rate progressively decreases until a plateau is reached when no more cell divisions can be observed (40 population-doublings). The number of population-doublings at each passage is defined as log N/log 2, where N is the number of harvested cells divided by the number of cells initially seeded. However, the use of cells with more than 20 population-doublings is not recommended in order to avoid any modification of cell characteristics.
10. To accurately estimate the shRNA-mediated silencing effect and evaluate the phenotype, it is necessary to obtain pure shRNA expressing cells. To meet this goal, lentiviral vectors expressing both the shRNA and the EGFP protein are used, thus allowing easy isolation of transduced cells by FACS. Clonal shRNA/$EGFP^+$ cell population could be obtained by the fastidious limited dilution method. However, we prefer to use FACS to obtain a pure knockdown cell population (up to 99% transduced cell population).

Acknowledgments

This work was supported by INSERM and the Association Vaincre les Maladies Lysosomales (VML). ER was supported by postdoctoral fellowships from VML and the Association Française contre les Myopathies (AFM). GD was supported by doctoral fellowship from Genzyme (France) and AFM.

References

1. Matzke, M.A., Birchler, J.A. (2005) RNAi-mediated pathways in the nucleus. *Nat Rev Genet* 6, 24–35.
2. Agrawal, N., Dasaradhi, P.V., Mohmmed, A., Malhotra, P, Bhatnagar, R.K., Mukherjee, S.K. (2003) RNA interference: biology, mechanism, and applications. *Microbiol Mol Biol Rev* 67, 657–685.
3. Naldini, L., Blömer, U., Gallay, P., Ory, D., Mulligan, R., Gage, F.H., Verma, I.M., Trono, D. (1996) In vivo gene delivery and stable transduction of nondividing cells by a lentiviral vector. *Science* 272, 263–267.
4. Zufferey, R., Dull, T., Mandel, R.J., Bukovsky, A., Quiroz, D., Naldini, L., Trono, D. (1998) Self-inactivating lentivirus vector for safe and efficient in vivo gene delivery. *J Virol* 72, 9873–9880.
5. Li, S., Kimura, E., Fall, B.M., Reyes, M., Angello, J.C., Welikson, R., Hauschka, S.D., Chamberlain, J.S. (2005) Stable transduction of myogenic cells with lentiviral vectors expressing a minidystrophin. *Gene Ther* 12, 1099–1108.
6. Ikemoto, M., Fukada, S., Uezumi, A., Masuda, S., Miyoshi, H., Yamamoto, H., Wada, M.R., Masubuchi, N., Miyagoe-Suzuki, Y., Takeda, S. (2007) Autologous transplantation of SM/C-2.6(+) satellite cells transduced with micro-dystrophin CS1 cDNA by lentiviral vector into mdx mice. *Mol Ther* 15, 2178–2185.
7. Richard, E., Douillard-Guilloux, G., Batista, L, Caillaud, C. (2008) Correction of glycogenosis type 2 by muscle-specific lentiviral vector. *In Vitro Cell Dev Biol Anim* 44, 397–406.
8. Rubinson, D.A., Dillon, C.P., Kwiatkowski, A.V., Sievers C., Yang, L., Kopinja, J., Rooney, D.L., Zhang, M., Ihrig, M.M., McManus, M.T., Gertler, F.B., Scott, M.L., Van Parijs, L. (2003) A lentivirus-based system to functionally silence genes in primary mammalian cells, stem cells and transgenic mice by RNA interference. *Nat Genet* 33, 401–406. Erratum in: (2003) *Nat Genet* 34, 231.
9. Tiscornia, G., Singer, O., Ikawa, M., Verma, I.M. (2003) A general method for gene knockdown in mice by using lentiviral vectors expressing small interfering RNA. Proc Natl Acad Sci USA 100, 1844–1848.
10. Hirschhorn, R., Reuser, A. (2001) Glycogen storage disease type II: acid alpha-glucosidase (acid maltase) deficiency. In *The Metabolic and Molecular Bases of Inherited Disease*, McGraw Hill, New York, pp 3389–3420.
11. Kishnani, P.S., Steiner, R.D., Bali, D., Berger, K., Byrne, B.J., Case, L.E., Crowley, J.F., Downs, S., Howell, R.R., Kravitz, R.M., Mackey, J., Marsden, D., Martins, A.M., Millington, D.S., Nicolino, M., O'Grady, G., Patterson, M.C., Rapoport, D.M., Slonim, A., Spencer, C.T., Tifft, C.J., Watson, M.S. (2006) Pompe disease diagnosis and management guideline. *Genet Med* 8, 267–288.
12. Raben, N., Nagaraju, K., Lee, E., Kessler, P., Byrne, B., Lee, L., LaMarca, M., King, C., Ward, J., Sauer, B, Plotz P. (1998) Targeted disruption of the acid alpha-glucosidase gene in mice causes an illness with critical features of both infantile and adult human glycogen storage disease type II. *J Biol Chem* 7, 53–62.
13. Bijvoet, AG., van de Kamp, E.H., Kroos, M.A., Ding, J.H., Yang, B.Z., Visser, P., Bakker, C.E., Verbeet, M.P., Oostra, B.A., Reuser, A.J., van der Ploeg, A.T. (1998) Generalized glycogen storage and cardiomegaly in a knockout mouse model of Pompe disease. *Hum Mol Genet* 7, 53–62.
14. Bruni, S., Loschi, L., Incerti, C., Gabrielli, O., Coppa, G.V. (2007) Update on treatment of lysosomal storage diseases. *Acta Myol* 26, 87–92.
15. Douillard-Guilloux, G., Raben, N., Takikita, S., Batista, L., Caillaud, C., Richard, E. (2008) Modulation of glycogen synthesis by RNA interference: towards a new therapeutic approach for glycogenosis type II. *Hum Mol Genet* 17, 3876–3886.
16. Mäkinen, P.I., Koponen, J.K., Kärkkäinen, A.M., Malm, T.M., Pulkkinen, K.H., Koistinaho, J., Turunen, M.P., Ylä-Herttuala, S. (2006) Stable RNA interference: comparison of U6 and H1 promoters in endothelial cells and in mouse brain. *J Gene Med* 8, 433–441.
17. Naito, Y., Yamada, T., Ui-Tei, K., Morishita, S., Saigo, K. (2004) siDirect: highly effective, target-specific siRNA design software for mammalian RNA interference. *Nucleic Acids Res* 32, W124–W129.
18. Ui-Tei, K., Naito, Y., Takahashi, F., Haraguchi, T., Ohki-Hamazaki H., Juni, A., Ueda, R., Saigo, K. (2004) Guidelines for the selection of highly effective siRNA sequences for mammalian and chick RNA interference. *Nucleic Acids Res* 32, 936–948.
19. Ohanna, M., Sobering, A.K., Lapointe, T., Lorenzo, L., Praud, C., Petroulakis, E., Sonenberg, N., Kelly, P.A., Sotiropoulos, A., Pende, M. (2005) Atrophy of S6K1(−/−) skeletal muscle cells reveals distinct mTOR effectors for cell cycle and size control. *Nat Cell Biol* 7, 286–294.

Part III

Methods for Muscle Gene Transfer in Large Animal Models

Chapter 15

Fetal Muscle Gene Therapy/Gene Delivery in Large Animals

Khalil N. Abi-Nader and Anna L. David

Abstract

Gene delivery to the fetal muscles is a potential strategy for the early treatment of muscular dystrophies. In utero muscle gene therapy can also be used to treat other genetic disorders such as hemophilia, where the missing clotting proteins may be secreted from the treated muscle. In the past few years, studies in small animal models have raised the hopes that a phenotypic cure can be obtained after fetal application of gene therapy. Studies of efficacy and safety in large animals are, however, essential before clinical application can be considered in the human fetus. For this reason, the development of clinically applicable strategies for the delivery of gene therapy to the fetal muscles is of prime importance. In this chapter, we describe the protocols for in utero ultrasound-guided gene delivery to the ovine fetal muscle in early gestation. In particular, procedures to inject skeletal muscle groups such as the thigh and thoracic musculature and targeting the diaphragm in the fetus are described in detail.

Key words: Fetus, Fetal gene therapy, Prenatal therapy, Gene therapy, Sheep, Ultrasound, Intramuscular, Intrapleural, Intraperitoneal, Muscle

1. Introduction

Gene delivery to the fetal muscles is a potential strategy to treat congenital diseases of the muscle such as the muscular dystrophies. Duchenne muscular dystrophy (DMD) is the most common form of muscular dystrophy, where absence of functional dystrophin leads to progressive muscle weakness beginning in early childhood and culminating in death secondary to respiratory or cardiac failure during the third decade of life. Prenatal gene transfer may offer advantages over postnatal treatment. Efficient gene delivery to several affected muscles groups is technically difficult postnatally, and fetal treatment may target a population of rapidly expanding muscle cells. It may also avoid the development of an immune response to the transduced dystrophin-expressing fibers (1). Importantly, dystrophin expression during fetal life has the ability

Dongsheng Duan (ed.), *Muscle Gene Therapy: Methods and Protocols*, Methods in Molecular Biology, vol. 709, DOI 10.1007/978-1-61737-982-6_15,

to correct some phenotypes that may be difficult to rescue with the postnatal approach (2).

Besides treating muscular dystrophies, in utero muscle gene therapy could also be used to treat life-threatening genetic disorders of other organs where the pathology begins before birth and where fetal application may be of advantage. For this, the ability of the muscle to release protein into the circulation can be exploited by assigning the actual site of production secondary importance as long as therapeutic plasma levels are realized. Thus fetal muscle gene transfer could be used to ameliorate conditions such as metabolic storage disorders or hemoglobinopathies for example, if sustained therapeutic levels of transgenic protein are achievable in utero.

Gene transfer to the fetal muscle has been investigated using adenovirus, adeno-associated virus (AAV), and lentivirus vectors in murine models (3–8). Long-term gene expression has been achieved by intramuscular delivery of adenovirus vectors or VSV-G pseudotyped lentivirus (such as equine infectious anemia virus, EIAV). In small animals such as rodents, gene delivery to the fetal muscle is commonly performed by hysterotomy where the uterus is opened and the fetus is exposed. This method, however, will be difficult to translate into clinical practice since it is associated with a high risk of maternal complications and preterm labor (9), and more sophisticated less invasive routes are needed.

When investigating prenatal gene therapy in large animals, in utero ultrasound-guided injection routes to deliver the therapy are available. These techniques have been mainly developed in the fetal sheep for a number of reasons (10). Sheep have a consistent gestation period of 145 days, and their physiology is similar to human pregnancy. They can provide mainly singleton pregnancies, the fetus is tolerant to in utero manipulations, and the development of the fetal immune system shows similarities with humans. The gestational age of the fetus can be accurately determined by ultrasound using fetal measurements (11, 12). This can all be achieved at a relatively low cost compared to other large animals such as the nonhuman primate. Although there are few sheep models of human genetic disease, these are becoming available with the application of genetic screening for pathological mutations or by cloning techniques (13).

In 2003, we showed that gene delivery to the hindlimb musculature of the early gestation fetal sheep using ultrasound-guided injection of adenovirus-hFIX and adenovirus-LacZ resulted in a highly efficient gene transfer to the muscle. In addition, therapeutic levels of hFIX protein were detected short term in the fetal plasma and there was a low complication rate (14). When a number of ultrasound-guided delivery routes were compared in early gestation fetal sheep, the highest circulating hFIX levels were achieved after intraperitoneal injection of the adenovirus hFIX vector as compared to intramuscular, intrahepatic or intraamniotic delivery (14). This would be anticipated from clinical practice since intraperitoneal fetal blood transfusion has a high success rate in treating fetal anemia and has a low complication rate. The intraperitoneal route also targeted the fetal diaphragm, a muscle that is important in the context of therapy for muscular dystrophy, although transduction was only at low level when compared to other organs such as the fetal liver, heart, adrenal, spleen, lung, and kidney that were preferentially transduced (14).

The respiratory muscles are another therapeutic target in muscular dystrophy treatment, but they are difficult to reach via systemic delivery or by direct injection. Creation of a hydrothorax into which adenovirus vectors are introduced can be achieved using ultrasound-guided injection in the preimmune sheep fetus and this results in efficient transduction of the intercostal muscles (15).

In human fetuses, there exists a potential therapeutic window of opportunity during the late first trimester and early second trimester during which exposure to a foreign protein might result in induction of tolerance and a real opportunity of long-term expression with the potential for cure (16). In the fetal sheep, when compared to late intrauterine gene transfer at >72 days of gestation (term = 145 days), early gestation delivery at 54–65 days takes advantage of multiple tolerogenic mechanisms promoting both central and peripheral tolerance to the transgene products (17). Thus delivery of gene therapy in early gestation delivery will most likely be the timing of choice.

Ultrasound guidance has also been used to deliver gene therapy to the fetal nonhuman primate by a number of other groups using a variety of routes (18–22). Recently, Tarantal showed that injection of the thoracic musculature of early second trimester fetal rhesus monkeys with AAV vectors resulted in transduction of the intercostal muscles, myocardium, and muscular component of the diaphragm (23), and bioluminescence imaging suggested that gene transfer was maintained long term. Delivery to other muscle groups is yet to be reported in the nonhuman primate. It is likely that nonhuman primates will have to be used for safety studies in immediate preparation for a clinical trial of fetal gene therapy, but their breeding conditions and high costs make them prohibitory for routine use for the development of novel techniques.

A number of canine models of genetic disease are available, the Golden Retriever muscular dystrophy and Canine X-linked muscular dystrophy dog being the most notable for muscular genetic diseases (24). Few studies of fetal gene transfer, however, have been performed in canine models. The procedures are complicated by the large numbers of pups, which makes laparotomy necessary, with injection of the pups under ultrasound guidance imaging through the uterine wall (25). Studies of fetal gene transfer to canine muscle are yet to be performed.

Here, we describe the protocols for in utero ultrasound-guided gene delivery to the ovine fetal muscle in early gestation (<70/145 days). In particular, procedures to perform intramuscular, intraperitoneal, and intrapleural fetal injections will be described. Because of the small fetal size in early gestation, intramuscular injection of the fetal muscles at this stage may be considered more as a regional rather than a local gene delivery method since the gene is delivered to a muscle group rather than to a specific muscle. The intramuscular route to target specific peripheral muscular groups combined with an intrapleural/intraperitoneal injection to target the respiratory muscles may be most appropriate for prenatal gene therapy of muscular dystrophies. As the vectors and therapeutic genes vary considerably depending on the experimental design and aims and, because our experience suggests that retrovirus- and lentivirus-mediated gene delivery to the early gestation sheep fetus is poor (unpublished), we will limit our description to the vector that we have used in proof-of-principle studies namely the adenovirus vector, which results in efficient short-term transgenic protein expression. We will also briefly describe time-mating and the anesthetic protocol that we have used and optimized in over 300 pregnant sheep over the past 10 years.

2. Materials

2.1. Generation of Time-Mated Sheep

1. Ewes and rams in a ratio of 10:1.
2. Vaccination to Chlamydia abortus (Enzovax®, Intervet UK) and toxoplasmosis (Toxovax®, Intervet UK).
3. Chronogest® CR sponges (Intervet UK Limited, Milton Keynes, UK) containing 20 mg Flugestone acetate.
4. Sponge applicator.
5. Ram Mating Raddle (Agrimark Limited, Isle of Man, UK).
6. Super Ewe & Lamb UFAS compound feed (J & W Atlee Ltd., Dorking, Surrey, UK).

2.2. Confirmation of Pregnancy and Gestational Age by Ultrasound

1. Ultrasound coupling gel (Electro Medical Supplies Ltd., Oxon, UK).
2. Clippers for fleece (Masterclip Duo, Outlandish Items Ltd., Leicestershire, UK).
3. Ultrasound scanner with a curved-array 3.0–5.0 MHz abdominal probe with real time image display, cine loop

facility, B-mode, M-mode pulse wave Doppler flow mode, and color flow mode, e.g., Logiq 400 (GE Medical Systems, Milwaukee, USA), or Acuson 128 XP10 ultrasound scanner (Siemens, Bracknell, UK).

4. Ultrasound charts in use for sheep (11, 12).

2.3. Sheep Anesthesia

1. Wood chips for bedding (GPM Shavings, East Angus, Canada).
2. A large sling with poles.
3. A balance to weigh the sheep (needs to measure up to 150 kg).
4. Clippers for fleece (Masterclip Duo, Outlandish Items Ltd., Leicestershire, UK).
5. 19 Gauge butterfly winged perfusion set (Terumo Europe NV, Leuven, Belgium).
6. BD Vacutainer EDTA, 9NC, and SST tubes (BD Vacutainer systems, Plymouth, UK).
7. 15 ml of 10% w/v thiopental sodium i.v. (Thiovet, Novartis Animal Health UK Ltd., Hertfordshire, UK).
8. 2–2.5% isoflurane (Isoflurane-Vet, Merial Animal Health Ltd., Essex, UK).
9. 100% oxygen.
10. An oral gastric tube.
11. A laryngoscope with a 30 cm blade (Penlon, UK).
12. A 9.0 mm cuffed endotracheal tube (Portex, UK).
13. A semiclosed anesthetic system (Ohmeda Anesthesia system, BOC Healthcare).
14. A stethoscope.
15. A capnograph.
16. A pulse oximeter (5250 RGM, Ohmeda).
17. A surgical theater table.

2.4. Ultrasound-Guided Procedures

1. Povidone iodine antiseptic cleaning solution (Grampian Pharmaceuticals Ltd., UK).
2. Chlorhexidine gluconate cleaning scrub (Hibiscrub, AstraZeneca Ltd., UK).
3. Povidone iodine antiseptic solution (Grampian Pharmaceuticals Ltd., UK).
4. A 3.0–5.0 MHz curved-array abdominal probe on an ultrasound scanner.
5. Sterile ultrasound coupling gel (Electro Medical Supplies Ltd., Oxon, UK).

6. A no.11 blade.
7. 20G, 15 cm long and 22G, 9 cm long echotip needles with a lancet tip (Cook Medical, UK).
8. Sterile normal saline (0.9% NaCl).
9. 1, 2, and 5 ml syringes.
10. Viricide agent, e.g., Virkon (Antec International Limited).

2.5. Vector Dose

1. 5×10^{11}–5×10^{12} particles/ml of first generation adenovirus vectors AdRSVβgal or AdTG9397 containing the β-galactosidase reporter gene and the human factor IX gene respectively (Transgene Laboratory, France).
2. Phosphate-buffered saline (PBS).
3. Diethylamnioethyl (DEAE) dextran (5 mg/ml) (Sigma-Aldrich).

3. Methods

3.1. Generation of Time-Mated Sheep

1. Vaccinate the selected experimental sheep for Chlamydia abortus and toxoplasmosis (see Note 1).
2. Induce ovulation by placing Chronogest® sponges (Flugestone acetate) containing 30 mg progesterone in the vagina of ewes for 2 weeks (see Note 2).
3. Two days after removal of the progesterone sponges, put ewes in a pen with the ram overnight. Mark the ram on its belly with Ram Raddle®, a colored powder mixed with liquid paraffin. The mark is transferred to the back of the ewe once she has been served. Marked ewes are presumed to have been tupped.
4. Put ewes that are unmarked or not marked clearly back for a "return service" with the ram 2 weeks later.
5. Use a spreadsheet to check on regular progress with tupping and to schedule experiments at specific gestational ages, with close liaison with the farmer.
6. When the sheep breeding season is advanced (March onward), superovulation may help. Place Chronogest® sponges vaginally for 12 days followed by an intramuscular injection of 1,000 IU Folligon®. Place the ewe with the ram 2 days later and return as before 14 days later. Superovulation often results in the generation of higher order multiples such as triplets and quadruplets, and the conception rate is not as high as that achieved using progesterone alone.

3.2. Confirmation of Pregnancy and Gestational Age by Ultrasound

1. Scan all ewes that have been tupped to confirm pregnancy, fetal number, and gestation age at approximately 40 days of age.
2. Hold ewes in a sitting position. Scan with a standard B-mode gray-scale ultrasound scanner and an abdominal probe. This is the same position as is used by farmers to clip the toenails and for shearing sheep. Sheep are generally placid and will happily remain in this position for at least 15 min.
3. Clip an area of the fleece close to the skin just above the udders extending 10 cm up the abdomen to allow a clear view of the whole uterus (see Note 3).
4. Apply the sonographer ultrasound coupling medium to the clipped area and scan the uterus using a 3.0–5.0 MHz probe held just above the udders. Scan across both horns of the uterus to examine for twins and triplet pregnancies (see Note 4).
5. Obtain still images of the occipito-snout length and biparietal diameter and measure the dimensions on the machine. Confirm gestational age according to standard tables (see Note 5).

3.3. Sheep Anesthesia (see Note 6)

1. Starve ewes for 12 h before surgery but allow free access to water and bedded on wood chips. This prevents bloating, which can occur when sheep under anesthesia are unable to belch (eructate) normally and release methane gas, which is a by-product of fermentation of their food.
2. Weigh animals to allow calculation of the correct dose for drugs.
3. Walk sheep to the theatre on the morning of surgery. There should be a dedicated theater space if possible that is kept clean, has anesthetic gas extraction facilities, and air-conditioning to maintain the temperature at 37°C.
4. Turn up the sheep with an assistant sitting comfortably behind. Clip the wool from the neck. Cannulate the jugular vein using a 19G butterfly winged perfusion set and secure the cannula with tape.
5. Take blood for baseline investigations of antibodies, hematology, and biochemistry parameters. Collect blood into BD Vacutainer tubes containing potassium EDTA for plasma and into plain tubes for serum (SST).
6. Administrate 15 ml of thiopental sodium (10% w/v) intravenously over 1 min for induction of anesthesia.
7. Once asleep, roll the ewes onto a mat with poles or handles to attach to a hoist. Lift ewes mechanically onto the operating table and lay them supine. With the sheep in supine position,

extend the neck and lift up the jaw using tape applied across the tongue and teeth. Pass a laryngoscope with a 30 cm blade down to the trachea to visualize the vocal cords. Pass a 9.0 mm cuffed endotracheal tube containing a stylet through the vocal cords. Remove the stylet and inflate the cuff with the required volume of air to prevent inhalation of regurgitated ruminal contents. Secure the tube with a tape tied around the lower jaw (see Note 7).

8. Connect the endotracheal tube to the anesthetic gas machine and check the correct placement of the endotracheal tube by (a) auscultating both lungs while the bag and valve mask are being compressed to force air into the lungs; (b) observing inflating and deflating of the bag and valve mask in the closed system; (c) feeling air movement within the endotracheal tube as the ewe breathes.
9. For lengthy experiments (over 20 min), pass an oral gastric tube into the sheep stomach to allow the release of methane gas.
10. Once the sheep is supine on the operating table and anesthetized, lower the head to allow drainage of saliva and ruminal fluid. Because access to the whole abdomen is needed for procedures, keep the animals in dorsal recumbency (see Note 8).
11. For long procedures (over 15 min), connect a capnograph to the endotracheal tube to monitor carbon dioxide saturation. Keep end-tidal CO_2 <5% if possible.
12. Maintain anesthesia by inhalation of 2–2.5% isoflurane in 100% oxygen run at 2–3 l/min for spontaneously breathing animals and at 6–8 l/min for animals on intermittent positive pressure ventilation (IPPV) to maintain an end-tidal CO_2 <5%, O_2 saturation >95%, and a respiratory rate of 12–14/min using a semi-closed anesthetic circuit.
13. Place a pulse oximeter on the ewe's ear or tongue to monitor oxygen saturation and pulse rate. Record oxygen saturation, pulse rate, respiration rate, carbon dioxide concentration, anesthetic gas concentration, and level of anesthesia on an anesthetic sheet every 15 min.
14. Turn off the anesthetic gas 1–2 min before the end of the experiment to allow recovery.
15. Hoist the ewe off the theater table to the floor and roll onto her sternum. Place her legs underneath to ensure that she can get upright as soon as she is awake. Extubate the ewe once strong swallowing reflexes return and then return the ewe to the sheep pen.
16. Carefully monitor the ewe to ensure she stands up within a few minutes of the procedure. Help the ewe to stand if she does not do so spontaneously after 20 min. Standing allows

recovery of ruminal function which can be affected by a long general anesthetic.

17. Assess the well-being of the ewe the day after the procedure. Assess fetal survival by ultrasound scan. Carefully examine the ultrasound-guided injection sites for infection.

3.4. Ultrasound-Guided Procedures

1. Once anesthetized, clip the wool from the ewe's abdomen from the udders up to the sternum and round to the sides (see Note 9).
2. Scrub the skin for 5 min using povidone iodine antiseptic cleaning solution (see Note 10).
3. Perform a detailed ultrasound examination of the uterus and its contents using a 3–5 MHz probe on the ultrasound scanner. Measure fetal femur length, abdominal circumference, biparietal diameter, and occipito-snout length to assess the gestational age (Fig. 1). Identify the appropriate site for injection (see Note 11).
4. Scrub the abdomen again for further 5 min using chlorhexidine glucuronate cleaning solution and make a final clean with povidone iodine solution over the injection site (see Note 10).
5. Wipe the ultrasound probe with 70% ethanol wipes and then spray again. Allow ethanol to evaporate.
6. Drape the abdomen now with sterile drapes to reveal a window for the ultrasound-guided injection (see Notes 12 and 13).
7. Use sterile ultrasound coupling medium (ultrasound gel) for ultrasound scanning (see Note 14).
8. Make a small skin incision using a no.11 blade to facilitate needle insertion and to avoid blunting the sharp needle tip.
9. Insert the needle under ultrasound guidance (with an angle of 45° relative to the ultrasound beam) through the maternal skin, the uterus, and into the amniotic cavity avoiding passage through the placentomes if possible. Keep the whole needle length and particularly the needle tip under ultrasound view as this view most accurately reflects the position of the needle, otherwise it is easy to be mislead about the true needle orientation (see Note 15).
10. Insert the needle through the fetal skin into the target by flicking it through the uterine wall and into the fetus. Once the needle tip is thought to be in the correct position, remove the stylet and inject the vector followed by 40 μl 0.9% saline to flush the vector from the dead space of the spinal needle (see Note 15).
11. Intramuscular injection. Obtain a sagittal ultrasonographic plane of the fetal thigh. Insert a 20 or 22G Echotip Lancet

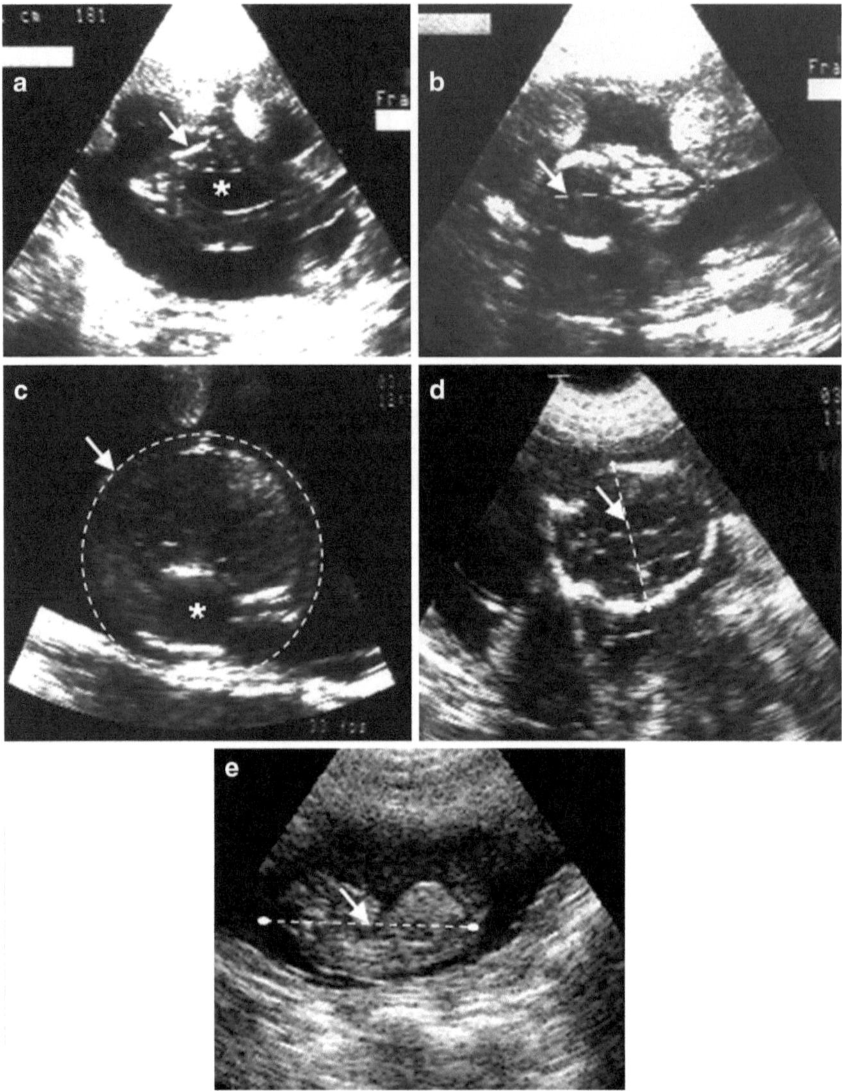

Fig. 1. Ultrasound measurements of fetal sheep. (**a**) Femur length (10.1 mm, *white arrow*) and fetal bladder (*white asterisk*). (**b**) Occipito-snout length (40.2 mm, *white arrow*) at 60 days of gestation. (**c**) Abdominal circumference (162 mm, *white arrow*) and fetal stomach (*white asterisk*) at 83 days of gestation. (**d**) Bi-parietal diameter (22 mm, *white arrow*) at 60 days of gestation. (**e**) Crown-rump length (27.4 mm, *white arrow*) at 36 days of gestation.

needle along the length of the femur and/or the buttock (Fig. 2). Check the position of the needle in longitudinal and transverse sections. Once the correct position of the needle is confirmed, remove the stylet. With a longitudinal view of the needle within the fetal muscle, attach the syringe containing the vector to the needle and as the needle is slowly withdrawn along the length of the muscle, inject the vector. Deliver 100–500 μl by preferably 1 (up to 4) injections depending on the viral titer. Observe echogenic foci within the muscle

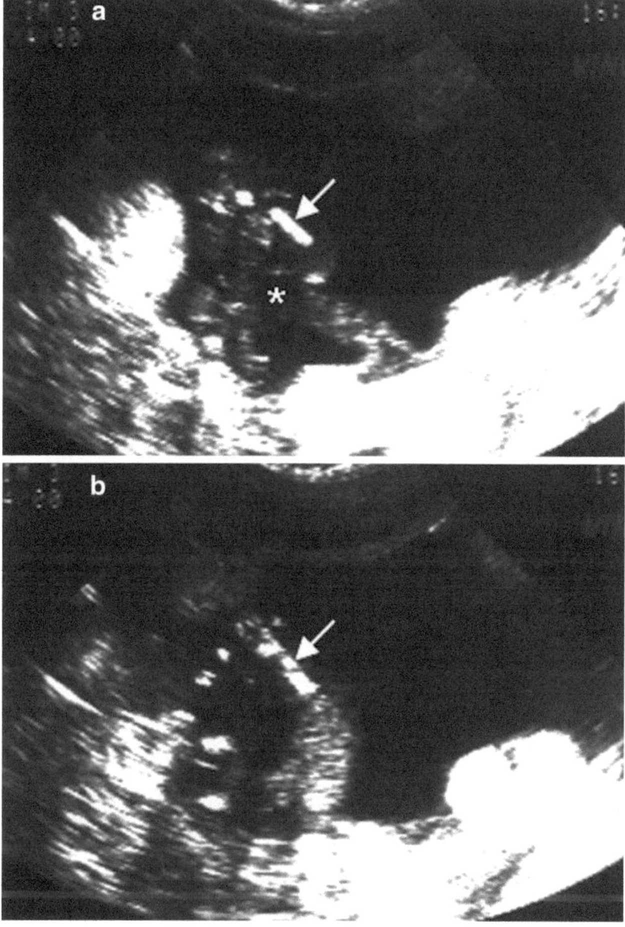

Fig. 2. Ultrasound-guided injection of fetal thigh muscle in early gestational fetal sheep. (**a**) The femur bone is seen in longitudinal view within the fetal thigh (*white arrow*). The fetal bladder is indicated (*white asterisk*). (**b**) The tip of a 22G (*white arrow*) is seen positioned just beside the femur within the fetal thigh muscle. The needle has been inserted along the muscle and parallel to the femur bone.

parenchyma to confirm vector placement. Check for bleeding after the removal of needle (see Note 16).

12. Intrapleural injection. Obtain a view of the fetal chest in longitudinal section. Create a hydrothorax by inserting a 20G Echotip Lancet needle through the thoracic musculature between the ribs and up to the margin of the lung parenchyma. Check the position of the needle in longitudinal and transverse sections of the fetal chest. Inject 500 μl PBS until a pool of fluid is seen by ultrasound (Fig. 3). This not only confirms the correct position of the needle, but also creates a hydrothorax into which the vector is delivered. Inject 100 μl vector into the pool of fluid. Check for bleeding after the removal of needle.

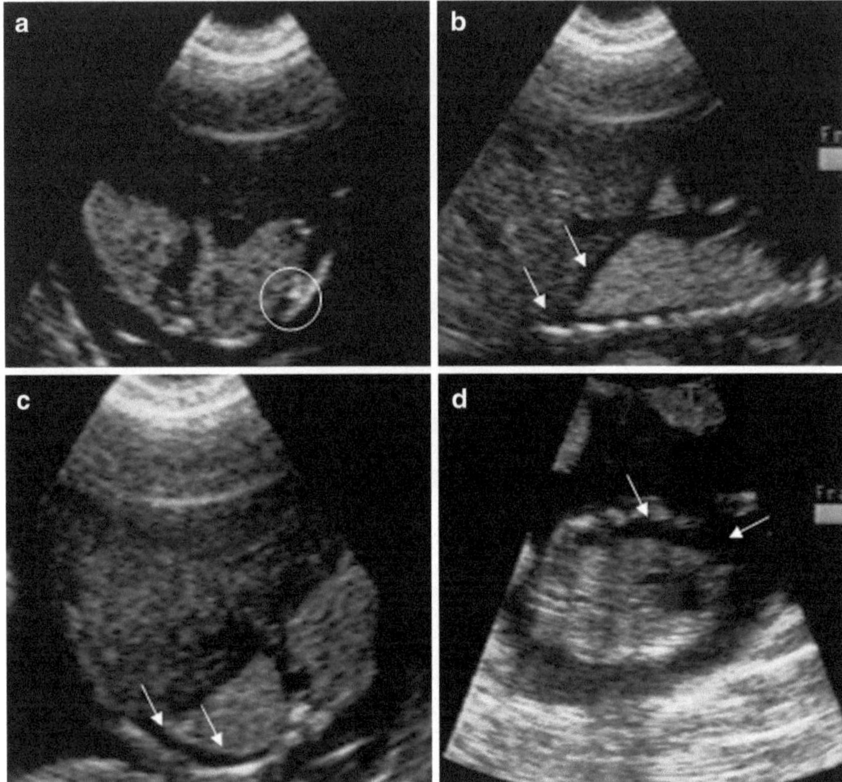

Fig. 3. Ultrasound-guided creation of hydrothorax. (**a**) Introduction of the needle tip through the thoracic musculature and up to the margin of the lung parenchyma. The needle tip is seen within the white circle. Visualization of hydrothorax after instillation of 500 microliters of phosphate-buffered saline (PBS in sagittal (**b**), coronal (**c**), and transverse (**d**) views. The white arrows indicate echolucent areas that contain the injected fluid.

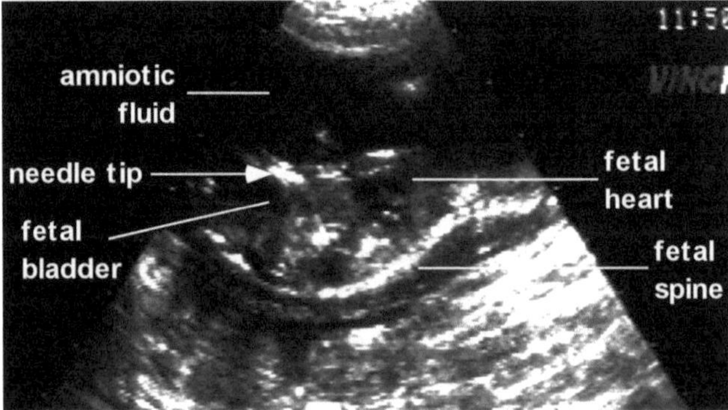

Fig. 4. Ultrasonogram of intraperitoneal injection of sheep fetus. A longitudinal view of the sheep fetus at 54 days is shown. The tip of the 22G needle (*white arrow*) is seen within the fetal peritoneal cavity just superior to the bladder.

13. Intraperitoneal injection. Obtain a mid-sagittal ultrasonographic plane of the fetal abdomen. Target the needle tip below the level and to the side of the cord insertion and just superior to the fetal bladder (Fig. 4). View the position of the needle to

confirm correct placement. Deliver the vector in volumes of 100–500 μl before 60 days, up to 1 ml between 60 and 70 days and 1–2 ml after 70 days of gestation (see Note 17).

3.5. Vector Dose

1. An adenoviral vector dose of around 5×10^{11}–5×10^{12} particles/ml is advised since higher doses may be toxic (see Note 18) (14).
2. For local delivery, complex the viral particles with DEAE dextran for better transfection efficiency (15). Prepare the adenovirus polycation by adding adenovirus particles to PBS containing 5 mg/ml DEAE. Incubate for 30 min at room temperature before injection.
3. Aliquot all vectors freshly using pipettes. Dilute in the needed volume and deliver to the fetus within 10–15 min.
4. After vector delivery, immediately place the contaminated needle into a sharps bin to avoid sharps injury. Flush the syringe with a viricide such as Virkon for 30 min to destroy the vector particles before being discarded.

4. Notes

1. Research in pregnant large animals such as the sheep requires considerable planning to ensure that time-mated ewes are presented to the researcher at the correct gestational age in the best of health. A dedicated animal research unit is required as well as good communication between the animal technicians looking after the flock before experiments and the researcher themselves. Time-mating or tupping of sheep usually begins in early autumn at the beginning of the sheep breeding season. A few breeds can be tupped all year round (e.g., Pol Dorset sheep), but they tend to be more expensive.
2. Large animals need to be moved to the experimental facility at least 1 week before experiments to ensure their acclimatization. Care needs to be taken with animal welfare so as to reduce stress, for example housing sheep in pairs, or close by so that they can see each other is recommended. The diet of pregnant may be different to nonpregnant large animals, and it is worth consulting a specialist for large animals to determine the best feeding strategy. For example in sheep, those that have a twin and other higher multiple conception may require a concentrated source of oil, protein, and vitamins found in special feed, which is fed at increasing amounts as gestation advances. This ensures adequate nutrition for the ewe and reduces the risk of toxemia caused by the large late

gestation fetus compressing the rumen and reducing the ability of the ewe to digest enough hay for nutrition. Attention to this level of detail will improve animal well-being and thereby reduce the miscarriage rate. Ewes that are allowed to come to birth should be vaccinated with Heptavac-P plus® (Hoechst Roussel Vet Limited, Dublin) 1 month before their delivery date to prevent Clostridium and Pasteurella infection in the lambs. Further, two doses are given to the lambs at 10 and 16 weeks of age.

3. It is useful to clip the fleece of the ewe just above the udders to improve the view of the fetal sheep during ultrasound assessment with the ewe awake and turned up. This improves the view and reduces the chance of missing a twin or triplet pregnancy.
4. It is important to confirm the gestational age by ultrasound before the induction of anesthesia in sheep particularly if the ewe had a "return service" 2 weeks after the first tup. The best assessment of gestational age in the first and second trimesters is provided by the fetal occipito-snout length. After approximately 100 days of gestation, the head is often deeply engaged in the pelvis so that measurement of occipito-snout length is not possible. The bi-parietal diameter can be used as an alternative in the third trimester.
5. Gestational age at injection is an important consideration. Before approximately 70–80 days of gestation, the fetal thigh musculature is relatively small compared to the size of the needle, making the injection technically more challenging. On the other hand, the relatively smaller muscle bulk means more is exposed to the vector, leading to better levels of gene transfer, and transduced muscle tissue is more easily identified on postmortem analysis.
6. Analgesia is not considered necessary for ultrasound-guided procedures since all procedures involve a single injection of a fine needle (20G or less) and all are done under general anesthesia. These procedures are routinely carried out in humans without the need for analgesia.
7. Alternatively, the ewe may also be intubated while she is held sitting up by an assistant behind her. Using a tape the upper jaw is lifted up to straighten the trachea, and while the tongue is gently pulled to one side with some gauze, the endotracheal tube is passed through the vocal cords.
8. Over long periods the rumen can compress the lungs causing dyspnea and the vena cava reducing venous return to the heart. This is not felt to be a problem in the case of in utero ultrasound-guided procedures because they are usually of sufficiently short duration.

9. It is vital to clean the skin/fleece of large animals adequately before embarking on surgery or ultrasound-guided procedures. Close clipping of the fleece or hairs from the skin helps remove dirt and a source of bacteria. Care is taken to ensure that even the finest wool is removed.
10. We routinely clean the abdomen three times prior to surgery while the animal is under general anesthesia. The first scrub is performed before the animal is scanned to remove the majority of animal dander, and we use povidone iodine antiseptic cleaning solution, which is worked into the skin/fleece using a scrubbing brush and then wiped clean. The second scrub is done after the ultrasound scan assessment of fetal size and gestational age. For this, we use chlorhexidine glucuronate cleaning solution, which is again worked into the skin/fleece using a scrubbing brush and then wiped clean. Finally, we dribble povidone iodine antiseptic cleaning solution onto the skin site over the incision or planned ultrasound-guided injection site. We have found that judicious attention to this aspect of care reduces the rate of miscarriage.
11. Ultrasound coupling medium (ultrasound gel) is used during scans. At this stage nonsterile gel can be used.
12. The operator performing the ultrasound-guided injection must have a comfortable view of the ultrasound screen since procedures may take some time to perform. Injecting the sheep fetus requires a lighter touch in comparison to the human fetus, since the rectus muscle are more lax, and the fetus is more superficial. Care must be taken not to compress the fetus within the uterus by leaning on the probe too much.
13. The operators (usually two) scrub up and don sterile gowns and gloves. All instruments (needles, syringes containing the vector and saline) are laid out on a small operating table draped with a sterile drape.
14. Small aliquots of ultrasound gel can be sterilized by irradiation for use during operative procedures although afterward they become quite liquid. Liquid paraffin is a good alternative that is used for human fetal medicine procedures. It can be sterilized by irradiation for use.
15. For each injection route a freehand technique is used. Needle guides can be used but reduce the flexibility of the approach, especially if the fetal position moves relative to the needle as it is inserted. The fetus does not move around once the ewe has been under general anesthesia for at least 10 min. Flushing after vector injection is important because in many experiments, small volumes of vector are often applied. If the position of the needle tip is not certain, a small volume (40 μl 0.9% saline) can be introduced to confirm it. A white flash

will be seen on the screen as turbulence created by microbubbles. Procedures are always videotaped so that after-injection review can be done to confirm the injection route.

16. The Injection sites include the fetal thigh muscles (quadriceps), buttocks, and biceps/triceps. The fetal thigh (quadriceps or hamstring) and buttock muscles are chosen for injection because these are the largest accessible muscle groups at this gestational age in the sheep fetus. The fetal thigh is best observed by obtaining a cross-sectional view of the fetal bladder and the femora are seen extending either side. A 22G is used before 70 days of gestation and 20G used after 70 days of gestation. Echotip needles are used because the tip is easier to see while next to the echogenic femur bone, when compared with other needles. The cross-sectional diameter of the fetal thigh or buttock in early gestation muscle is as wide as the bevel of the Echotip needle. Thus insertion of the needle tip into the muscle in a transverse section may result in immediate loss of the vector into the amniotic fluid from the needle tip. To avoid this and to ensure a longer needle track, the needle is placed deeply within the muscle parallel to the bone. In later gestations (100 days of gestation and beyond), the muscle bulk is larger and a more angled approach to the fetal bone is available without the risk of vector leakage. If necessary to achieve the required vector dose, a number of intramuscular injections on the same fetus can be done, on the same leg or on the opposite leg, if accessible.

17. A 20 or 22G Echotip Lancet needle is used, depending on gestational age. 20G is used after 80 days of gestation. Care should be taken to prevent damage to the umbilical cord insertion site or the umbilical arteries as they flow down medio-laterally along the posterior aspect of the anterior fetal abdominal wall. Color or power Doppler is helpful in delineating the vascular anatomy if the operator is in doubt. Needle placement can also be verified by injecting 100 μl saline and observing turbulence as microbubbles spread throughout the peritoneal cavity. In addition, an echodense area surrounding the fetal bowel usually develops at the injection site.

18. The lethal toxic level for adenovirus vectors in fetal sheep is approximately 8×10^{12} particles/kg (26). The dose given to the fetus is calculated according to the approximate weight of the fetal sheep determined from standard charts of fetal size (11, 12). We have found adenovirus vectors to provide short-term and highly efficient muscular gene transfer to fetal sheep. The vectors are tolerated to a high dose with minimal observed muscular damage on histological analysis (14, 15, 27). Other vector systems have not proved so useful. Gene transfer to fetal sheep muscle in vitro is achievable using retrovirus

vectors (28). We have found, however, that retrovirus vectors based on the Moloney Leukemia Virus (MLV) give poor gene transfer to the quadriceps of late gestation fetal sheep (120 days of gestation). This may reflect the relatively low dose that was applied (1×10^8 particles/kg fetus) to a relatively large muscle bulk at that gestation. Lentivirus vectors also seem to give poor gene transfer in the fetal sheep. Vectors based on human immunodeficiency (HIV) or equine immune anemia virus (EIAV) and pseudotyped with VSV-G or Mokola gave no significant gene transfer to the fetal sheep muscle in early gestation when injected at a dose of 1×10^8 particles/kg fetus.

References

1. Chamberlain JS. (2002) Gene therapy of muscular dystrophy. *Hum Mol Genet* 11, 2355–2362.
2. Ahmad A, Brinson M, Hodges BL, Chamberlain JS, Amalfitano A. (2000) *Mdx* mice inducibly expressing dystrophin provide insights into the potential of gene therapy for Duchenne muscular dystrophy. *Hum Mol Genet* 9, 2507–2515.
3. MacKenzie TC, Kobinger GP, Kootstra NA, Radu A, Sena-Esteves M, Bouchard S et al. (2002) Efficient transduction of liver and muscle after *in utero* injection of lentiviral vectors with different pseudotypes. *Mol Ther* 6, 349–358.
4. MacKenzie TC, Kobinger GP, Louboutin JP, Radu A, Javazon EH, Sena-Esteves M et al. (2005) Transduction of satellite cells after prenatal intramuscular administration of lentiviral vectors. *J Gene Med* 7, 50–58.
5. Bilbao R, Srinivasan S, Reay D, Goldberg L, Hughes T, Roelvink PW et al. (2003) Binding of adenoviral fiber knob to the coxsackie-adenovirus receptor is crucial for transduction of fetal muscle. *Hum Gene Ther* 14, 645–649.
6. Bilbao R, Reay DP, Hughes T, Biermann V, Volpers C, Goldberg L et al. (2003) Fetal muscle gene transfer is not enhanced by an RGD capsid modification to high-capacity adenoviral vectors. *Gene Ther* 10, 1821–1829.
7. Bilbao R, Reay DP, Li J, Xiao X, Clemens PR. (2005) Patterns of gene expression from *in utero* delivery of adenoviral-associated vector serotype 1. *Hum Gene Ther* 16, 678–684.
8. Gregory LG, Waddington SN, Holder MV, Mitrophanous KA, Buckley SMK, Mosley KL et al. (2004) Highly efficient EIAV-mediated in utero gene transfer and expression in the major muscle groups affected by Duchenne muscular dystrophy. *Gene Ther* 11, 1117–1125.
9. Kitano Y, Flake AW, Crombleholme TM, Johnson MP, Adzick NS. (1999) Open fetal surgery for life-threatening fetal malformations. *Semin Perinatol* 23, 448–461.
10. David AL, Peebles D. (2007) Gene therapy for the fetus: is there a future? *Best Pract Res Clin Obstet Gynaecol* 22, 203–218.
11. Barbera A, Jones OW, Zerbe GO, Hobbins JC, Battaglia FC, Meschia G. (1995) Ultrasonographic assessment of fetal growth: comparison between human and ovine fetus. *Am J Obstet Gynecol* 173, 1765–1769.
12. Kelly RW, Newnham JP. (1989) Estimation of gestational age in Merino ewes by ultrasound measurement of fetal head size. *Aus J Agric Res* 40, 1293–1299.
13. Porada CD, Sanada C, Long CR, Wood JA, Desai J, Frederick N et al. (2010) Clinical and molecular characterization of a re-established line of sheep exhibiting hemophilia A. *J Thromb Haemost* 8, 276–285.
14. David AL, Cook T, Waddington S, Peebles D, Nivsarkar M, Knapton H et al. (2003) Ultrasound guided percutaneous delivery of adenoviral vectors encoding the beta-galactosidase and human factor IX genes to early gestation fetal sheep *in utero*. *Hum Gene Ther* 14, 353–364.
15. Weisz B, David AL, Gregory LG, Perocheau D, Ruthe A, Waddington SN et al. (2005) Targeting the respiratory muscles of fetal sheep for prenatal gene therapy for Duchenne muscular dystrophy. *Am J Obstet Gynecol* 193, 1105–1109.
16. Touraine J-L, Plotnicky H, Roncarolo MG, Bachetta R, Gebuhrer L. (2007) Immunological lessons learnt from patients transplanted with fully mismatched stem cells. *Immunol Res* 38, 201–209.

17. Colletti E, Lindstedt S, Park PJ, Almeida-Porada G, Porada CD. (2008) Early fetal gene delivery utilizes both central and peripheral mechanisms of tolerance induction. *Exp Hematol* 36, 816–822.
18. Larson JE, Morrow SL, Delcarpio JB, Bohm RP, Ratterree MS, Blanchard JL et al. (2002) Gene transfer into the fetal primate: evidence for the secretion of transgene product. *Mol Ther* 2, 631–639.
19. Tarantal AF, McDonald RJ, Jimenez DF, Lee CI, O'Shea CE, Leapley AC et al. (2005) Intrapulmonary and intramyocardial gene transfer in rhesus monkeys (Macaca mulatta): safety and efficiency of HIV-1-derived lentiviral vectors for fetal gene delivery. *Mol Ther* 12, 87–98.
20. Bennett M, Galan H, Owens G, Dewey R, Banks R, Hobbins J et al. (2001) *In utero* gene delivery by intraamniotic injection of a retroviral vector producer cell line in a nonhuman primate model. *Hum Gene Ther* 12, 1857–1865.
21. Garrett DJ, Larson JE, Dunn D, Marrero L, Cohen JC. (2003) In utero recombinant adeno-associated virus gene transfer in mice, rats, and primates. *BMC Biotechnol* 3, 16.
22. Lai L, Davison BB, Veazey RS, Fisher KJ, Baskin GB. (2002) A preliminary evaluation of recombinant adeno-associated virus biodistribution in rhesus monkeys after intrahepatic inoculation in utero. *Hum Gene Ther* 13, 2027–2039.
23. Tarantal AF, Lee C. (2010) Long-term Luciferase expression monitored by bioluminescence imaging after AAV-mediated fetal gene delivery in rhesus monkeys (Macaca mulatta). *Hum Gene Ther* 21, 143–148.
24. Wells DJ, Wells KE. (2005) What do animal models have to tell us regarding Duchenne muscular dystrophy? *Acta Myol* 24, 172–180.
25. Meertens L, Zhao Y, Rosic-Kablar S, Li L, Chan K, Dobson H et al. (2002) In utero injection of alpha-L-iduronidase-carrying retrovirus in canine mucopolysaccharidosis type I: infection of multiple tissues and neonatal gene expression. *Hum Gene Ther* 13, 1809–1820.
26. Themis M, Schneider H, Kiserud T, Cook T, Adebakin S, Jezzard S et al. (1999) Successful expression of b-galactosidase and factor IX transgenes in fetal and neonatal sheep after ultrasound-guided percutaneous adenovirus vector administration into the umbilical vein. *Gene Ther* 6, 1239–1248.
27. David AL, Peebles DM, Gregory L, Waddington SN, Themis M, Weisz B et al. (2006) Clinically applicable procedure for gene delivery to fetal gut by ultrasound-guided gastric injection: toward prenatal prevention of early-onset intestinal diseases. *Hum Gene Ther* 17, 767–779.
28. John HA. (1994) Variable efficiency of retroviral-mediated gene transfer into early-passage cultures of fetal lamb epithelial, mesenchymal, and neuroectodermal tissues. *Hum Gene Ther* 5, 283–293.

Chapter 16

Electroporation of Plasmid DNA to Swine Muscle

Angela M. Bodles-Brakhop, Ruxandra Draghia-Akli, Kate Broderick, and Amir S. Khan

Abstract

For plasmid-mediated gene therapy applications, a major limitation to scale up from rodents to large animals is the low expression level of injected plasmid DNA. The electroporation technique, which results in the passage of foreign material through the cell membrane, is one method that has been shown to be effective at improving local plasmid uptake and consequently, expression levels. Previous studies have determined that optimized electroporation parameters (such as electric field intensity, number of pulses, lag time between plasmid injections and electroporations, and optimal plasmid formulation conditions) are dependent on the target muscle type and individual species. Here, we provide a detailed protocol to optimize conditions for the successful intramuscular electroporation of plasmid DNA to swine, a large animal model. Our results suggest that the technique is safe and effective for veterinary applications. Furthermore, these results provide evidence for the feasibility of upcoming human applications.

Key words: Electroporation, Gene therapy, Muscle, Plasmid DNA, Swine

1. Introduction

Unfortunately, positive results from preclinical studies in mice do not always translate to larger animals and the same is true in the field of gene therapy. Electroporation (EP) is one physical method that is currently being investigated to overcome the issue of efficacy for delivery of plasmids for gene therapy purposes to large animals and ultimately to humans. Intramuscular (IM) EP provides a convenient and accessible site for the production of therapeutic or antigenic proteins. It provides a system for in vivo manufacture of the desired gene product and, in the case of therapeutic proteins, long-term expression. Clearly, the optimization of plasmid constructs plays an important role in the success of

Dongsheng Duan (ed.), *Muscle Gene Therapy: Methods and Protocols*, Methods in Molecular Biology, vol. 709,
DOI 10.1007/978-1-61737-982-6_16,

gene therapy (1); however, it has been shown that expression of naked plasmid DNA can be enhanced with EP when compared to injection alone (2, 3).

The ability to use the technique of EP for gene therapy applications in a large animal model, such as swine, not only provides valuable information for the transition of the technology to a human clinical setting but also presents the veterinarian and the farm production industry with the necessary tools to improve the health and well-being of companion and production animals. The optimization of EP parameters is necessary to achieve maximum plasmid expression while also minimizing muscle damage (4). EP parameters are dependent on species, individual tissue-specific resistance, and DNA formulation (5, 6). It has been shown that effective expression is achievable with optimized EP at both the intact (whole) muscle and the cellular (single fiber) levels while maintaining the contractile function of skeletal muscles (7). Successful EP of large animals has previously been reported for dogs (8–11), horses (12), pigs (13, 14), and cattle (15, 16) for the treatment of several different ailments such as cancer-associated anemia, renal disease, and arthritis-like conditions, as well as to improve the quality of life and survival, and to increase production parameters.

In swine, we have previously shown that IM injection of a growth hormone-releasing hormone (GHRH) plasmid DNA followed by EP in piglets resulted in an increase in body weight compared to placebo-injected controls (13, 17). Similarly, the IM EP of GHRH to gilts at day 85 of gestation resulted in piglets that were larger at birth and weaning compared to controls (18). A myogenic plasmid encoding for GHRH can be successfully and safely delivered to the skeletal muscle of sows in late gestation. In a large study of gestating sows ($n = 997$), we have shown that IM EP is a safe, viable alternative or supplement method to conventional treatments for improving life expectancy and increasing production parameters over three successive pregnancies (14). A different version of this plasmid-mediated GHRH technology (LifeTide™ SW5) gained regulatory approval in Australia in early 2008 for the purpose of reducing piglet mortality and marks the first nonviral gene therapy product to be approved by a regulatory agency for use in food animals delivered by EP.

In this protocol, we describe administration of a muscle-specific plasmid encoding for secreted embryonic alkaline phosphatase (SEAP) to the muscle of young pigs. SEAP expression is determined over a 2-week period posttreatment as an immune-mediated clearance of the protein occurs after this time point. We provide the strategies to the optimal EP parameters for optimal expression of plasmid DNA in swine muscle (see Note 1).

In particular, we have compared EP lag time between injection and onset of EP (4 and 80 s) and electric field intensity (0.1 and 0.5 A).

2. Materials

2.1. Plasmid Preparation

1. Plasmid DNA, pAV5012. Plasmid DNA is formulated into a volume of between 0.5 and 1 mL per injection site. Virtually, any plasmid can be used (see Note 2). However, for optimal expression after intramuscular (IM) injection, we recommend using the muscle-specific promoters (17).
2. Sterile water.

2.2. Intramuscular Injection and Electroporation Procedure

1. 3-week-old pigs, weighing between 8 and 11 kg.
2. Approved animal experimental protocols (in accordance with the NIH, USDA, and Animal Welfare Act).
3. CELLECTRA® constant current electroporator (Inovio Pharmaceuticals, Blue Bell, PA, USA) (Fig. 1).
4. Intramuscular applicator (Inovio Pharmaceuticals) (Fig. 1).
5. Blood collection equipment including syringes/needles, tubes for storing sera, centrifuge for isolating sera, and 1 cc syringes with 21G 2″ needles (Becton-Dickinson, Franklin Lakes, NJ, USA).
6. Phospha-light chemiluminescent reporter assay kit (Applied Biosystems, Foster City, CA, USA). A detection kit for SEAP in sera.

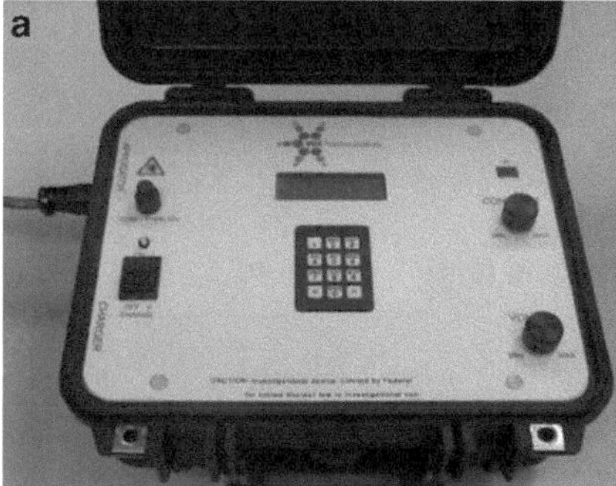

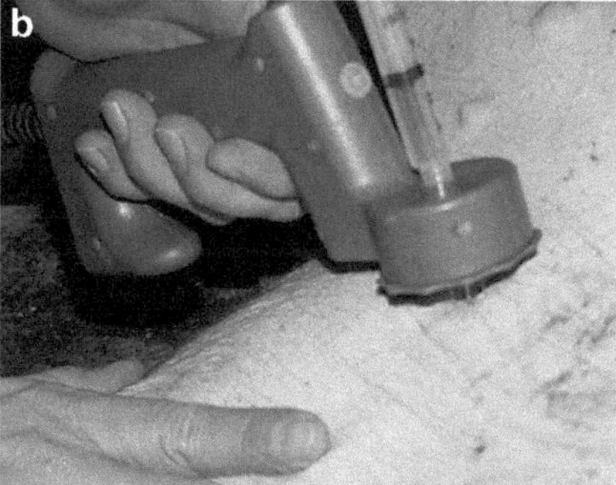

Fig. 1. Injection and electroporation into swine muscle. (**a**) CELLECTRA® constant current electroporator. (**b**) Intramuscular applicator.

3. Methods

3.1. Plasmid Preparation

1. Dilute plasmid preparations (0.5 mg SEAP plasmid) in 1 mL sterile water.
2. Prepare syringes with the volume of plasmid to be injected.

3.2. Intramuscular Injection and Electroporation Procedure

1. Group house animals with ad lib access to food and water throughout the study. Randomly assign pigs to different groups ($n = 5$/group) (Table 1). Acclimate for 4 days.
2. Anesthetize animals with ketamine (22 mg/kg) and acepromazine (1.1 mg/kg). Maintain light anesthesia with isoflurane (or equivalent) when necessary (see Note 3).
3. Carefully choose the injection and electroporation site (semimembranosus muscle) to avoid all surface vessels.
4. Clean the injection site with ethanol and ensure the skin is completely dry before proceeding.
5. Initiation of electroporation and injection of DNA (Fig. 1). Insert the electrode needle array firmly into the preferred injection site of the muscle ensuring the applicator is pressed firmly against the skin. Insert the injection needle in the center of the electrode array. Check that the injection needle is not in a blood vessel by gently pulling back the piston of the syringe. If blood flows back into the syringe the device should be removed, repositioned, and the check repeated. Inject the plasmid by slowly emptying the contents of the syringe into the muscle. After injection remove the syringe and dispose of in a Sharps container (see Note 4). Press the initiation button on the applicator to start the countdown to electroporation. Hold the applicator firmly as electroporation

Table 1
Treatment groups and electroporation conditions

Groups	Plasmid (mg)	Ampere	Lag time (s)
1	SEAP, 0.5	0.1	4
2	SEAP, 0.5	0.1	80
3	SEAP, 0.5	0.5	4
4	SEAP, 0.5	0.5	80

pulses will cause involuntary muscle contractions. The electroporator counts down from 4 s and then following audible beeps the device delivers the electroporation pulses. Check the display immediately following the procedure for any errors (see Note 5). After a successful procedure, wait a few seconds before gently removing the array from the muscle. Dispose of the array in a Sharps/Biohazard container. Closely monitor animals for 24 h to ensure a full recovery from anesthesia.

6. Collect 1 mL of blood from the jugular vein on study day 0 (immediately prior to treatment) and on days 2, 5, 7, and 9. Weigh animals prior to each bleed. Centrifuge the blood samples at 3,000 × *g* for 10 min. Aliquote serum to at least two prelabeled tubes per sample. Immediately place tubes on ice. Freeze at –80°C.
7. Measure SEAP activity with the Phospha-light chemiluminescent reporter assay kit (see Note 6) (Figs. 2 and 3).
8. Perform statistical analysis using the Student's *t*-test program in the Microsoft Excel statistics package (see Notes 7 and 8).

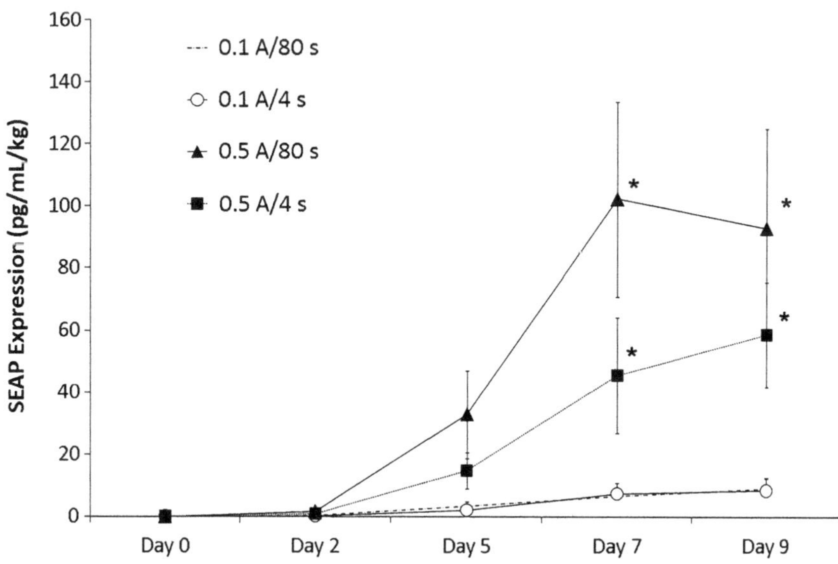

Fig. 2. Expression of SEAP (pg/mL/kg) over 9 days. SEAP expression at day 0, 2, 5, 7, and 9 indicated a significant difference (<0.05) between the electric field intensity (0.1 A compared to 0.5 A), but there was no difference noted for the 80 vs. 4 s lag time.

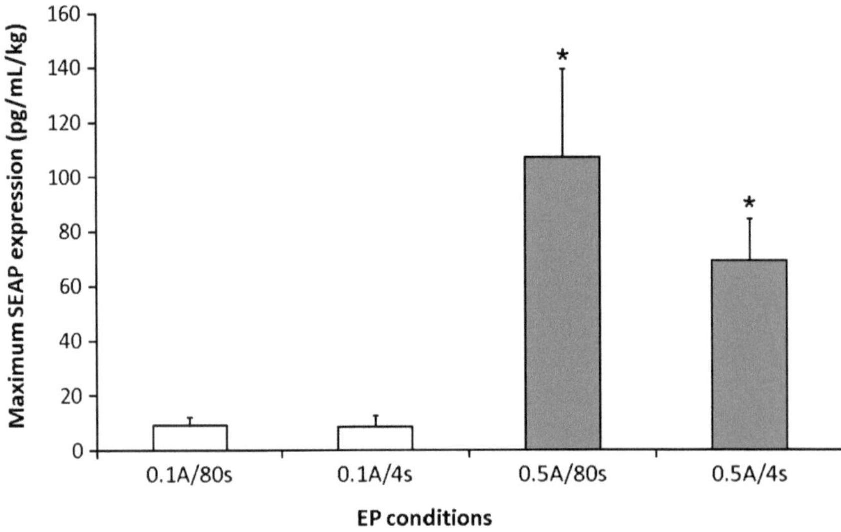

Fig. 3. Maximum SEAP expression over days 5–9. The higher electric field intensity of 0.5 A resulted in significantly higher maximum SEAP expression ($P < 0.05$).

4. Notes

1. The voltage necessary to ensure a certain current intensity varies with animal size and age (which will influence muscle and fiber size, fat and collagen content of muscle tissues, etc.). Therefore, it is important to always perform a pilot study to determine the optimal EP conditions.
2. DNA vaccines or other therapeutic plasmids can be used in place of the SEAP plasmid.
3. Large animals, such as pigs, may require an anesthetic during the procedure to avoid injury to the operator, animals or both. A licensed veterinarian should be consulted regarding the anesthesia protocol.
4. Inject plasmid slowly to prevent the solution seeping out.
5. The CELLECTRA® device continuously measures and records the tissue resistance, varying the voltage to maintain a constant current in the tissue throughout the procedure. If the correct amount of current was delivered, the original operating parameters will be displayed. If less than the desired current is delivered, an “Electroporation Error” message will be displayed. If less than the desired output was achieved, the EP procedure can be repeated once without removing the array.
6. Samples may require dilution to keep them within the detection limits of the assay. If a dilution is done, a correction factor must be added during the analysis to correctly compare results.

7. Data must be normalized for body weight.
8. Due to interanimal variability, SEAP expression will not peak on exactly the same days. Therefore, the maximal expression for a given animal between days 5 and 9 was used to calculate the SEAP expression and compare among treatment groups to identify the most optimal EP conditions.

Acknowledgments

The authors would like to thank Dr. Patricia Brown for her help as an attending veterinarian to oversee the animal anesthesia and perform the electroporation animal procedures. This work was supported by Inovio Pharmaceuticals (Blue Bell, PA).

References

1. van Gaal, E.V., Hennink, W.E., Crommelin, D.J., Mastrobattista, E. (2006) Plasmid engineering for controlled and sustained gene expression for nonviral gene therapy. *Pharm Res* 23,1053–1074.
2. Draghia-Akli, R., Khan, A.S., Cummings, K.K., Parghi, D., Carpenter, R.H., Brown, P.A. (2002) Electrical enhancement of formulated plasmid delivery in animals. *Tech Cancer Res Treat* 1, 365–371.
3. Miyazaki, S., Miyazaki, J. (2008) In vivo DNA electrotransfer into muscle. *Dev Growth Differ* 50, 479–483.
4. Khan, A.S., Smith, L.C., Abruzzese, R.V., Cummings, K.K., Pope, M.A., Brown, P.A., Draghia-Akli, R. (2003) Optimization of electroporation parameters for the intramuscular delivery of plasmids in pigs. *DNA Cell Biol* 22, 807–814.
5. Khan, A.S., Pope, M.A., Draghia-Akli, R. (2005) Highly efficient constant-current electroporation increases in vivo plasmid expression. *DNA & Cell Biology* 24, 810–818.
6. Draghia-Akli, R., Khan, A.S., Brown, P.A., Pope, M.A., Wu, L., Hirao, L., Weiner, D.B. (2008) Parameters for DNA vaccination using adaptive constant-current electroporation in mouse and pig models. *Vaccine* 26, 5230–5237.
7. Schertzer, J.D., Plant, D.R., Lynch, G.S. (2006) Optimizing plasmid-based gene transfer for investigating skeletal muscle structure and function. *Mol Ther* 13, 795–803.
8. Tone, C.M., Cardoza, D.M., Carpenter, R.H., Draghia-Akli, R. (2004) Long-term effects of plasmid-mediated growth hormone releasing hormone in dogs. *Cancer Gene Ther* 11, 389–396.
9. Draghia-Akli, R., Cummings, K.K., Khan, A.S., Brown, P.A., Carpenter, R.H. (2003) Effects of plasmid-mediated growth hormone releasing hormone supplementation in young healthy Beagle dogs. *J Anim Sci* 81, 2301–2310.
10. Bodles-Brakhop, A.M., Brown, P.A., Pope, M.A., Draghia-Akli, R. (2008) Double-blinded, placebo-controlled plasmid GHRH trial for cancer-associated anemia in dogs. *Mol Ther* 16, 862–870.
11. Brown, P.A., Bodles-Brakhop, A.M., Pope, M.A., Draghia-Akli, R. (2009) Gene therapy by electroporation for the treatment of chronic renal failure in companion animals. *BMC Biotechnol* 9, 4.
12. Brown, P.A., Bodles-Brakhop, A., Draghia-Akli, R. (2008) Plasmid growth hormone releasing hormone therapy in healthy and laminitis-afflicted horses-evaluation and pilot study. *J Gene Med* 10, 564–574.
13. Draghia-Akli, R., Ellis, K.M., Hill, L.A., Malone, P.B., Fiorotto, M.L. (2003) High-efficiency growth hormone releasing hormone plasmid vector administration into skeletal muscle mediated by electroporation in pigs. *FASEB J* 17, 526–528.
14. Person, R., Bodles-Brakhop, A.M., Pope, M.A., Brown, P.A., Khan, A.S., Draghia-Akli, R. (2008) Growth hormone-releasing hormone plasmid treatment by electroporation decreases offspring mortality over three pregnancies. *Mol Ther* 16, 1891–1897.
15. Brown, P.A., Davis, W.C., Draghia-Akli, R. (2004) Immune enhancing effects of growth hormone releasing hormone delivered by plasmid injection and electroporation. *Mol Ther* 10, 644–651.

16. Brown, P.A., Bodles-Brakhop, A.M., Draghia -Akli, R. (2008) Effects of plasmid growth hormone releasing hormone treatment during heat stress. *DNA & Cell Biology* 27, 629–635.

17. Draghia-Akli, R., Fiorotto, M.L., Hill, L.A., Malone, P.B., Deaver, D.R., Schwartz, R.J. (1999) Myogenic expression of an injectable protease-resistant growth hormone-releasing hormone augments long-term growth in pigs. *Nat Biotechnol* 17, 1179–1183.

18. Khan, A.S., Fiorotto, M.L., Cummings, K.K., Pope, M.A., Brown, P.A., Draghia-Akli, R. (2003) Maternal GHRH plasmid administration changes pituitary cell lineage and improves progeny growth of pigs. *Am J Physiol Endocrinol Metab* 285, E224–E231.

Chapter 17

Local Gene Delivery and Methods to Control Immune Responses in Muscles of Normal and Dystrophic Dogs

Zejing Wang, Stephen J. Tapscott, and Rainer Storb

Abstract

Adeno-associated viral vector (AAV)-mediated gene transfer represents a promising gene replacement strategy for treating Duchenne muscular dystrophy (DMD). However, recent studies demonstrated cellular immunity specific to AAV capsid proteins in animal models, which resulted in liver toxicity and elimination of transgene expression in a human trial of hemophilia B. We have recently developed immunosuppressive strategies to prevent such immunity for successful long-term transgene expression in dog muscle. Here, we describe in detail the immunosuppressive regimens employed in both normal and DMD dogs and provide methods for evaluating the efficiency of the regimens following intramuscular injection of AAV in dogs.

Key words: Adeno-associated virus, AAV, Duchenne muscular dystrophy, DMD, Cellular immunity, Muscle, Dog, *cxmd*, Immunosuppression, CSP, MMF, ATG

1. Introduction

Duchenne muscular dystrophy (DMD) is a fatal, X-linked, recessive muscle disease caused by lack of dystrophin due to mutations in the dystrophin gene (1–3). The clinical course of DMD is severe and progressive, and the muscle tissue is eventually replaced by connective tissue and adipose cells (4). Despite its well-understood pathogenesis, no curative therapies for DMD exist at present. Viral vector-mediated gene delivery represents a promising gene replacement strategy for treatment because DMD is caused by a single gene mutation (5–7).

One major hurdle in using viral vectors for in vivo gene delivery is the development of host immune responses to both the virus and the transgene product, especially in the setting of preexisting inflammatory lesions in dystrophic muscle. Vectors based on the

Dongsheng Duan (ed.), *Muscle Gene Therapy: Methods and Protocols*, Methods in Molecular Biology, vol. 709, DOI 10.1007/978-1-61737-982-6_17, © Springer Science+Business Media, LLC 2011

adeno-associated virus (AAV) have shown promise in preclinical studies and clinical trials for a number of acquired and inherited diseases (8–12). However, we found that local intramuscular injection of AAV2 or AAV6 vectors into wild-type dogs induced robust T cell-mediated immune responses against AAV capsid proteins that peaked 4 weeks after vector injection and limited long-term transgene expression (13). In line with this finding, cellular immunity specific to AAV capsid proteins has also been observed in mouse and nonhuman primate models, and moreover, resulted in liver toxicity and elimination of transgene expression in a human trial of hemophilia B (14–16). Taken together, these data suggest that immune modulation might be necessary to achieve successful long-term transgene expression in these species.

Antigen-specific T cells are mediators of allo-specific immune responses and, therefore, have been targeted by immunosuppression regimens employed in hematopoietic cell transplantation (HCT) and solid organ transplantations with the aims of preventing graft rejection and inducing immunological tolerance (17, 18). We have used a combination of three immunosuppressive drugs: cyclosporine (CSP), a calcineurin inhibitor; mycophenolate mofetil (MMF), an antimetabolite; and rabbit anti-dog thymocyte globulin (ATG), a mixture of polyclonal antibodies, as an initial strategy for averting immune responses to AAV capsid and the transgene product. The combination of CSP and MMF is synergistic both for enhancing engraftment and controlling graft-versus-host disease (GVHD) in a dog model of HCT and is effective in

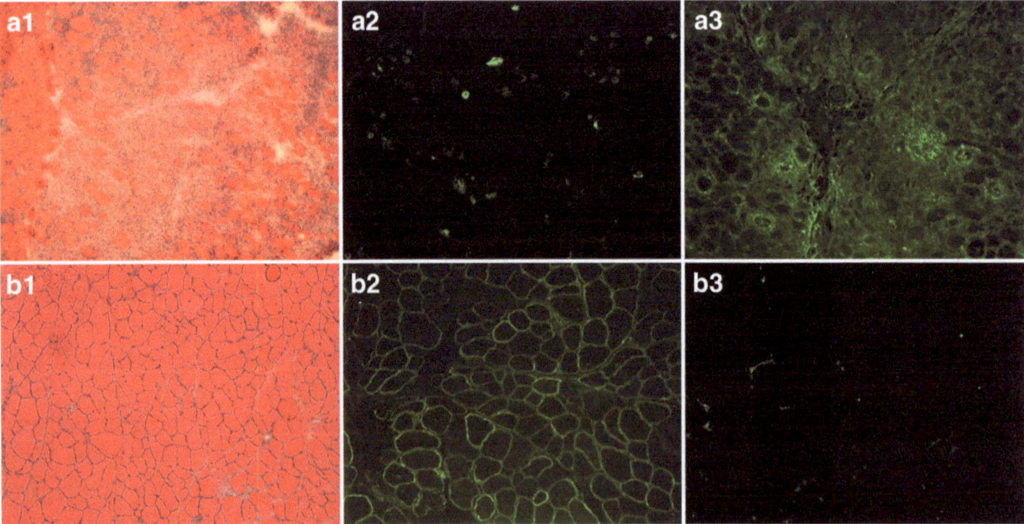

Fig. 1. Effectiveness of immunosuppression and intramuscular injection of AAV vectors in dogs. A1–A3, injection of AAV6-CMV-cFIX in muscles of normal dog without immunosuppression. *A1*, H&E staining; *A2*, cFIX expression in *green*; *A3*, CD8 staining in *green*. B1–B3, injection of AAV6-CMV-c-μ-dys in muscles of *cxmd* dog with CSP/MMF/ATG. *B1*, H&E; *B2*, expression of c-μ-dys in *green*; *B3*, CD8 staining in *green*.

preventing rejection of solid organ transplants (18–20). This drug combination has been translated successfully to the human HCT setting (21, 22). ATG has been used in dogs and humans as part of conditioning regimens for HCT (23, 24), to treat GVHD (25, 26), and to treat patients with aplastic anemia (27). We demonstrated that a brief course of immunosuppression with CSP/MMF/ATG was sufficient to avert the cellular immune responses against AAV6 vectors and permitted robust long-term expression of a canine microdystrophin (c-μ-dys) transgene in the skeletal muscle of dogs with DMD, *cxmd* (28) (Fig. 1).

Here, we outline the immunosuppressive regimen for preventing cellular immune responses to AAV vectors following intramuscular injection in both normal and dystrophic dogs and describe methods for monitoring cellular infiltration and transgene expression.

2. Materials

2.1. Immunosuppression Regimen

1. Soft gelatin CSP capsule (IVAX Pharmaceuticals, North Wales, PA, USA).
2. CellCept Intravenous MMF (Roche, Nutley, NJ, USA).
3. Rabbit anti-canine ATG (produced in collaboration with SangStat, Fremont, CA, stored at –80°C, concentration varies).

2.2. In Vivo Evaluation of Effectiveness of Immunosuppression

1. AAV vectors carrying CMV-canine factor IX (cFIX), CMV-human-microdystrophin (hu-μ-dys), or CMV- canine-μ-dys are provided by Dr. Jeffrey Chamberlain (University of Washington, Seattle, WA). The vectors are purified either by affinity purification through a HiTrap heparin column (Amersham, Piscataway, NJ, USA) or by double CsCl gradient centrifugation followed by dialysis into Hank's buffered salt solution (HBSS, Invitrogen, Carlsbad, CA, USA).
2. HBSS (Invitrogen).
3. 31 G syringes (Becton-Dickinson, Franklin Lakes, NJ, USA).
4. Anesthesia reagents: acepromazine (Fort Dodge Animal Health, Fort Dodge, IA, USA); atropine (Baxter Health Corporation, Deerfield, IL, USA); butorphanol (Fort Dodge Animal Health); glycopyrrolate (American Regent, Shirly, NY, USA); propofol (Abbott Laboratories, North Chicago, IL, USA); and lidocaine hydrochloride, injectable-2% at the site (VEDCO, St. Joseph, MO, USA).
5. Surgical instruments and sutures: sterile forceps and scissors (World Precision Instruments, Sarasota, FL, USA), 4-0 Maxon (US Surgical, Tyco Healthcare LP, Norwalk, CT, USA), 4-0

Prolene nonabsorbable suture, blue microfilament (Ethicon, Somerville, NJ, USA), scalpel #10 blade (Medline Industries, Mundelein, IL, USA), and skin stapler (US Surgical, Tyco Healthcare LP, Norwalk, CT, USA).

6. Tissue-Tek O.C.T. medium (Tissue-Tek, Hatfield, PA, USA).
7. Tongue depressor (Fisher Scientific, Waltham, MA, USA).
8. 2-methylbutane (Sigma, St. Louis, MO, USA).

2.3. Evaluation of Cellular Immunity and Gene Expression

2.3.1. Quantifying CD3+ Cells by Flow Cytometry

1. Heparin sulfate (APP Pharmaceuticals, LLC Schaumburg, IL, USA).
2. Hemolytic buffer (store at 25°C): 0.15 M ammonium chloride, 0.01 M sodium bicarbonate, 0.13 M EDTA in distilled H_2O (dH_2O).
3. Fetal bovine serum (Gibco, Carlsbad, CA, USA).
4. (HBSS, Gibco).
5. Mouse anti-canine CD3 (CA17.6F9, IgG2b) (provided by Dr. Peter Moore, School of Veterinary Medicine, University of California, Davis; produced, purified, and FITC conjugated by the Biologics Production Facility at FHCRC) and FITC-conjugated mouse IgG2b (Caltag, Bangkok, Thailand).
6. Sorvall Lengend Heraeus Centrifuge, 750064446 B (Fisher Scientific).
7. Fluorescence-activated flow cytometer (Becton-Dickinson).

2.3.2. Examine Muscle Histology by Hematoxylin and Eosin–Phloxine Staining (H&E)

1. Gills #3 hematoxylin (Fisher Scientific).
2. Eosin–phloxine: 5% of 1% eosin Y, 1% of 1% phloxine B, 39% of 95% alcohol, and 0.2% glacial acetic acid.
3. $MgSO_4$–$NaHCO_2$ tap water: 0.08 M magnesium sulfate and 0.02 M sodium bicarbonate.
4. Methanol (Fisher Scientific).
5. Xylene (Fisher Scientific).
6. Alcohol (Decon Lab, King of Prussia, PA, USA).
7. VectaMount (Vector laboratories, Burlingame, CA, USA).
8. Slides and slide cover (Fisher Scientific).

2.3.3. Detection of T cells and Expression of Canine Factor IV and Dystrophin by Immunofluorescence

1. Bovine serum albumin (Sigma). Store at 4°C.
2. 1× Dulbecco's phosphate-buffered saline (PBS, Gibco).
3. 4', 6-Diamidino-2-phenylindole (DAPI, Sigma).
4. Primary antibodies: Rabbit anti-mouse dystrophin (N-terminus) polyclonal antibody (Jeff Chamberlain, University of Washington); mouse anti-human dystrophin (C-terminus) monoclonal antibody (Vector laboratories); mouse anti-canine CD4 and CD8 monoclonal antibodies

(IE4 and JD3, respectively, provided by Dr. Peter Moore, UC Davis, and produced by Fred Hutchinson Cancer Research Center Biologics Production Facility); and rabbit anti-canine factor IX (Affinity Biologicals, Ontario, Canada).

5. Secondary antibodies: goat anti-mouse FITC-conjugated or donkey anti-rabbit FITC-conjugated secondary antibody (Jackson laboratory, West Grove, PA, USA).
6. Normal mouse and rabbit isotope antibodies (Invitrogen).
7. Vectashield (Vector Laboratories).
8. Slides and slide cover (Fisher Scientific).

3. Methods

3.1. Immunosuppressive Regimens

Immunosuppression in normal dogs consists of oral CSP and subcutaneous MMF (see Note 1). In dystrophic *cxmd* dogs, the preexisting inflammatory process is a major feature of the pathology of dystrophin-deficient muscle and, therefore, requires a more potent immunosuppression. Adding a canine-specific ATG for transient T-cell depletion to the combination of CSP and MMF is very efficient in *cxmd* dogs for preventing cellular immunity against AAV vector following injection into dog muscle.

1. For intramuscular injection of AAV in normal dog: Start oral administration of CSP 3 days before (day 3) AAV injection (day 0) (Fig. 2), at 7.5 mg/kg, twice daily; and continue for

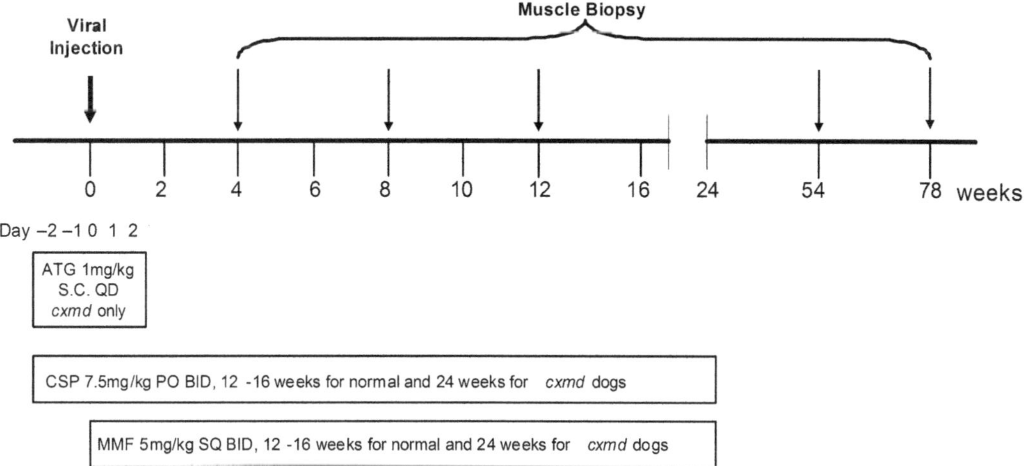

Fig. 2. Immunosuppression regimens in normal and dystrophic *cxmd* dogs receiving intramuscular AAV injection. S.C., subcutaneous injection; *QD,* once daily; *PO,* oral administration; *BID,* twice daily; *cxmd*, canine X-linked muscular dystrophy.

12–16 weeks, twice daily at 7.5 mg/kg (see Note 2). Start subcutaneous (S.C.) MMF on the day of AAV injection at 5 mg/kg, twice daily. Continue for 12–16 weeks, twice daily at 5 mg/kg, S.C. Check blood CSP level on day 3, 7, and weekly thereafter. Complete blood counts on day 2 and day 0 and weekly thereafter, including white blood cell count, red blood cell count, platelet count, white blood cell differential, hematocrit, and hemoglobin.

2. For intramuscular injection of AAV in dystrophic *cxmd* dog: Start ATG 2 days before AAV injection; on day 2, 1 mg/kg, S.C. once. Continue once daily from day 1 to day 2 at 1 mg/kg, S.C. Start oral administration of CSP 3 days before (day 3) AAV injection (day 0) (Fig. 2), at 7.5 mg/kg, twice daily; and continue for 24 weeks, twice daily at 7.5 mg/kg (see Note 2). Start MMF on the day of AAV injection at 5 mg/kg, S.C., twice on day 0; and continue for 24 weeks, twice daily at 5 mg/kg, S.C. Check blood CSP level on day 3, 7, and weekly thereafter. Complete blood counts on day 2 and day 0 and weekly thereafter, including white blood cell count, red blood cell count, platelet count, white blood cell differential, hematocrit, and hemoglobin.

3.2. In Vivo Evaluation of Effectiveness of Immunosuppression

All research experiments performed on dogs follow the *Guide for Laboratory Animal Facilities and Care* prepared by the National Academy of Sciences, National Research Council, and after approval by the Institutional Animal Care and Use Committee. All dogs are immunized for leptospirosis, distemper, hepatitis, papillomavirus and parvovirus (see Note 3), dewormed, and observed for disease for at least 2 months before being entered in studies. Three muscles in hind limbs are chosen for intramuscular AAV injection for easy access: biceps femoris, semitendinosus, and semimembranosus.

1. Intramuscular AAV vector injection. Shave injection sites. For normal dogs, apply general anesthesia. For *cxmd* dogs, sedate with acepromazine 0.025 mg/kg, atropine 0.025 mg/kg, butorphanol 0.1 mg/kg IV, and continuous propofol at 0.025 mg/kg/min during procedure (see Note 4). Place dogs in a lateral decubitus position. Infiltrate incision site with 2% lidocaine hydrochloride S.C. Make a 4–6 cm incision in the skin along the longitudinal axis of the hind limb to expose the selected muscles. Place nonabsorbable 4-0 sutures in the belly of the muscle to mark each injection site. Slowly inject 1×10^{11} to 1×10^{12} vector genome of rAAV vectors in 250 μl of HBSS per site into the muscle belly right underneath each suture using 31 G syringes. Close skin and monitor animals daily for recovery.
2. Muscle biopsy. Perform muscle biopsies at 4, 8, 12, 54, and 78 weeks after AAV injection, or at a desired time of your

study. Take muscle biopsy samples from uninjected sites as negative controls. Prepare dogs as described above. Make an incision as described above. Locate nonabsorbable sutures. Dissect muscle sample at a size about 1 cm^3 around the suture. Lay tissue on a tongue presser. Embed tissue in Tissue-Tek O.C.T. medium. Immediately snap freeze in liquid nitrogen-cooled 2-methylbutane. Store samples at –80°C until use. Close muscle and skin. Monitor animal daily for recovery.

3.3. Evaluation of Cellular Immunity and Gene Expression

To examine the effectiveness of peripheral blood T-cell depletion by ATG, blood samples are collected at different time points for quantifying absolute lymphocyte counts using flow cytometry: before initiating immunosuppression (day 3) as baseline, and on days 0, 3, and 7, and weekly until week 8. Muscle biopsy samples are collected at different time points as described above using histology and immunofluorescence analysis.

3.3.1 Quantifying CD3+ Cells by Flow Cytometry

1. Collect 10 ml of blood containing 10% heparin as anticoagulant.
2. Transfer blood into 50 ml Corning tube, fill tube with hemolytic buffer.
3. Mix blood sample with buffer and incubate at room temperature (RT) for 10 min or until sample is clear and becomes dark red.
4. Centrifuge at 365 ×*g* for 5 min.
5. Discard supernatant and wash cells with cold HBS supplemented with 2% heat-inactivated horse serum, 2× 5 min at 365 ×*g*. Resuspend cells in 5% FBS/PBS.
6. Add fluorescein isothiocyanate (FITC)-conjugated canine CD3 antibody at concentration of 4 μg/ml to 100 μl of cellular suspension containing approximately one million cells per milliliter, and incubate for 30 min at 4°C in the dark; FITC-conjugated mouse IgG2b as control.
7. Wash the cells with HBS for 3× 5 min at 365 ×*g*, and resuspend in HBS/2% heat-inactivated horse serum supplemented with propidium iodide for excluding dead cells.
8. Analyze on a fluorescence-activated flow cytometer.

3.3.2. Examine Muscle Histology by Hematoxylin and Eosin–Phloxine Staining (H&E)

1. Cut 10 μm cryostat sections.
2. Completely dry sections at RT for 15 min.
3. Fix sections in methanol for 5 min at RT.
4. Wash in distilled water (dH_2O) for 30 s.
5. Incubate in Gills #3 hematoxylin for 3 min.
6. Dip in dH_2O for 4 times.
7. Dip in tap H_2O for 10 times.
8. Rinse $MgSO_4$–$NaHCO_2$ tap H_2O for 1 min.
9. Rinse in running H_2O for 2 min.

10. Incubate in eosin–phloxine for 30 s.
11. Dehydrate in 95% alcohol, 2× 2 min.
12. Dehydrate in 100% alcohol, 2× 2 min.
13. Dehydrate in xylene, 2× 3 min.
14. Mount in VectaMount (Vector laboratories), and dry slides overnight in hood.

3.3.3. Detection of T cells and Expression of Canine Factor IV and Microdystrophin by Immunofluorescence

1. Dry sections for 15 min at RT.
2. Block sections in rabbit anti-mouse dystrophin (N-terminus) polyclonal antibody with 4% BSA, 0.2% tween 20 for 15 min.
3. Incubate for 60 min at RT with the following primary antibodies diluted in 1× PBS with 4% BSA: mouse anti-canine CD4 and CD8 monoclonal antibody (1:100 dilution of 0.97 mg/ml and 1.2 mg/ml stock, respectively), rabbit anti-canine (1:100 dilution), rabbit anti-mouse dystrophin (N-terminus) polyclonal antibody (1:600 dilution), and mouse anti-human dystrophin (C-terminus) monoclonal antibody (1:100 dilution) (see Note 5). Normal mouse and rabbit isotope antibodies are used as negative controls.
4. Rinse sections in 1× PBS, 3× 10 min each.
5. Incubate for 30 min at RT with goat anti-mouse FITC-conjugated (for CD4 and CD8) or donkey anti-rabbit FITC-conjugated secondary antibody.
6. Sections are counterstained with DAPI in 1× PBS (1:1,000 dilution from 1 mg/ml stock) for 15 min.
7. Rinse sections in 1× PBS, 3× 15 min each, and mount in Vectashield.
8. Examine staining using a Nikon Eclipse 800 fluorescence microscope.

4. Notes

1. The administration of CSP in current immunosuppressive regimen in *cxmd* dogs is required for 24 weeks. Dogs react strongly to intravenous CSP, with significant gastrointestinal symptoms including vomiting and diarrhea. It is not practical to apply CSP twice daily via intravenous route for long-term purpose. *Cxmd* dogs do not swallow well in general; however, by closely monitoring blood CSP levels and adjusting CSP dosage accordingly, we are able to ensure blood CSP level to be within the range of its therapeutical level, 100–350 ng/ml.

2. Potential toxicities associated with the use of ATG, CSP, and MMF are broad. ATG toxicities include fevers, skin rashes, and serum sickness for which patients will receive corticosteroids. Toxicities associated with long-term use of MMF and CSP include the risk of renal toxicity, hypertension, tremors, magnesium wasting, and central nervous system toxicities, among others. The dosages used here for drug are relatively low; however, blood CSP level should be closely monitored on days 3, 7, and weekly thereafter. The therapeutic range for CSP levels in dogs is from 100 to 350 ng/ml. The dosages for CSP and MMF should be adjusted when CSP level is outside of the range as the two drugs have synergistic effect. Affected *cxmd* dogs are more susceptible to intussusceptions; therefore, the incidence of intussusceptions should be closely monitored, and should it happen, surgery should be performed to remove it. Standard supportive care including oral antibiotics with fluid support, platelet, and blood transfusions should be provided when needed.
3. Canine Parvovirus capsid protein has between 10 and 60% of identity to that of AAV6; however, we have found that vaccination with canine parvovirus in these dogs did not induce a humoral immune response to AAV capsid proteins (Wang, Tapscott, and Storb, unpublished data). Furthermore, there was limited/no homology between canine parvovirus in the epitopes of AAV6 capsid protein that were identified to be immunogenic in our ongoing study (Wang, Tapscott, and Storb, unpublished data).
4. DMD is caused by loss of dystrophin, a critical component of the dystrophin–glycoprotein complex (DGC) at the muscle membrane (2, 3). The disease affects muscles throughout the body, including cardiac muscle. Cardiomyopathy is a leading cause of death in DMD patients and dogs (1). While normal dogs are under general anesthesia in this protocol, *cxmd* dogs are only under sedation to avoid causing potential stress on cardiac muscle.
5. Canine microdystrophin is constructed based on sequence homology with human and mouse μ-dys, which contains essentially the same domains that are present in the human ΔR4-R23/ΔCT μ-dys, without a large portion of the central rod domain and the full C-terminus (28, 29). Commercial available antibody to N-terminus of dystrophin (Vector laboratories) does not cross-react in canine muscle. Antibody to C-terminus from Vector does cross-react and can be used to detect expression of endogenous dystrophin and, therefore, to distinguish expression of endogenous dystrophin from that of AAV-delivered microdystrophin.

Acknowledgements

We thank Dr. Jeffery Chamberlain for providing vectors used in the protocols and Dr. Kathy High for providing the c-FIX plasmid. We thank E. Zellmer, Diana Jensen, Jenee O'Brian, and E. Finn for technical assistance, and A. Joslyn, B. Steinmetz, and their teams, and M. Spector, DVM, and J Duncan, DVM, for their care of the dogs. We further thank S. Carbonneau, H. Crawford, B. Larson, K. Carbonneau, Vermeulen, and D. Gayle for administrative assistance and manuscript preparation. This work was supported by NIH U54-HD47175 and the Seattle Senator Paul D. Wellstone Muscular Dystrophy Cooperative Research Center, NIH CA15704, NIH AR056949, and by Career Development Award for Z. Wang from the Muscular Dystrophy Association (MDA 114979).

References

1. Finsterer J, Stollberger C (2003) The heart in human dystrophinopathies *Cardiol* 99, 1–19.
2. Muntoni F, Torelli S, Ferlini A (2003) Dystrophin and mutations: one gene, several proteins, multiple phenotypes (Review). *Lancet Neurol* 2, 731–740.
3. Tyler KL (2003) Origins and early descriptions of "Duchenne muscular dystrophy". *Muscle Nerve* 28, 402–422.
4. Foidart M, Foidart JM, Engel WK (1981) Collagen localization in normal and fibrotic human skeletal muscle. *Arch Neurol* 38, 152–157.
5. Duan D (2006) Challenges and opportunities in dystrophin-deficient cardiomyopathy gene therapy. *Hum Mol Genet* 15 (Spec. No. 2), R253–R261.
6. Gregorevic P, Allen JM, Minami E, Blankinship MJ, Haraguchi M, Meuse L, *et al.* (2006) rAAV6-microdystrophin preserves muscle function and extends lifespan in severely dystrophic mice. *Nat Med* 12, 787–789.
7. Wang Z, Chamberlain JS, Tapscott SJ, Storb R (2009) Gene therapy in large animal models of muscular dystrophy. *ILAR J* 50, 187–198.
8. Athanasopoulos T, Fabb S, Dickson G (2000) Gene therapy vectors based on adeno-associated virus: characteristics and applications to acquired and inherited diseases. *Int J Mol Med* 6, 363-375.
9. Sun B, Zhang H, Franco LM, Young SP, Schneider A, Bird A, *et al.* (2005) Efficacy of an adeno-associated virus 8-pseudotyped vector in glycogen storage disease type II. *Mol Ther* 11, 57–65.
10. Yue Y, Ghosh A, Long C, Bostick B, Smith BF, Kornegay JN, *et al.* (2008) A single intravenous injection of adeno-associated virus serotype-9 leads to whole body skeletal muscle transduction in dogs. *Mol Ther* 16, 1944–1952.
11. Athanasopoulos T, Graham IR, Foster H, Dickson G (2004) Recombinant adeno-associated viral (rAAV) vectors as therapeutic tools for Duchenne muscular dystrophy (DMD). *Gene Ther* 11 Suppl 1, S109–S121.
12. Warrington KH, Jr., Herzog RW (2006) Treatment of human disease by adeno-associated viral gene transfer. *Hum Genet* 119, 571–603.
13. Wang Z, Allen JM, Riddell SR, Gregorevic P, Storb R, Tapscott SJ, *et al.* (2007) Immunity to adeno-associated virus-mediated gene transfer in a random-bred canine model of Duchenne muscular dystrophy. *Hum Gene Ther* 18, 18–26.
14. Sabatino DE, Mingozzi F, Hui DJ, Chen H, Colosi P, Ertl HC, *et al.* (2005) Identification of mouse AAV capsid-specific CD8+ T cell epitopes. *Mol Ther* 12, 1023–1033.
15. Gao G, Lu Y, Calcedo R, Grant RL, Bell P, Wang L, *et al.* (2006) Biology of AAV serotype vectors in liver-directed gene transfer to nonhuman primates. *Mol Ther* 13, 77–87.
16. Manno CS, Pierce GF, Arruda VR, Glader B, Ragni M, Rasko JJ, *et al.* (2006) Successful transduction of liver in hemophilia by AAV-Factor IX and limitations imposed by the host

immune response (erratum appears in Nat Med 12, 592). *Nat Med* 12, 342-347.

17. Maris M, Storb R (2003) The transplantation of hematopoietic stem cells after non-myeloablative conditioning: A cellular therapeutic approach to hematologic and genetic diseases. *Immunol Res* 28, 13-24.
18. Kirk AD (2003) Crossing the bridge: large animal models in translational transplantation research. *Immunol Rev* 196, 176–196.
19. Storb R, Yu C, Wagner JL, Deeg HJ, Nash RA, Kiem H-P, *et al.* (1997) Stable mixed hematopoietic chimerism in DLA-identical littermate dogs given sublethal total body irradiation before and pharmacological immunosuppression after marrow transplantation. *Blood* 89, 3048–3054.
20. Yu C, Seidel K, Nash RA, Deeg HJ, Sandmaier BM, Barsoukov A, *et al.* (1998) Synergism between mycophenolate mofetil and cyclosporine in preventing graft-versus-host disease among lethally irradiated dogs given DLA-nonidentical unrelated marrow grafts. *Blood* 91, 2581–2587.
21. McSweeney PA, Niederwieser D, Shizuru JA, Sandmaier BM, Molina AJ, Maloney DG, *et al.* (2001) Hematopoietic cell transplantation in older patients with hematologic malignancies: replacing high-dose cytotoxic therapy with graft-versus-tumor effects. *Blood* 97, 3390–3400.
22. Maris MB, Niederwieser D, Sandmaier BM, Storer B, Stuart M, Maloney D, *et al.* (2003) HLA-matched unrelated donor hematopoietic cell transplantation after nonmyeloablative conditioning for patients with hematologic malignancies. *Blood* 102, 2021–2030.
23. Storb R, Floersheim GL, Weiden PL, Graham TC, Kolb H-J, Lerner KG, *et al.* (1974) Effect of prior blood transfusions on marrow grafts: Abrogation of sensitization by procarbazine and antithymocyte serum. *J Immunol* 112, 1508–1516.
24. Buckley JD, Chard RL, Baehner RL, Nesbit ME, Lampkin BC, Woods WG, *et al.* (1989) Improvement in outcome for children with acute nonlymphocytic leukemia. A report from the Childrens Cancer Study Group. *Cancer* 63, 1457–1465.
25. Storb R, Kolb HJ, Graham TC, Kolb H, Weiden PL, Thomas ED (1973) Treatment of established graft-versus-host disease in dogs by antithymocyte serum or prednisone. *Blood* 42, 601–609.
26. Storb R, Gluckman E, Thomas ED, Buckner CD, Clift RA, Fefer A, *et al.* (1974) Treatment of established human graft-versus-host disease by antithymocyte globulin. *Blood* 44, 57–75.
27. Doney K, Leisenring W, Storb R, Appelbaum FR, for the Seattle Bone Marrow Transplant Team (1997) Primary treatment of acquired aplastic anemia: outcomes with bone marrow transplantation and immunosuppressive therapy. *Ann Intern Med* 126, 107–115.
28. Wang Z, Kuhr CS, Allen JM, Blankinship M, Gregorevic P, Chamberlain JS, *et al.* (2007) Sustained AAV-mediated dystrophin expression in a canine model of Duchenne muscular dystrophy with a brief course of immunosuppression. *Mol Ther* 15, 1160–1166.
29. Phelps SF, Hauser MA, Cole NM, Rafael JA, Hinkle RT, Faulkner JA, *et al.* (1995) Expression of full-length and truncated dystrophin mini-genes in transgenic mdx mice. *Hum Mol Genet* 4, 1251–1258.

Chapter 18

Gene Transfer to Muscle from the Isolated Regional Circulation

Mihail Petrov, Alock Malik, Andrew Mead, Charles R. Bridges, and Hansell H. Stedman

Abstract

Vector transport across the endothelium has long been regarded as one of the central "bottlenecks" in gene therapy research, especially as it pertains to the muscular dystrophies where the target tissue approaches half of the total body mass. Clinical studies of gene therapy for hemophilia B revealed the limitations of the intramuscular route, compelling an aggressive approach to the study of scale-independent circulatory means of vector delivery. The apparent permeability of the microvasculature in small animals suggests that gravitational and/or inertial effects on the circulation require progressive restriction of fluid and solute flow across the capillary wall with increasing body size. To overcome this physiological restriction, we initially used a combined surgical and pharmacological approach to temporarily alter permeability within the isolated pelvic limb. Although this was successful, new information about the cell and molecular biology of histamine-induced changes in microvascular permeability suggested an alternative approach, which substituted pressure-induced transvenular extravasation. Here we outline the details of our surgical approaches in the rat. We also discuss the modifications that are appropriate for the dog.

Key words: AAV, Adenovirus, Muscular dystrophy, Hemophilia, Gene therapy, Regional or isolated limb perfusion, Histamine, Large animal (canine) models

1. Introduction

The recessively inherited Duchenne and limb-girdle muscular dystrophies are considered, with some important exceptions (1), to be targets for gene replacement therapy that do not require finely tuned control of the level of gene expression. There is ample evidence from model systems to support the idea that much higher than wild type levels of transgene expression in postmitotic myocytes will not result in cellular toxicity (2, 3). In small animal models, regional and more recently, systemic transfer of AAV -vectors

Dongsheng Duan (ed.), *Muscle Gene Therapy: Methods and Protocols*, Methods in Molecular Biology, vol. 709, DOI 10.1007/978-1-61737-982-6_18, © Springer Science+Business Media, LLC 2011

encoding delta-sarcoglycan and microdystrophins has resulted in dramatic amelioration of the cellular pathology and functional consequences (3–5). Direct extension of these promising results to clinical trials is risky when issues of scaling and immune response are taken into consideration. Fortunately, the existence of a homologous genetic disease in the dog affords a unique opportunity to minimize these risks (6). This lesson was learned painfully some 40 years ago when the unsuccessful results of early human studies, based almost exclusively on studies in small animal models, mandated a return to the lab to work out critical details of bone marrow transplantation, with a particular emphasis on studies in the canine model (7). Despite the long hiatus, success in the large animal studies enabled a dramatic improvement of results in the clinic, as now widely recognized in historical accounts of the scientific foundation for the modern-day approach to a wide range of hematological diseases (8).

In our earliest studies of adenovirus-mediated gene transfer to skeletal muscle, we naively thought that simple intravenous administration of the vectors would result in widespread transduction of muscle. When this failed, our studies of biodistribution suggested that most of the vector was taken up by the liver across the fenestrated endothelium. Transient vascular exclusion of the liver did not measurably increase the transduction of muscle nor did direct injection of the femoral artery, although a detailed analysis of this experiment provided an important clue. The histological appearance of the microvasculature revealed highly efficient transduction of the capillary endothelium and the virtual absence of gene transfer to the myocytes (4). The identical vectors produced the reciprocal transduction pattern when injected intramuscularly, i.e., efficient (but highly localized) myocyte transduction and the virtual absence of endothelial staining for marker transgene expression. This suggested that either the endothelial cells themselves, the adjacent basement membrane, or both were preventing vector egress from the microvascular lumen to the muscle interstitium. In the adult rat and hamster hindlimb, this pattern was maintained even in the setting of applied high-pressure arterial side infusion.

Histamine has long been known to produce transient alterations in microvascular permeability. We reasoned that using histamine to prime the circulation of the isolated limb might provide a window of enhanced permeability to adenovirus vector. To minimize the risk to the central organs, we extensively flushed out the histamine with saline prior to the release of hindlimb tourniquet. To further enhance its effect, we added papaverine to dilate the arterioles controlling access to the peripheral capillary beds. This approach provided a dramatic increase in the level of myocyte transduction throughout the hindlimb. It also proved to be equally successful with the delta-sarcoglycan null, cardiomyopathic

BIO14.6 hamster model for limb-girdle muscular dystrophy 2F, when we substituted an AAV vector encoding human delta-sarcoglycan under the control of a constitutive promoter (4). Importantly, this AAV2-based vector was equally restricted in biodistribution in the absence of histamine. On the basis of this success, we became interested in an early opportunity to address the scale dependence of this effect in a canine model for hemophilia B. We first undertook a series of control studies using readily quantified marker transgenes. Then, in collaboration with the laboratories of Katherine High and Timothy Nichols and using a previously developed AAV2 vector encoding canine factor IX, we scaled this approach for use in the point mutational canine model for hemophilia B under transient immunosuppression to minimize the risk of an immune response to the transgene product. This too proved successful, as gauged by the level and longevity of recombinant factor IX detected in these dogs (9). Nonetheless, translation of this approach to gene therapy into the clinical realm is complicated by the unpredictability of the biological response to histamine (10–12).

At about this time, new data emerged about the mechanism and exact site of histamine's action on the microvasculature. Compelling photomicrographs revealed that the primary endothelial localization of the peripheral histamine receptors was on the postcapillary venules, not the capillaries per se (13). Moreover, the endothelial cells at this location generally do not show the extensive tight junctional architecture of the capillaries, suggesting that radial and or axial forces might reversibly displace or fenestrate these cells, thereby providing a low resistance and low reflectance path for fluid and solute transport, respectively, from lumen to interstitium. We addressed this possibility in the context of adenovirus and AAV2 vectors by cannulating the venous side of the circulation, and infusing a saline suspension of vector under pressure against a hindlimb tourniquet. The result was a pattern of muscle transduction that resembled, in the absence of histamine, that previously seen only in the presence of this mediator. In scaling to the larger animal, we were initially concerned that the extensive network of one way valves in the limb would limit the pattern of transduction. However, careful review of the literature and some cadaveric dissections confirmed that in the dog and the human, valves are limited essentially to the named veins and their large branches, providing a pathway for orthograde flow from distal to proximal and from superficial to deep, with locally retrograde flow from larger veins to smaller branches that drain the individual muscles. This approach was thus further modified for application in the mature dog, with the finding that pressurized infusion of adenovirus, AAV1 or AAV2 vector into the distal saphenous vein of the dog would result in virtually universal, uniform transduction of the hindlimb musculature (14).

At no time did the dogs exhibit any clinically apparent functional pathology in the limbs following this procedure. In contrast to our previous experience with the histamine-dependent approach, there were no measurable acute hemodynamic alterations after tourniquet release. The additional advantages from a clinical safety perspective included the absence of the need for proximal exposure or cannulation of the femoral vessels, papaverine, and systemic heparinization. There remains a theoretical risk of venous endothelial injury, but we have yet to see a clinically apparent problem in over 30 dogs studied in this way, and all veins have remained patent at the level of the ankle. As noted below, we eliminate the risk of focal leak near the catheterization site by inserting the catheter under direct visualization, ligating the vessel distally, and snaring it proximally. Fluid is administered under pressure both to drive transvenular extravasation and to overcome reflex venoconstriction which would otherwise restrict access to the microvascular beds.

In the protocols that follow, we provide a detailed account of the sequence of steps we have used to achieve widespread vector-mediated gene transfer into the hindlimbs of the adult rat and dog. Many studies are currently underway in the maturing large animal to extend this technology for use in the regional circulation for various diseases and in the systemic circulation for muscular dystrophy. It is important to note that even in the childhood onset muscular dystrophies, the diagnosis is usually not made until the child is 2–4 years of age, after the observation of delayed motor milestones. Thus, any theoretical neonatal window of reduced microvascular permeability, due to the developmental progression of microvascular parameters in going from weightless intrauterine environment to skeletal maturity, will be lost by this time. Among terrestrial vertebrates, it is unclear where humans fit on the spectrum from mouse to giraffe in terms of the influence of size and posture on the development of these parameters. For instance, an adult mouse can accommodate the infusion, within seconds, of a volume of saline equivalent to at least its own blood volume administered by tail vein injection, as most immediately leaks into the general interstitium. An identical maneuver would likely result in critical illness in the dog or human, as the restricted peripheral permeability serves to shunt most of the edema fluid into the lungs in large mammals (15, 16). In another illustrative example, a newborn horse must be able to stand and run within hours of birth, and the transmural pressure gradient in the extremity veins must immediately withstand that of the aortic wall in utero. This magnitude of pressure change never happens in the small animal, illustrating the possibility that critical cellular mechanisms for cardiovascular homeostasis have been lost or gained in distinct evolutionary lineages (e.g., the absence of valves from the limb veins of mice).

2. Materials

1. Adult Fisher 344 rats (Jackson Laboratory, Bar Harbor, ME, USA) (see Note 1).
2. 2-0 Polypropylene suture, tapered needle (Ethicon, Somerville, NJ, USA).
3. Suitable magnification, preferably binocular operating microscope.
4. PE 10 tubing (Becton-Dickinson, Sparks, Maryland, USA).
5. 27 and 30 G needles (Becton-Dickinson).
6. Microsurgical instruments: Scissors, forceps, and needle driver (Fine Science Tools, Foster City, CA, USA).
7. Microcautery device (GOMAC Healthcare, El Sobrante, CA, USA).
8. Hibiclens or equivalent for decontaminating skin.
9. Sterile towel drapes or fenestrated sterile plastic drape.
10. Heparin 1,000 U/mL (Abraxis Pharmaceutical Products, Schaumburg, IL, USA).
11. Papaverine (Sigma, St Louis, MO, USA).
12. Histamine (Sigma).
13. Phosphate buffered saline (PBS).
14. Isoflurane vaporizer and scavenging system (Kent Scientific Cooperation, Torrington, CT, USA) (see Note 2).
15. Microvascular clamps (Accurate Surgical & Scientific Instruments, Westbury, NY, USA).

3. Methods

1. Obtain proper authorization from the relevant Institutional Animal Use and Care Committee prior to undertaking these studies. Make sure all appropriate personnel training is completed and Institutional Biosafety Regulations are in place before experimental animals are made available to the investigators.
2. Anesthetize adult Fisher 344 rats by isoflurane inhalation to achieve an adequate plane of surgical anesthesia. Specifically, anesthetize rats in a chamber and maintain anesthesia with a mask at 1–3% of isoflurane in room air, with adequate local scavenge to suitable filtered suction (see Note 2).

3. Place the animal in dorsal recumbency. Shave the entire hind limb, lower abdomen, flank, and gluteal area. Prep and drape in a routine sterile fashion to afford access to the extremity for vascular access and tourniquet placement. Secure the selected hindlimb in an abducted position with small, sterilized rubber bands placed gently around the mid foot and secured through the drapes to an underlying cork board with a sterilized thumbtack.
4. Orthograde infusion/perfusion. Make a 2–3-cm groin incision. Mobilize the vessels of the groin fat pad. Catheterize smaller vessels to allow access to the common femoral artery and vein through a side branch of sufficient caliber to allow perfusion, while eliminating the risk of repair of the main vessels that would otherwise be required following direct catheterization. Dissect the femoral vessels to the degree required for clamp placement between the inguinal ligament and the bifurcation of the common into deep and superficial femoral branches. Place a pair of 2-0 polypropylene sutures and tapered (noncutting) needles just deep to the common femoral artery and vein to encompass all of the soft tissues at the proximal thigh (see Note 3). Place Rammel snares for later use to tighten the tourniquet.

   Catheterize the fat pad vessels after vasodilating with topical papaverine solution. Attach heat tapered PE 10 polyethylene tubes to 27 or 30 G needles that are prefilled with PBS (see Note 4).

   Systemically heparinize the rats before tightening the tourniquet by infusion of 100 U/kg body weight heparin. Three minutes later, tighten the tourniquet and clamp the common femoral vessels (see Note 5).

   Prime the hindlimb circulation by arterial infusion of either PBS or 150 μg of papaverine in 500 μL of 10 mM histamine/PBS, pH 7.4. Five minutes later, infuse 2.5×10^{11} adenovirus particles/kg or 3×10^{12} particles of AAV/kg body weight in 500 μL of either PBS or 10 mM histamine/PBS followed by a "chase" of PBS at 10 mL/kg (see Note 6). After perfusion/infusion, flush the limb circulation with 10 mL/kg of PBS with drainage through the groin fat pad venous cannula to prevent spillage into the open wound. Cauterize the fat pad vessels. Remove the catheters. Open the clamps and withdraw from the femoral vessels. Release the tourniquet (see Note 7). Close the incision in two layers with resorbable sutures. Allow the rats to recover from anesthesia with routine monitoring and environmental temperature control(see Notes 8–10).
5. Retrograde infusion. Follow the same procedure as above except the groin fat pad dissection step. Instead, catheterize the anatomic equivalent of the greater saphenous vein just proximal to the ankle joint. Make a 1-cm counterincision in

the distal limb. Ligate the vein distally and place a snare around the heat-tapered catheter tip. Perform retrograde infusion using a comparable volume of PBS, with or without histamine (see Note 11).

4. Notes

1. The protocol described here is readily modified for use in larger mammals such as dogs.
2. Alternatively, ketamine (100 mg/mL) and xylazine (20 mg/mL) or comparable injectable anesthetic combination can be used in rats at the dose of ketamine 75–100 mg/kg and xylazine 5–10 mg/kg, mixed in physiological saline. For studies in the dog, preanesthetic sedation is achieved with dexmedetomidine IM at the manufacturer's recommended dose, followed by placement of an IV catheter and induction with propofol IV, followed by endotracheal intubation and spontaneous or mechanical ventilation using 1–3% isoflurane with the balance oxygen. For dogs, we recommend routine use of a recirculating temperature control blanket and indwelling rectal or esophageal temperature monitor. We also recommend capnography, pulse oximetry, and noninvasive blood pressure monitoring on a routine basis for any procedure requiring a surgical plane of anesthesia.
3. This is an essential step because in our experience with alternative tourniquet configurations, there is very high risk of a mediator leak into the central circulation when the perfusion pressure is increased. The tapered shape of the thigh incites distal slippage of the tourniquet, and this problem is prevented by the internal anchoring scheme.
4. A soldering iron is suitable for heat tapering while rolling the tips under the dissecting microscope.
5. Three minute delay is to ensure systemic distribution of the heparin.
6. The histamine and papaverine are allowed 5 min for receptor activation.
7. If the flush is adequate, there is at most a few minute period of hyperpnea noted in the rats following tourniquet release, while normal pulse is maintained as visualized in the femoral artery under the microscope.
8. Special cautions should be taken during catheterization in the dog's regional limb. For orthograde flow, it is appropriate to directly catheterize the common femoral vessels. This will require the use of a standard vascular surgical instrument set

and an atraumatic catheter with a bulbous tip to allow it to be secured in the artery after arteriotomy. A Javid or equivalent carotid endarterectomy shunt works well for this, with the open end snugly fit into tygon tubing of a size appropriate for the animal under consideration. For smaller dogs, a Webster cannula is appropriate. For flow from the venous side distally, we have found that it is best, especially in the dystrophic dogs in which the vascular smooth muscle is affected by the dystrophin deficiency, to place the catheters under direct visualization. Access is gained through a 1-cm incision over the bifurcation of the saphenous vein on the dorsum of the foot. After a 1-cm length of vein is carefully dissected free of the surrounding soft tissues, two 4-0 silk ties are used to encircle the vein proximally and distally. A transverse venotomy is made, the lumen is dilated with a fine-tipped hemostat, an introducer is used to secure access to the venotomy, and the largest diameter catheter that will fit is introduced. The silk ties are tightened and the catheter is secured in place with an additional suture placed through the skin at the level of the catheter hub. After the infusion and tourniquet release, the catheter is withdrawn and the vein proximal to the venotomy is ligated.

9. There are also differences in tourniquet placement when performing this procedure in dogs. An appropriately sized pneumatic tourniquet may be used for distal hindlimb infusions (i.e., to transduce muscles below the knee). However, for access to the thigh musculature, an insulated indistensible wire tourniquet is best wrapped tightly 4–5 times around the limb at the level of the groin. If this is appropriately placed, tightened, and monitored, it will occlude blood flow across a proximal plane that encompasses the hip joint and prevent the systemic spread of the vector. We have attached the tourniquets to harnesses to prevent migration distally on the limb during the period of occlusion.

10. Infusion pressure should be monitored in dogs. Many blood pressure transducers record a maximal pressure of 200 torr, so it is important to have in the circuit a device that can withstand and record higher pressures. We routinely use a dial from a blood pressure cuff attached to a sterile saline/air interface and then to a three-way stopcock between the syringe and the catheter. We use a pressure of 300 torr in dystrophic and nondystrophic dogs, with the vector infused in the first 1/4 of the total infused volume of PBS. As with the adult rats, if the tourniquet is placed at the level of the hip, we infuse 10 mL/kg body weight.

11. Heparin may be used as above, but no clotting of the visible vessels was noted in its absence. If histamine is used, the flush

effluent can be contained and recovered by making a small nick in the saphenous artery and collecting the PBS with a pipette tip attached to a biohazard vacuum suction trap. The catheterized vein is cauterized, the catheter withdrawn, and the incisions closed as above. A high-impedance infusion pressure monitoring transducer can be used in conjunction with a three-way stopcock placed between the syringe and the needle tip to provide a record of the pressure achieved during the vector infusion.

References

1. Zhu, X., Hadhazy, M., Groh, M. E., Wheeler, M. T., Wollman, R., and McNally, E. M. (2001) Overexpression of {gamma}-sarcoglycan induces severe muscular dystrophy: Implications for the regulation of sarcoglycan assembly. *J Biol Chem 276*, 21785–21790.
2. Cox, G., Cole, N., Matsumura, K., Phelps, S., Hauschka, S., Campbell, K., Faulkner, J., and Chamberlain, J. (1993) Overexpression of dystrophin in transgenic mdx mice eliminates dystrophic symptoms without toxicity. *Nature 364*, 725–729.
3. Zhu, T., Zhou, L., Mori, S., Wang, Z., McTiernan, C. F., Qiao, C., Chen, C., Wang, D. W., Li, J., and Xiao, X. (2005) Sustained whole-body functional rescue in congestive heart failure and muscular dystrophy hamsters by systemic gene transfer. *Circulation 112*, 2650–2659.
4. Greelish, J., Su, L., Lankford, E., Burkman, J., Chen, H., Konig, S., Mercier, I., Desjardins, P., Mitchell, M., Zheng, X., Leferovich, J., Gao, G., Balice-Gordon, R., Wilson, J., and Stedman, H. (1999) Stable restoration of the sarcoglycan complex in dystrophic muscle perfused with histamine and a recombinant adeno-associated viral vector. *Nat Med 5*, 439–543.
5. Gregorevic, P., Allen, J. M., Minami, E., Blankinship, M. J., Haraguchi, M., Meuse, L., Finn, E., Adams, M. E., Froehner, S. C., Murry, C. E., and Chamberlain, J. S. (2006) rAAV6-microdystrophin preserves muscle function and extends lifespan in severely dystrophic mice. *Nat Med 12*, 787–789.
6. Cooper, B. J., Winand, N. J., Stedman, H., Valentine, B. A., Hoffman, E. P., Kunkel, L. M., Scott, M. O., Fischbeck, K. H., Kornegay, J. N., Avery, R. J., and et al. (1988) The homologue of the Duchenne locus is defective in X-linked muscular dystrophy of dogs. *Nature 334*, 154–156.
7. Ladiges, W. C., Storb, R., and Thomas, E. D. (1990) Canine models of bone marrow transplantation. *Lab Anim Sci 40*, 11–15.
8. Baron, F., and Storb, R. (2004) Allogeneic hematopoietic cell transplantation as treatment for hematological malignancies: a review. *Semin Immunopathol 26*, 71–94.
9. Arruda, V. R., Stedman, H. H., Nichols, T. C., Haskins, M. E., Nicholson, M., Herzog, R. W., Couto, L. B., and High, K. A. (2004) Regional intravascular delivery of AAV-2-F.IX to skeletal muscle achieves long-term correction of hemophilia B in a large animal model. *Blood 105*, 3458–3464.
10. Baier, H., Rodriguez, J. L., Chediak, A., and Wanner, A. (1988) Tracheal narrowing during histamine-induced bronchoconstriction. *J Appl Physiol 64*, 1223–1228.
11. Agarwala, S. S., Glaspy, J., O'Day, S. J., Mitchell, M., Gutheil, J., Whitman, E., Gonzalez, R., Hersh, E., Feun, L., Belt, R., Meyskens, F., Hellstrand, K., Wood, D., Kirkwood, J. M., Gehlsen, K. R., and Naredi, P. (2002) Results from a randomized phase III study comparing combined treatment with histamine dihydrochloride plus interleukin-2 versus interleukin-2 alone in patients with metastatic melanoma. *J Clin Oncol 20*, 125–133.
12. Hornyak, S. C., Orentas, D. M., Karavodin, L. M., and Gehlsen, K. R. (2005) Histamine improves survival and protects against interleukin-2-induced pulmonary vascular leak syndrome in mice. *Vascul Pharmacol 42*, 187–193.
13. Simionescu, M., Gafencu, A., and Antohe, F. (2002) Transcytosis of plasma macromolecules in endothelial cells: a cell biological survey. *Microsc Res Tech 57*, 269–288.
14. Su, L. T., Gopal, K., Wang, Z., Yin, X., Nelson, A., Kozyak, B. W., Burkman, J. M., Mitchell, M. A., Low, D. W., Bridges, C. R., and Stedman, H. H. (2005) Uniform scale-independent gene transfer to striated muscle after transvenular extravasation of vector. *Circulation 112*, 1780–1788.
15. Brauer, L. P., Svensen, C. H., Hahn, R. G., Kilicturgay, S., Kramer, G. C., and Prough, D. S.

(2002) Influence of rate and volume of infusion on the kinetics of 0.9% saline and 7.5% saline/6.0% dextran 70 in sheep. *Anesth Analg 95*, 1547–1556.

16. Narayan, S. M., Drinan, D. D., Lackey, R. P., and Edman, C. F. (2007) Acute volume overload elevates T-wave alternans magnitude. *J Appl Physiol 102*, 1462–1468.

Chapter 19

AAV-Mediated Gene Therapy to the Isolated Limb in Rhesus Macaques

Louise R. Rodino-Klapac, Chrystal L. Montgomery, Jerry R. Mendell, and Louis G. Chicoine

Abstract

The development of a nonhuman primate (NHP) model for vascular delivery of therapeutic transgenes with adeno-associated viral (AAV) vectors is crucial for successfully treating muscular dystrophies. Current animal models for Duchenne muscular dystrophy (DMD) gene therapy have species limitations related to assessing function, immune response, and distribution of the micro- and minidystrophin transgenes in a clinically relevant manner. In addition, there are many forms of muscular dystrophy for which there are no available disease models. NHPs provide the ideal model to optimize vector delivery across a vascular barrier and provide accurate dose estimates for local or broadly targeted gene therapy studies. The vascular anatomy NHPs more clearly parallels humans providing an appropriate substrate for translational experiments. Here we outline the development of a rhesus macaque isolated focal limb perfusion (IFLP) protocol targeting the vascular bed of the gastrocnemius. This protocol serves as a model with broad implications for other muscle diseases along with the capability of targeting multiple muscle groups. To overcome the partial homogeneity between portions of the human microdystrophin transgene and those of the NHP dystrophin gene, we utilized a FLAG tag for tracking distribution of microdystrophin. We also provide methods for assessing transduction efficiency of microdystrophin.FLAG following the IFLP vascular delivery protocol.

Key words: Adeno-associated virus (AAV), Gene therapy, Muscle disease, Duchenne muscular dystrophy, Microdystrophin, *mdx* Mice, Nonhuman primates, Isolated focal limb perfusion

1. Introduction

Duchenne muscular dystrophy (DMD) is the most common devastating muscle disease of childhood. The disease is caused by monogenic mutations in the dystrophin gene, making this condition potentially amenable to gene therapy. Several gene replacement strategies are under investigation and progress towards clinical gene therapy with mini- and microdystrophin constructs

Dongsheng Duan (ed.), *Muscle Gene Therapy: Methods and Protocols*, Methods in Molecular Biology, vol. 709,
DOI 10.1007/978-1-61737-982-6_19, © Springer Science+Business Media, LLC 2011

delivered by adeno-associated virus (AAV) has gained momentum (1, 2). Proof of principle studies in the *mdx* mouse with microdystrophin have demonstrated reversal of the dystrophic process with reduced central nucleation, improvement in tetanic force measures, and increased resistance to eccentric contractions (3–5).

Gene transfer to multiple muscle groups will be required to produce clinically meaningful outcomes. To this end, both arterial and venous approaches to gene therapy have demonstrated success using AAV6, AAV8, and AAV9 in preclinical studies with *mdx* mice and dystrophic canines (5–7). However, these important models fail to replicate some critical obstacles related particularly to dosing and anatomical distribution of blood vessels that dominate the clinical environment. To overcome these obstacles, we utilized the rhesus macaque where the arterio-venous circulation of the muscle presents anatomical similarities with DMD boys. In addition, the nonhuman primate (NHP) endothelial capillary junctions, the interior wall pressures, and the size ratio of capillaries to muscle fibers more closely simulate the clinical setting. We developed a clinically relevant arterial approach that enables gene delivery to a muscle-specific vascular bed that is scalable to include multiple beds as the clinical condition requires. Specifically, we developed a rhesus macaque isolated focal limb perfusion (IFLP) protocol targeting the vascular bed of the gastrocnemius muscle. This approach provides delivery of the transgene throughout the gastrocnemius and is accomplished with a relatively small delivery volume that does not induce muscle injury.

A limitation in utilizing NHP in gene therapy studies is the homogeneity of the endogenous and transgene protein products, especially when a disease model is not available. This lack of a biomarker prevents the precise assessment of vector-mediated gene expression and distribution, both at the microcellular and tissue levels. Thus, we utilized an 8 amino acid FLAG tag for tracking distribution of microdystrophin using AAV8 and the muscle-specific promoter – muscle creatine kinase (MCK) – to restrict expression for safety in preparation for clinical application.

In this protocol, we outline the procedures used to generate our human microdystrophin.FLAG construct and subsequent assessment of its bioactivity. We also describe the methodology for detecting the transgene and finally describe in detail the methodology for the vascular administration of the viral vector.

2. Materials

2.1. Generation of Microdystrophin. FLAG Vector

1. pUC57.Human Microdystrophin cDNA (GenScript Corporation, Piscataway, NJ, USA). The human microdystrophin codon-optimized (human) cDNA-containing (R4-R23/Δ71-78) domains was in vitro synthesized and

cloned into a pUC57 vector with flanking Not I restriction enzyme sites.

2. pAAV.MCK plasmid: AAV2 ITRs using flanking Xba I restriction enzyme sites in a plasmid derived from pCMVβ (Clontech, Mountain View, CA, USA). The plasmid also includes a consensus Kozak sequence, an SV40 chimeric intron, and synthetic polyadenylation site (53 bp).
3. FLAG primers: 25 nM, Standard Desalting (Integrated DNA Technologies, Coralville, IA, USA).
4. Pfu Ultra™ II Fusion HS DNA Polymerase (Stratagene, Cedar Creek, TX, USA).
5. dNTPS master mix 25 mM Stock (Denville Scientific, Metuchen, NJ, USA).
6. Not I restriction enzyme (New England BioLabs, Ipswich, MA, USA).
7. BSA (New England BioLabs).
8. QIAquick gel extraction kit (Qiagen, Valencia, CA, USA).
9. UltraPure Agarose (Invitrogen, Carlsbad, CA, USA).
10. Zero Blunt Topo PCR cloning kit (Invitrogen).
11. Quick T4 DNA Ligase (New England BioLabs).
12. One Shot® TOP10 Chemically Competent *E. coli* (Invitrogen).
13. LB-Amp Selection Media: Lennox L Broth Base (Invitrogen) with100 mg/mL, Ampicillin. Store at 4°C.
14. S.O.C. medium (Invitrogen). Store at room temperature.
15. Amp selection LB agar plates (100 mg/mL, Ampicillin). Store at 4°C.
16. 14-mL Polypropylene round bottom tubes (Becton Dickinson, Franklin Lakes, NJ, USA).
17. Sequencing Primers: 25 nM, Standard Desalting (Integrated DNA Technologies).

2.2. Biotesting of Microdystrophin. FLAG Vector

1. Chinese rhesus macaques (Covance Research Products, Denver, PA, USA) are being used for these experiments. There is no DMD primate disease model, thus dystrophin expression is present in this strain.
2. Anesthetic reagents and instruments for local muscle injection: Telazol (5 mg/kg), Buprenorphine (0.01 mg/kg), 30 G 3/10 cc Ultra-Fine insulin syringe (Becton Dickinson), Sterile forceps (World Precision Instruments, Sarosota, FL, USA), needle holders (Miltex, York, PA, USA), 4.0 Vicryl suture (Owens and Minor, Mechanicsville, VA, USA).
3. Mouse monoclonal antibody for the N-terminus of dystrophin: Dys-2 (clone Dy8/6C5, IgG1), an antibody that

reacts with the C-terminal domain of human dystrophin (Novocastra, Newcastle, UK).

4. Polyclonal antibody to recognize the FLAG epitope: F7425 – rabbit anti-FLAG antibody, affinity purified (Sigma-Aldrich, St Louis, MO, USA).
5. Gum Tragacanth: 7 g gum tragacanth (Sigma), 2 mg Thymol (Sigma) in 100 mL dH_2O.
6. 2-Methylbutane (Fisher Scientific, Pittsburgh, PA, USA).
7. Superfrost/Plus Slides (Fisher Scientific).
8. PBS: 10× (Fisher Scientific).
9. 10% Goat serum: Normal goat serum (Gibco-BRL, Grand Island, NY, USA) diluted with PBS.
10. Triton X-100 (Biorad, Hercules, CA, USA).
11. Alexa 568-conjugated goat antirabbit antibody (Molecular Probes, Carlsbad, CA, USA).
12. Alexa 488-conjugated goat antimouse antibody (Molecular Probes).
13. Vectashield Mounting Medium for Fluorescence (Vector Laboratories, Burlingame, CA, USA).
14. 1.5 mL Microfuge tubes (Fisher Scientific).

2.3. IFLP to Target the Gastrocnemius of the Macaque

1. Surgical suite or clean room.
2. Fluoroscope: GE OEC9900 Elite C-ARM.
3. 12 mL Luer Lock syringe X3 (Kendall Monoject, Mansfield, MA, USA).
4. Suture, Silk Blk BR 3-0 18 SH (Ethicon, Somerville, NJ, USA).
5. Suture, Vicryl UD BR CT 4-0 (Ethicon).
6. Sterile angiography drape pack (Medline: Custom Pack): sterile gowns, cover fluoroscope blue (44 × 36), table cover, OR towels (6–12), 32 oz plastic bowl (×2), syringe 10 mL luer lock with green plunger, syringe 10 mL luer lock (×5, Merit Medical), needles (18 G × 1.5 and 23 G × 1.5, Becton-Dickinson), No. 11 scalpel blade and holder, tape scissors, small scissors Metzenbaum (Fine Science Tools, Foster City, CA, USA), straight Kelly forceps (5.5 × 2, Roboz, Gaithersburg, MD, USA), curved Kelly forceps (5.5 × 2, Roboz), needle driver med (Roboz), Povidone 1% iodine paint sponge sticks (Aplicare, Meridien, CT, USA), ethyl alcohol.
7. 3 Fr sheath VCF-3.0-18-6-J Introducer (Cook, Bloomington, IN, USA).

8. 3 Fr catheter, 50 cm long, P3.0-NT-50-P-NS-JB1 (Cook).
9. Glidewire/Heart wire 0.018, 180 length Runthrough NS (Terumo, Elkton, MD, USA).
10. Dermaprene surgical sterile gloves PF (Generic).
11. Ioversol injection 51% (50 mL) diluted to 50% with normal saline (Tyco, Princeton, NJ, USA).
12. Pouch, self seal 3.5″×9″ (Medical Action Industries, Brentwood, NY, USA).
13. Sterilization pouch 3.5″×5.25″ self seal (Medical Action Industries).
14. N/WVN 4×4 N/S sponges (Kendall Monoject, Mansfield, MA, USA).
15. N/WVN 4×4 N/S guaze roll (Kendall).
16. Viral stock (2×10^{12} viral particles per kg body weight in 2.5 mL normal saline per kg body weight).
17. Sterile normal saline for flushes (Baxter, Deerfield, IL, USA).
18. Heparin sodium injection 1,000 USP Units/mL (does: 50 U/kg).
19. Vacular pick (Becton-Dickinson).

3. Methods

3.1. Generation of Microdystrophin. FLAG Vector

Generation of a microdystrophin.FLAG fusion protein can be constructed with the FLAG tag placed at either the N or C terminus of the protein. For this construct, the tag was placed at the C terminus, but for other transgenes both N- and C-terminal tags may need to be tested for transgene potency and potential immunogenicity.

1. *Synthesis of human microdystrophin cDNA*. Have the cDNA-containing (R4-R23/Δ71-78) domains (3) synthesized by GenScript using species-specific codon optimized for human. Engineer Not I sites or appropriate restriction sites flanking the cDNA to facilitate subsequent excision and transfer. Clone cDNA into a pUC57 or appropriate shuttle vector.
2. *pAAV.MCK ITR-containing entry vector*. Use pCMVβ as the vector backbone plasmid containing AAV2 ITRs. Synthesize the MCK promoter, chimeric SV40 intron, consensus Kozak sequence, synthetic polyadenylation site (53 bp), and a mouse microdystrophin cDNA (5) (Fig. 1). Remove the mouse microdystrophin cDNA by Not I digestion.

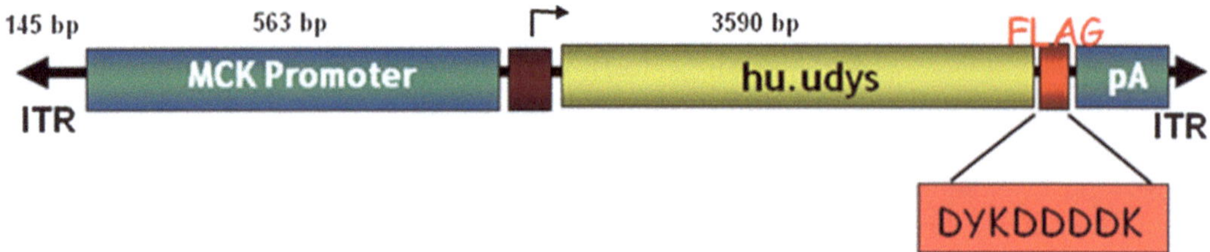

Fig. 1. Human microdystrophin.FLAG cassette. The FLAG epitope tag allows for the assessment of microdystrophin distribution across normal muscle primate muscle. An eight amino acid tag was placed in frame at the C terminus of human microdystrophin. The cassette includes a 563-bp MCK promoter, a chimeric intron (*purple*) to augment RNA processing, and a synthetic polyadenylation site (53 bp) flanked by two AAV2 ITRs.

3. *PCR amplification of C-terminal FLAG tag*. Design primers to amplify human microdystrophin. The reverse primer includes sequence encoding the FLAG epitope (DYKDDDDK) sequence (see Note 1) in frame with the microdystrophin C terminus. The forward and reverse primers contain Not I restriction sites to facilitate subsequent cloning followed by 10 random nucleotides for stability. PCR amplification protocol consists of final concentrations of 50 ng human microdystrophin template, 2 mM primers, 0.5 mM dNTPs, 1 U Pfu Ultra proofreading polymerase, and 1× Pfu Ultra Buffer. Conditions: (1) 94°C 10 min; (2) 40 cycles: 92°C-30 s, 58°C-30 s, 72°C-4 min ; and (3) 72°C-10 min. Insert PCR product into the Blunt Topo vector for long-term use.
4. *Cloning of microdystrophin.FLAG cDNA into AAV2 ITR containing entry plasmid*. Digest the microdystrophin.FLAG insert and the entry plasmid with Not I enzyme (2 h at 37°C). Separate components on 1% agarose gel. Gel purify vector backbone and microdystrophin.FLAG insert with the Gel Extraction kit. Quantify products and proceed to ligation. Using Quick Ligase kit, ligate insert into the entry vector at a ratio of 3:1. Transform Top10 chemically competent cells by heatshocking at 42°C for 45 s. Shake with SOC for 1 h at 34°C and plate on LB-Amp plates. Screen colonies with Not I digest and PCR using forward primer from vector backbone and reverse primer from insert. Sequence the entire plasmid to insure no mutations have been induced by PCR.

3.2. Biotesting of Microdystrophin.FLAG Vector in Rhesus Macaques

For all animal experiments, get approval from the institute Animal Care and Use Committee following NIH guidelines. The tibialis anterior (TA) muscle was used to evaluate the bioactivity of human microdystrophin.FLAG.

1. *Intramuscular gene delivery to the TA muscle*. Anesthetize the animal and expose the belly (central/thickest portion of muscle) with a 1.5-cm incision. Inject vector (~5×10^{12} vg in 1 mL) into three areas 0.5 cm apart in a linear fashion parallel to the incision using an Ultra-fine 30 G insulin syringe inserted the

entire length of the needle (12.7 mm). Suture the wound and use India ink to mark the three injection sites over the skin.

2. *Necropsy.* Euthanize the animal with sodium pentobarbital (1 mL/10 lbs). Expose the TA and remove the entire muscle (treated and control) using a scalpel and forceps. Cut the entire muscle into 0.75 × 0.75 cm blocks from site of proximal to distal insertion working in layers. Assign a number to each block for subsequent sample tracking.
3. *Muscle tissue embedding.* Mount each tissue sample with muscle cross-section exposed using gum tragacanth as adhesive to wooden dowel rod (12 mm diameter, 1.5 cm height). Cool 2-methylbutane in metal cup over liquid nitrogen until white crystals form. Using large forceps grasping wooden dowel, insert tissue into the cooled 2-methylbutane and freeze for 35 s. Transfer to dry ice and allow to dry for 20 min with subsequent long-term storage at −80°C.
4. *Isolation of organs for biodistribution study.* Collect vital organs including gonads, heart, lung, spleen, kidney, liver, and lymph nodes. Using a new sterile set of forceps and scissors for each organ, isolate a biopsy from each organ and place in sterile microfuge tubes. Flash freeze in liquid nitrogen.
5. *Cryo-sectioning muscle samples.* Remove samples from storage (−80°C) and place in cryostat maintained at −20°C. Allow samples to acclimate for 20 min and then collect 12 mm sections onto slides. Dry slides on a 37°C hotplate for 30 min. Perform Hematoxylin and Eosin staining on one slide and proceed to FLAG antibody staining with another slide.
6. *Immunostaining with polyclonal FLAG antibody.* Block slides with PBS/20% goat serum/0.1% triton X-100 for 1 h at room temperature. Apply primary antibody (1:50 dilution in blocking solution) and incubate for 1 h at room temperature. Wash with 1× PBS (3× 20 min each). Block again for 30 min prior to secondary antibody application. Detect signal with an Alexa 568-conjugated goat antirabbit secondary antibody (1:250 dilution in blocking solution). Wash with 1× PBS (3× 20 min each). Coverslip with Vectashield mounting media and store at −20°C.
7. *Quantification of muscle fiber transduction.* Photograph each muscle section. Capture four 10× images per section to cover entire cross-sectional area. Capture images from contralateral control side in the same manner.
8. *Count sarcolemma positively stained fibers.* Count the number of muscle fibers with positive sarcolemmal staining (Fig. 2) along with the total number of fibers per section. Calculate transduction efficiency using the formula: % Transduction = total number of positive fibers/total number of fibers.

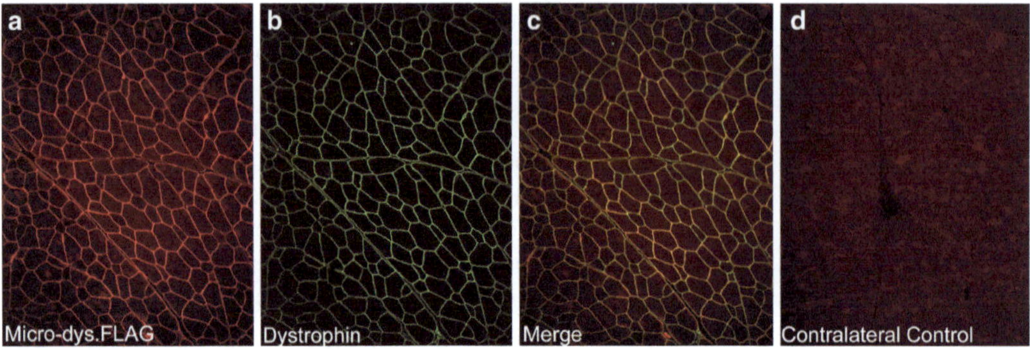

Fig. 2. Microdystrophin.FLAG Bioactivity following intramuscular injection. (**a**) Muscle biopsy from the injected TA from NHP stained with an anti-FLAG antibody. (**b**) C-terminal antibody to endogenous dystrophin. (**c**) Merged image demonstrating colocalization. (**d**) Contralateral control muscle.

3.3. IFLP to Target the Gastrocnemius of the Macaque

1. Perform complete physical examinations and screening for tuberculosis, simian retrovirus, simian T cell leukemia virus and simian immunodeficiency virus for Rhesus macaques upon arrival to the research facility. Apply a 6-week quarantine prior to entry into the study. The care of these animals is under the supervision of a DVM who has NHP clinical expertise.
2. Prior to removal from their enclosure and taken to the surgery preparation room, sedate the Rhesus macaque with an IM dose of telazol (3–6 mg/kg).
3. Shave the groin, including the femoral triangle, with electric clippers.
4. Intubate the animal with a size-appropriate endotracheal tube which is tied in place with umbilical tape. Carry the sedated and intubated animal to the surgical suite and place the animal on a heated pad and anesthetize with isoflurane (1.5–2% balance O_2) titrated for effect.
5. Place the animal in supine position with the lower extremities placed in a relaxed frog leg position. Place surgical towels beneath the knees for support. Prep the area of the femoral triangle and groin with Povidine scrub. Save the contralateral side as a backup site should vascular injury prevent cannulation of the targeted vascular bed. After a few minutes, wipe off the Povidine with ethanol soaked in 4×4 gauze. Cover the animal and work area with surgical towels just leaving the surgical site exposed.
6. Make a small incision (0.5–1.5 cm) within the femoral triangle, outlined by the inguinal ligament, sartorius, and adductor, parallel and over the palpated femoral artery pulse (see Note 2).
7. Isolate the femoral bundle by blunt dissection and carefully dissect the femoral artery free of the surrounding structures and secure it with both proximal and distal ligatures. Use a 21-G needle to puncture the femoral artery. Apply gentle tension to the ligatures to stop bleeding.

8. Place a 3.0 Fr introducer sheath via the Seldinger method by passing the preflushed sheath over a wire previously placed in the artery. Advance the sheath only a few centimeters and secure it in place with a 3.0 braided silk suture.
9. Administer heparin at the dose of 50 U/kg body weight via the sheath and clear the sheath with normal saline.
10. Generate a road map of the vasculature by administering a few milliliters of contrast agent through the sheath and capture the image by a fluoroscopy.
11. Place a 3 Fr, 50 cm long catheter into the introducer sheath and advance it a few centimeters. Place the guide wire (0.018 in. diameter) through the catheter. Under fluoroscopic guidance, advance it to the sural arteries which perfuse the two heads of the gastrocnemius.
12. Once the catheter is correctly positioned, isolate the vascular bed of the gastrocnemius with the placement of proximal and distal tourniquets (see Note 3). Place the proximal tourniquet above the knee and just proximal to the catheter tip. Assess optimal placement of the proximal tourniquet by partial tourniquet tightening and visualization of a small volume (few milliliters) of injected contrast agent (see Note 4). Once the relationship of the proximal tourniquet to catheter tip is established, flush the contrast from the limb with normal saline and position the distal tourniquet just below the gastrocnemius. This second tourniquet provides compartmentalization of the gastrocnemius (Fig. 3).
13. Once the two tourniquets are well positioned, start dosing with a preflush volume (2.5 mL/kg) of normal saline delivered over 60 s with the tourniquets pulled snug, but not tight (see Note 5). While the final volume is administered, pull the tourniquets tight to occlude blood flow (see Note 6).
14. With the tourniquets pulled tight, administer the vector (2×10^{12} viral genomes per kg in 2.5 mL per kg volume) over 60 s. Allow 10 min dwell time as the tourniquets are left tight (see Note 7).
15. Following the 10 min dwell and with the tourniquets still tight and occluding blood flow, administer a postvolume of normal saline (2.5 mL/kg) over 60 s (see Note 8).
16. Remove the tourniquets. Apply direct pressure to the wound for 10 min to control bleeding. Close the wound with a continuous subcuticular 4.0 Vicryl suture. Apply a pressure dressing to the site and keep it in place until the animal awakes from anesthesia, then pull it off. Usually this takes about one to two hours. Allow the animal to recover from general anesthesia in a warmed environment and return the animal to their cage when warmed and stable.

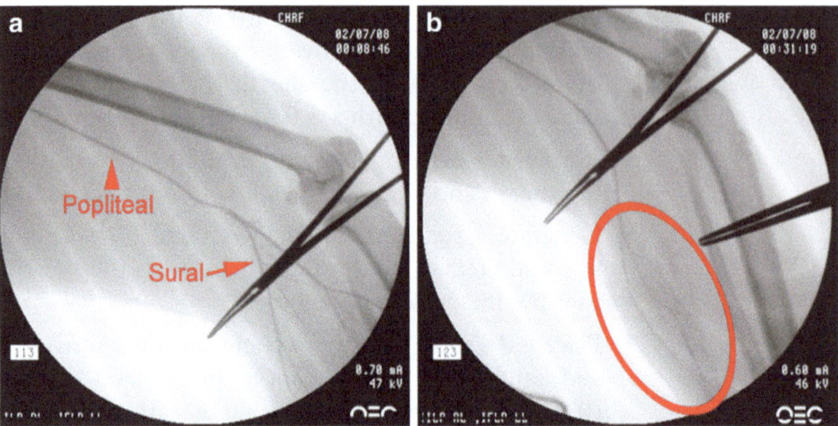

Fig. 3. Fluoroscopy-guided targeting of gastrocnemius. (**a**) Image of sural branch of the femoral artery. (**b**) Catheterization of sural artery and the use of tourniquets compartmentalizes the gastrocnemius muscle.

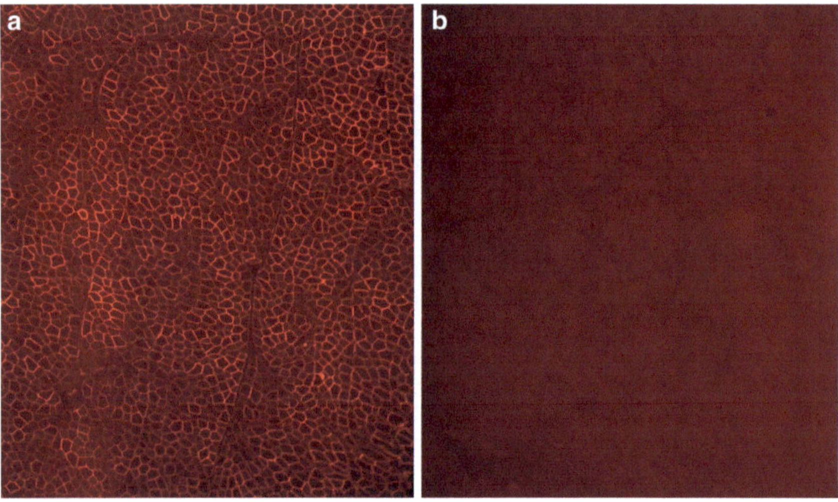

Fig. 4. Vascular delivery of microdystrophin.FLAG to the gastrocnemius muscle. (**a**) Microdystrophin.FLAG antibody staining in the treated gastrocnemius. (**b**) Contralateral control.

17. Necropsy at such a time as defined by the experimental conditions and conduct necropsy as described above (see Subheading 3.2, Step 2) (Fig. 4).

4. Notes

1. Primer sequence for C-terminal FLAG tag: 5′-TGGA GACCGATACAATGGACTACAAAGACGATGACGACAA GTAACTAGTGCGGCCGCtattct-3′.
2. We chose the direct cut-down approach as an alternative to the percutaneous approach as the study animals were small

relative to the introducer sheath and we had concerns of maintaining vascular integrity in using a blind stick. Furthermore, we did not cannulate the contralateral artery and advance the catheter up and over the aortic bifurcation as the sharp angel encountered at the bifurcation was not permitting. This approach has a role when studying larger animals (>2–3 kg).

3. Compartmentalization of the vascular bed provides several benefits to the procedure. First, it enhances the targeting of the vascular bed. Second, it reduces risk to our patients by minimizing off-target transgene exposure. Third, it provides a protective local environment from preexisting neutralizing antibody and serum components (complement) that could be a significant barrier to successful transduction (4, 5).
4. Appropriate placement of the proximal tourniquet just above the catheter tip enhances success of IFLP by preventing retrograde flow up the artery that can lead to off-target gene transfer.
5. The preflush provides a mechanism to remove potential binding elements from the targeted vascular compartment. For example, binding or neutralizing antibodies.
6. The amount of tension needed to occlude the vascular bed can be assessed by ultrasound evaluation. The now isolated compartment is ready to receive the vector.
7. The 10-min dwell provides more than ample time for virus attachment and uptake into the target cells.
8. The final volume clears the catheter and partially clears the vasculature of remaining vector.

Acknowledgments

We thank the Viral Vector Core at Nationwide Children's for vector production. Research supported by the Children's Hospital Foundation (J.R.M.); National Institutes of Health U54 (1U54NS055958-01A1), Muscular Dystrophy Association (J.R.M.); Jesse's Journey Foundation for Gene and Cell Therapy (J.R.M.), Ruth L. Kirschstein NRSA postdoctoral fellowship (1F32AR055008 to L.R.K.).

References

1. Rodino-Klapac, L. R., Chicoine, L. G., Kaspar, B. K., and Mendell, J. R. (2007). Gene therapy for duchenne muscular dystrophy: expectations and challenges. *Arch Neurol* 64, 1236–1241.
2. Wang, Z., Chamberlain, J. S., Tapscott, S. J., and Storb, R. (2009). Gene therapy in large animal models of muscular dystrophy. *ILAR J* 50, 187–198.
3. Harper, S. Q., et al. (2002). Modular flexibility of dystrophin: implications for gene therapy of Duchenne muscular dystrophy. *Nat Med* 8, 253–261.
4. Liu, M., Yue, Y., Harper, S. Q., Grange, R. W., Chamberlain, J. S., and Duan, D. (2005). Adeno-associated virus-mediated microdystrophin expression protects young mdx muscle from contraction-induced injury. *Mol Ther* 11, 245–256.
5. Rodino-Klapac, L. R., et al. (2007). A translational approach for limb vascular delivery of the micro-dystrophin gene without high volume or high pressure for treatment of Duchenne muscular dystrophy. *J Transl Med* 5, 45.
6. Gregorevic, P., et al. (2006). rAAV6-microdystrophin preserves muscle function and extends lifespan in severely dystrophic mice. *Nat Med* 12: 787–789.
7. Yue, Y., Ghosh, A., Long, C., Bostick, B., Smith, B. F., Kornegay, J. N., et al. (2008) A single intravenous injection of adeno-associated virus serotype-9 leads to whole body skeletal muscle transduction in dogs. *Mol Ther* 16, 1944–1952.

Chapter 20

Antisense Oligo-Mediated Multiple Exon Skipping in a Dog Model of Duchenne Muscular Dystrophy

Toshifumi Yokota, Eric Hoffman, and Shin'ichi Takeda

Abstract

Exon skipping is currently one of the most promising molecular therapies for Duchenne muscular dystrophy (DMD). We have recently developed multiple exon skipping targeting exons 6 and 8 in dystrophin mRNA of canine X-linked muscular dystrophy (CXMD), an animal model of DMD, which exhibits severe dystrophic phenotype in skeletal muscles and cardiac muscle. We have induced efficient exon skipping both in vitro and in vivo by using cocktail antisense 2'*O*-methyl oligonucleotides (2'OMePS) and cocktail phosphorodiamidate morpholino oligomers (morpholinos, or PMOs) and ameliorated phenotype of dystrophic dogs by systemic injections. The multiple exon skipping (double exon skipping) shown here provides the prospect of choosing deletions that optimize the functionality of the truncated dystrophin protein for DMD patients by using a common cocktail that could be validated as a single drug and also potentially applicable for more than 90% of DMD patients.

Key words: Multiple exon skipping, Morpholinos (phosphorodiamidate morpholino oligomers), 2′*O*-methylated antisense oligomers (phosphorothioate), Dystrophic dogs (canine X-linked muscular dystrophy), Duchenne/Becker muscular dystrophies

1. Introduction

Duchenne muscular dystrophy (DMD), a progressive and fatal X-linked myopathy, and its milder form, Becker muscular dystrophy (BMD), are caused by mutations in the *DMD* gene (1). Exon skipping using antisense oligonucleotides (AOs) is currently one of the most promising molecular therapies for DMD (2–4). Synthetic derivatives of nucleic acids have been designed and synthesized, where the backbone of RNA and DNA is replaced with other chemistries. One uses a morpholino backbone phosphorodiamidate morpholino oligomers (morpholinos, or PMOs) developed by AVI BioPharma, Portland, Oregon.

Dongsheng Duan (ed.), *Muscle Gene Therapy: Methods and Protocols*, Methods in Molecular Biology, vol. 709,
DOI 10.1007/978-1-61737-982-6_20,

Recently, we have successfully induced dystrophin expression by using morpholino-mediated systemic multiple exon skipping and ameliorated dystrophic pathology in dogs (5). Another antisense chemistry 2'*O*-methylated phosphorothioate (2'OMePS) has been also shown to effectively induce dystrophin expression systemically in mice in vivo (6).

The canine X-linked muscular dystrophy (CXMD) model contains a point mutation within the acceptor splice site of exon 7. This leads to exclusion of exon 7 from the mRNA transcript (7, 8). To restore the open reading frame, at least two further exons (exons 6 and 8) must be skipped (multiple exon skipping, or multiexon-skipping). Therefore, it is more challenging to rescue dystrophic dogs with exon-skipping strategy. Here, we summarize the method and protocol of antisense-mediated exon skipping in vitro and in vivo in dystrophic CXMD dogs.

2. Materials

2.1. Design of Antisense Oligos

1. Web sites for exonic splicing enhancer ESE targeting. ESE finder[http://rulai.cshl.edu/cgi-bin/tools/ESE3/esefinder.cgi?process=home] and Rescue ESE [http://genes.mit.edu/burgelab/rescue-ese/].

2.2. Transfection of Antisense 2'OMePS into Dog Myoblasts

1. Dulbecco's modified Eagle's medium (DMEM) (Gibco, Bethesda, MD, USA) supplemented with 10% fetal bovine serum (FBS, HyClone, Ogden, UT, USA).
2. 0.25% Trypsin and 1 mM ethylenediamine tetraacetic acid (EDTA) (Gibco).
3. Teflon cell scrapers (Fisher, Waltham, MA, USA).
4. Ham's F-10 nutrient mixture with HEPES (Gibco) (9).
5. Fetal calf serum (FCS) (Gibco).
6. Human recombinant basic fibroblast growth factor (bFGF) (Sigma-Aldrich, Natick, MA, USA).
7. Penicillin (200 U/mL) and streptomycin (200 μg/mL) (Sigma-Aldrich).
8. AOs (2'OMePS) (Eurogentec, Liège, Belgium) against exons 6 and 8 of the canine dystrophin gene. Ex6A (GUU GAUUGUCGGACCCAGCUCAGG), Ex6B (ACCUAUGA CUGUGGAUGAGAGCGUU), and Ex8A (CUUCCUGG AUGGCUUCAAUGCUCAC).
9. Lipofectin (Invitrogen, Carlsbad, CA, USA).
10. 2% Horse serum (Gibco).
11. Six-well plates (IWAKI, Funabashi, Japan).

12. Opti-MEM (Gibco).
13. Culture dish (10, 15 cm noncoat and 10, 15 cm collagen coat) (IWAKI).
14. Phosphate buffer saline (PBS).
15. Human recombinant insulin (10 mg/mL) (Sigma-Aldrich).
16. Proliferation medium: Nutrient Mixture F-10 Ham (Ham's F-10; developed by Ham et al. for mammalian cell proliferation (9)) supplemented with 200 U/mL penicillin, 200 μg/mL streptomycin, 2.5 ng/mL bFGF, and 20% FBS.
17. Differentiation medium: DMEM supplemented with 200 U/mL penicillin, 200 μg/mL streptomycin, and 10 μg/mL insulin.

2.3. Intramuscular Injections of Antisense Oligos in Dogs

1. CXMD dogs and wild type littermates.
2. Antisense morpholinos (Gene-tools, Philomath, OR, USA) against exons 6 and 8 of the dog dystrophin gene. Ex6A (GTTGATTGTCGGACCCAGCTCAGG), Ex6B (ACCTATGACTGTGGATGAGAGCGTT), and Ex8A (CTTCCTGGATGGCTTCAATGCTCAC) (see Note 1).
3. Saline (Ohtsuka-Pharmaceutical, Tokyo, Japan).
4. 27G Needles (TERUMO, Tokyo, Japan).
5. Thiopental sodium (Mitsubishi Tanabe Pharma, Osaka, Japan).
6. Isoflurane (Abbott laboratories, Chicago, IL, USA).
7. Butorphanol tartrate (Bedford Laboratories, Bedford, OH, USA).
8. Gauze (Johnson and Johnson, New Brunswick, NJ, USA).
9. Pledget (Johnson and Johnson).
10. Veterinary surgical instruments: forceps, scalpels, scissors, suture needles, threads, and needle holders (Mizuho, Narashino, Japan).
11. Povidone iodine (Meiji Seika, Tokyo, Japan).
12. Heparin sodium (Fuji Pharmaceutical, Tokyo, Japan).
13. Surgical glove (Ansell, Red bank, NJ, USA).
14. Surgical drape (Nagai Leben, Tokyo, Japan).
15. Sepham antibiotics (Cefamezine or Syncl) (Astellas, Tokyo, Japan, or Asahi-kasei, Tokyo, Japan).

2.4. Systemic Injections of Antisense Morpholinos

1. CXMD dogs and wild-type littermates.
2. Syringe infusion pump (Muromachi, Tokyo, Japan).
3. 22G Indwelling needles (TERUMO).
4. 50 mL syringe (TERUMO).

5. Antisense morpholinos (Gene-tools)against exons 6 and 8 of the dog dystrophin gene. Ex6A (GTTGATTGTCGGA CCCAGCTCAGG), Ex6B (ACCTATGACTGTGGATGA GAGCGTT), and Ex8A (CTTCCTGGATGGCTTCAATG CTCAC).

2.5. RNA Extraction

1. Eppendorf tubes (Eppendorf, Hamburg, Germany).
2. Trizol (Invitrogen).
3. Chloroform (Sigma-Aldrich).
4. Isopropanol (Sigma-Aldrich).
5. 75% Ethanol (Sigma-Aldrich).

2.6. RT-PCR

1. One-Step RT-PCR kit (Qiagen, Venlo, The Netherlands).
2. Forward primer in exon 5: CTGACTCTTGGTTTGA TTTGGA (Invitrogen).
3. Reverse primer in exon 10: TGCTTCGGTCTCTGTCAATG (Invitrogen).
4. RNAsin (Promega, Madison, WI, USA).

2.7. cDNA Sequencing

1. Gel extraction kit (Qiagen).
2. BigDye® Terminator v3.1 Cycle Sequencing Kit (Applied Biosystems, Foster City, CA, USA).
3. ExoSap-IT® (USB, Santa Clara, CA, USA).
4. MicroAmp® Reaction Plates (Applied Biosystems).
5. Qiagen gel extraction kit (Qiagen).
6. Hidi-formamide (Applied Biosystems).
7. ABI 3130 Genetic Analyzer (Applied Biosystems).

2.8. Muscle Sampling from Necropsy of Dogs

1. Tragacanth gum (Sigma-Aldrich).
2. Isopentane (Sigma-Aldrich).
3. Liquid nitrogen.
4. Cork disks (Iwai-kagaku, Tokyo, Japan).
5. Dry ice.

2.9. Immunostaining for Dog Muscles

1. Poly-l-lysine–coated slides (Fisher, Hampton, NH, USA).
2. Cover glasses (Fisher).
3. Cryostat Microsystem cm1900 (Leica, Wetzlar, Germany).
4. Dystrophin antibodies including DYS1 and DYS2 (Novocastra, Newcastle, UK).
5. Alexa 594 goat antimouse IgG_1, Alexa 594 goat antimouse IgG_2 highly cross-absorbed (Invitrogen).
6. DAPI containing mounting agent (Invitrogen).
7. Goat serum (Invitrogen).

8. Moisture chamber (Scientific Devise Laboratory, Des Plaines, IL, USA).
9. Chamber slide (Lab-tek, Naperville, IL).
10. 4% Paraformaldehyde (PFA).

2.10. Western Blotting from Dog Muscles

1. Lysis buffer: 75 mM Tris–HCl (pH 6.8), 10% SDS, 10 mM EDTA, and 5% 2-mercaptoethanol.
2. Bradford reagent (Bio-Rad, Hercules, CA, USA).
3. Bovine serum albumin (BSA) (Sigma-Aldrich).
4. 2× Laemmli SDS-loading buffer: 0.1 M Tris–HCl (pH 6.6), 2% (w/v) SDS, 2% (0.28 M) beta-mercaptoethanol, 20% glycerol, 0.01% bromophenol blue.
5. Ready-made 5% resolving SDS gels (Bio-Rad).
6. PVDF membrane (GE, Fairfield, CT, USA).
7. Transfer buffer (10×): 250 mM of Tris-Base, 1,920 mM of Glycine.
8. Transfer buffer (1×): 10% 10× buffer, 20% methanol.
9. Dystrophin antibodies including DYS1 and DYS2 (Novocastra) and desmin antibody (Abcam, Cambridge, MA, USA).
10. ECL plus kit (GE).
11. ECL and autography film (GE).
12. ImageJ software (NIH, Bethesda, MD, USA).

2.11. Clinical Grading of Dogs

1. Video camera.
2. Stop watch.

3. Methods

3.1. Design of Antisense Oligos

1. Identify ESE sites in exons using Rescue ESE and ESEfinder.
2. Design antisense sequences to target ESEs of exon 6 (Ex6A) and exon 8 (Ex8A), or exon/intron boundary between exon 6 and intron 6 (Ex6B), or between exon 8 and intron 8 (Ex8B) (see Note 2).
3. Select antisense oligonucleotide chemistries. 2'OMePS is preferred for myoblast experiment (see Note 3). PMOs are used for in vivo studies (see Note 1).

3.2. Transfection of Antisense 2'OMePS into Dog Myoblasts

1. Use standard preplating method to obtain primary myoblast cells from neonatal CXMD dogs (10).
2. Culture WT or CXMD myoblasts (1.5×10^5 cells) in growth medium containing F-10, FCS (20%), bFGF (2.5 ng/mL),

penicillin (200 U/mL), and streptomycin (200 μg/mL) for 72 h, on six-well plates.

3. Dilute lipofectin to a total of 100 mL in opti-MEM media at a ratio of 2:1 for lipofectin: RNA (Use 10 mL lipofectin for 5 mg RNA).
4. Allow to stand still at RT for 30–45 min, then dilute AOs to a final volume of 100 μL in opti-MEM media.
5. Combine diluted lipofectin and AOs and mix gently.
6. Incubate at RT for 10–15 min.
7. Remove serum-containing medium from cells and wash them with opti-MEM reduced serum media.
8. Add 0.8 mL opti-MEM media to the tube containing the lipofectin DNA complexes.
9. Mix gently and overlay the complex onto the cells.
10. Return cells to the incubator and after 3 h replace opti-MEM media with differentiation medium and wait 3–10 days until they differentiate into myotubes.

3.3. Intramuscular Injections of Antisense Oligos in Dogs

1. Induce general anesthesia by 20 mg/kg of thiopental sodium injections and maintain by isoflurane inhalation (2.0–3.0%).
2. Cut skin above tibialis anterior (TA) muscle with scalpel.
3. Stitch the fascia of TA muscles at two different points at 2 cm intervals as markers; i.e., inner side distal/outer side proximal.
4. Bend needles (10°) to inject PMOs horizontally using 27 G needle, inject PMO solutions slowly into muscles and wait 1 min before removing the needle to prevent leakage.
5. Inject butorphanol tartrate (0.2 mg/kg) before and after procedure.
6. Administer sepham antibiotics (Cefamezine or Syncl) for three days after surgical procedures.

3.4. Systemic Injections of Antisense Morpholinos

1. Dissolve 120–200 mg/kg of morpholinos Ex6A, Ex6B, and Ex8A at 32 mg/mL in saline.
2. Inject them into saphenous vein of a dog using 22 G indwelling needles for each injection using infusion pumps to inject at 50 mL/20 min.
3. Inject morpholinos for 5–11 times at weekly or biweekly intervals.

3.5. RNA Extraction from Myotubes

1. Remove medium.
2. Put 1 mL Trizol for each well of six-well plates.
3. Wait 10 min.

4. Add 200 μL of chloroform (for RNA)
5. Shake well.
6. Wait 2 min.
7. You can see three layers including the RNA layer (top), DNA layer (middle), and protein layer (bottom).
8. Centrifuge at 12,000×*g* for 15 min at 4°C.
9. Take 400 mL carefully from top layer. Remove supernatant from the top layer, and put in another tube.
10. Add 500 mL isopropanol.
11. Keep in –80°C for O/N.
12. Centrifuge at 12,000×*g*, 10 min, 4°C.
13. Decant fluid. You can see a pellet of RNA in bottom.
14. Wash with 75% EtOH.
15. Centrifuge at 8,000×*g*, 5 min, 4°C.
16. Dry up, keep upside down for 15 min or O/N.
17. Add 15–30 μL water, then quantify RNA concentration.

3.6. RT-PCR

1. Make Reaction mix containing 1.5 μL 10 mM forward primer, 1.5 μL 10 mM reverse primer, 1 μL dNTP, 5 μL one-step PCR kit buffer, 0.7 μL RNAsin, 1 μL enzyme mixture from one-step PCR kit, and 200 ng RNA and add water to the total of 25 μL.
2. Perform RT-PCR in the thermocycler with 1 cycle of 50°C 30 min, 1 cycle of 95°C 15 min, 35 cycles of 94°C 1 min, 60°C 1 min and 72°C 1 min. Finally add 1 cycle of 72°C 10 min and then store PCR product in 4°C.

3.7. cDNA Sequencing

1. Use Qiagen gel extraction kit to excise the band of interest for subsequent cDNA sequencing according to manufacturer's instructions. Exon 6-9 skipped band (101 bp) is identified by electrophoresis using 2% agarose gel (Fig. 1).
2. Use BigDye® Terminator v3.1 cycle sequencing kit for cDNA sequencing with the same primers following manufacturer's instructions (Fig. 1).

3.8. Muscle Sampling from Necropsy of Dogs

1. Inject with thiopental sodium for induction of general anesthesia, then maintain anesthetic status by isoflurane.
2. Euthanize dogs by bleeding from the carotid artery.
3. Collect following muscles by necropsy of dogs 2 weeks after final injection of oligos. These muscles include TA, extensor digitorum longus (EDL), Gastrocnemius, soleus, biceps femoris, rectus femoris, biceps brachii, triceps brachii, deltoid, extensor carpi ulnaris (ECU), extensor carpi radialis (ECR),

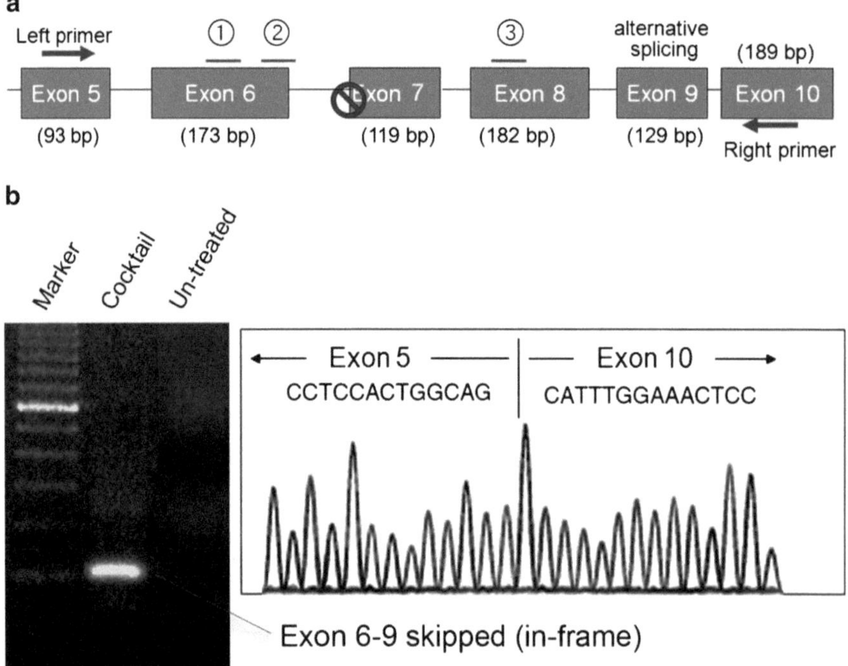

Fig. 1. Multiple exon skipping in dystrophic dogs. (**a**) Schematic outline of the protocol. A splice site mutation in intron 6 leads to deletion of exon 7 at mRNA level in dystrophic dogs. To restore the reading frame, two additional exons (exon 6 and exon 8) need to be skipped (removed) by three oligo cocktail of antisense. Exon 9 is known as alternative splice site. (**b**) RT-PCR and cDNA sequencing after exon skipping in dystrophic dogs. *Left panel*; RT-PCR reveals exon 6–9 skipped in-frame products (101 bp) in dystrophic dogs after the treatment of cocktail oligos. Alternative splice site Exon 9 is also mostly removed from the resulting mRNA. *Right panel*; Exon-skipping patterns are further confirmed by cDNA sequencing.

flexor carpi ulnaris (FCU), flexor carpi radialis (FCR), gracilis, intercostal, abdominal muscles, diaphragm, lateral dorsi, esophagus, sternocleidomastoid, and the heart.

4. Dissect muscles into small portion to stand on cork disks (1.2 cm diameter) labeled with the ID of the animal and muscle name on the back side.
5. Mix a portion of tragacanth gum (10–20 mL) well with equal amount of water until it becomes soft and sticky. Put them into 10 mL or 25 mL syringes. Unused gum in the syringe can be stored in freezer.
6. Put tragacanth gum to fix the muscle specimen on cork disks.
7. Put liquid nitrogen in a metal container and isopentane in a smaller metal container.
8. Lower the isopentane with the container into the liquid nitrogen. Wait for a couple of minutes until it becomes slushy and ready for freezing.
9. Put a portion of gum on the cork.

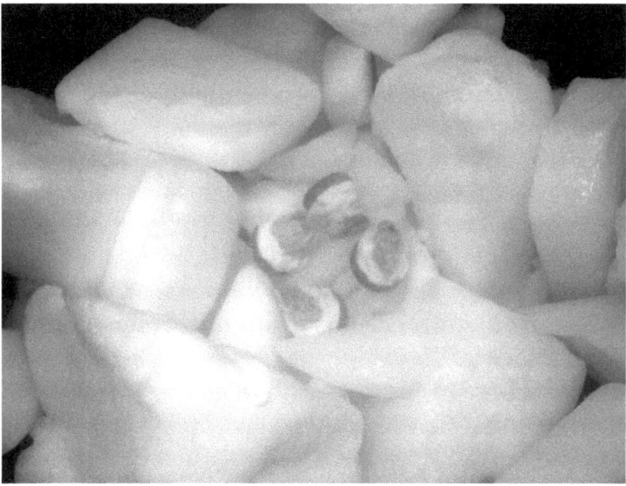

Fig. 2. Frozen muscle samples from dystrophic dogs.

10. Dissect out muscles and put it on the cork at RT. Place it on the cork longitudinally and put some gum blob around the bottom of them so that the longitudinal axis of the muscle is perpendicular to the cork and stable.
11. Place the muscle on cork into the cooled isopentane and shake vigorously for 1 min.
12. Place it on dry ice (Fig. 2).
13. Put samples in a glass vials, and store at –80°C.

3.9. Immunostaining for Dog Muscles

1. Set up cryostat for sectioning. The working temperature should be –25°C. Set the section thickness at 8 μm for immunohistochemistry and 12 μm for HE staining. Put in a blade.
2. Place muscle blocks on dry ice for transportation.
3. Label slide glasses in pencil with animal IDs, cut date, and muscle name.
4. Mount cork with muscle sample block and fix in place with water. Attach the chuck with tissue specimen onto the holder.
5. Start slicing the muscle until approximately one fourth of the way in the muscle.
6. Touch and transfer individual sections onto RT slide glass and leave at RT to dry.
7. Place every sixth section on the same slide (sections 1, 6, 11 on slide #1; sections 2, 7, 12 on slide #2) and cut them at interval of every 200 mm until you have five sections collected per slide. Keep sections clustered as closely as possible to reduce the amount of antibody solutions required.

8. When finished, allow slides to dry at RT for at least an additional 90 min. Slides can be stored at –80°C.
9. For immunohistochemistry, put slides in moisture chamber (and dry them for 30 min if they were stored in a freezer).
10. Blocking; 2 h in PBS with 15% goat serum at RT.
11. Incubate with a primary antibody; antidystrophin rod (DYS-1) or C-terminal monoclonal antibody (DYS-2) for dog dystrophin staining (1:150 dilutions) for overnight at 4°C.
12. Wash with PBS 5 min × 3 times.
13. Incubate with a secondary antibody, Alexa 594 goat antibody against mouse IgG_1 or IgG_2 (highly cross-absorbed) (1:2,500) for 30 min at RT.
14. Wash with PBS 5 min × 5 times.
15. Wipe off excess liquid and mount with DAPI-containing mounting agent for nuclear staining and then put cover glasses.
16. Count the number of positive fibers for DYS1 under fluorescent microscope and compare in sections where their biggest number of the positive fibers were as previously described (see Note 4) (11).
17. Immunohistochemistry is also applicable for myotubes. Use slide glasses with chambers to culture them and fix them by 4% PFA for 10 min (Fig. 3).

3.10. Western Blotting from Dog Muscles

1. Collect the 30–40 of cryo-sections of 15 μm in 1.5 mL tube on dry ice.
2. Add 150 mL of sample buffer and homogenize on ice.
3. Boil them for 3–5 min and centrifuge for 15 min at 16,500 × *g*.
4. Collect supernatant and keep the aliquot at –70°C.

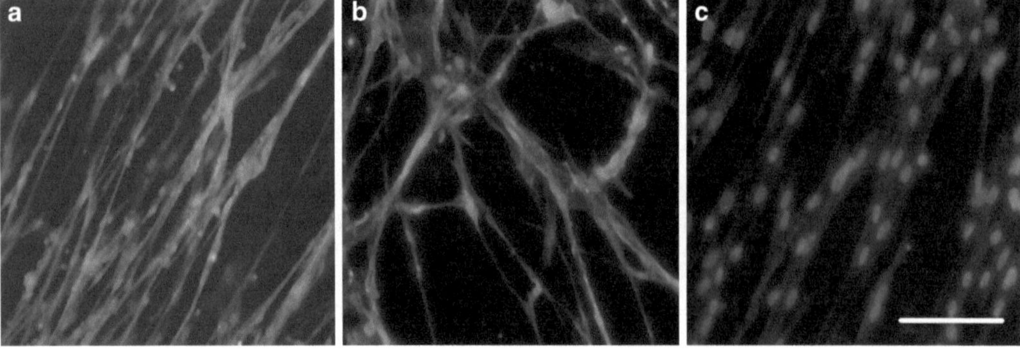

Fig. 3. Recovery of dystrophin expression after cocktail antisense transfection in dystrophic dog myoblast. Dystrophin expression and DAPI nuclear double staining of wild-type myotubes (**a**), cocktail of Ex6A, Ex6B, and Ex8A 2′OMePS transfected myotubes (**b**), and nontreated CXMD myotubes (**c**). Dystrophin C-terminal antibody DYS-2 is used. Bar: 50 μm.

5. Dilute an aliquot of protein 100-fold with distilled water to reduce final SDS concentration less than 0.1%. Measure protein concentration of the diluted protein sample with Bradford protein assay. Specifically, record the absorbance at 570 nm using a photospectrometer and calculate the concentration from standard curve.
6. For SDS-PAGE, set the glass plates for readymade mini-gel (5%).
7. Mix the samples with 2× Laemmli SDS-loading buffer.
8. Boil samples for 3 min, then load 20 mg of samples in each lane.
9. Run the gel at 150 V for approximately 3 h.
10. After running the gel, incubate the gel for 20 min in transfer buffer + 0.1 % SDS (optional for transferring high molecular weight proteins).
11. Wet four pieces of sponge and Whatman paper with ddH_2O, then soak them in the transfer buffer, and soak PVDF membrane using methanol for 1 min to prewet it, and then pour-off methanol and add H_2O, make sure that the membrane does not float. Leave it in water for 3 min.
12. Set the gel and membrane as shown in the manual.
13. Run 40–50 V o/n in cold room.
14. For blotting, prepare 2,000 mL of 0.05% PBS/Tween 20 (PBST). Wash the membrane briefly with 20 mL PBS.
15. Prepare 100 mL of PBST/5% milk powder, and incubate in 50 mL PBST/5% milk powder for 2 h.
16. Incubate the blot with primary antibody in the appropriate dilution with PBS/5% milk powder (1:100 dilution for Dys1 dystrophin antibody) for 1 h or O/N.
17. Wash the blot for 15 min each with 3× 100 mL PBST, then incubate the blot with the HRP conjugated secondary antibody.
18. Wash the blot for 20 min each with 3× 200 mL PBST.
19. Use ECL plus kit for detection. Mix two solutions at 40:1, and incubate with membrane for 1 min. Then use film and developer for the detection.
20. To preserve for the spare blot, rinse the blot with PBST, and store it in PBST at 4°C for a few weeks. Desmin antibody is used to normalize intersample loading amount. Signals are analyzed and quantified using Adobe Photoshop and ImageJ software (Fig. 4).

3.11. Clinical Grading of Dogs

1. Let a dog walk and evaluate gait disturbance: grade 1 = none, grade 2 = sitting with hind legs extended, grade 3 = bunny

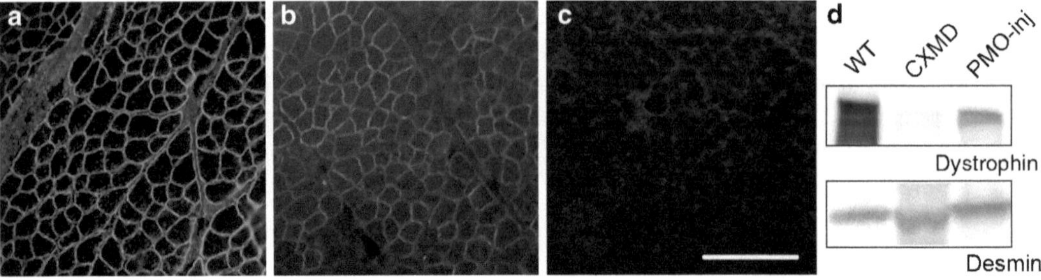

Fig. 4. Recovery of dystrophin expression after 7 × 200 mg/kg intravenous cocktail morpholino injections. Dystrophin expression of wild-type dog muscle (**a**), cocktail of Ex6A, Ex6B, and Ex8A PMOs injected dog muscle (**b**), nontreated CXMD dogs (**c**), Western blotting analysis with dystrophin antibody (**d**). Bar: 100 μm.

hops with hind legs, grade 4 = shuffling walk, and grade 5 = unable to walk (12).

2. Evaluate mobility disturbance: grade 1 = none, grade 2 = lying down more than normal, grade 3 = cannot jump on hind legs; grade 4 = increasing difficulty moving around, and grade 5 = unable to get up and move around.
3. Palpate limb or temporal muscle atrophy: grade 1 = none, grade 2 = suspect hardness, grade 3 = can feel hardness or apparently thin, grade 4 = between grades 3 and 5, and grade 5 = extremely thin or hard.
4. Evaluate drooling: grade 1 = none, grade 2 = occasionally dribbles saliva when sitting, grade 3 = some drool when eating and drinking, grade 4 = strings of drool when eating or drinking, and grade 5 = continuous drool.
5. Evaluate macroglossia: grade 1 = none, grade 2 = slightly enlarged, grade 3 = extended outside dentition, grade 4 = enlarged and slightly thickened, and grade 5 = enlarged and thickened.
6. Evaluate dysphagia: grade 1 = none; grade 2 = takes time and effort in taking food, grade 3 = difficulty in taking food from plate, grade 4 = difficulty in chewing, swallowing, or drinking, and grade 5 = unable to eat.
7. Add up the total score.
8. For running test, encourage each dog to run one time for 15 m, and record elapsed time.

4. Notes

1. Alternatively, one can also use 2'O-MePs (Eurogentec) against exons 6 and 8 of the dog dystrophin gene. These include Ex6A (GUUGAUUGUCGGACCCAGCUCAGG),

Ex6B (ACCUAUGACUGUGGAUGAGAGCGUU), and Ex8A (CUUCCUGGAUGGCUUCAAUGCUCAC).

2. The efficacy of antisense oligos is highly unpredictable, and hence several antisense oligos should be designed for each target exon. A preferred antisense sequence contains 40–60% of GC, does not have more than three consecutive guanine, and does not lead to self dimers or hetero dimers when injected as a cocktail. We have designed more than ten antisense sequences against exon 6 and exon 8 of dogs, and optimized the most efficient combination of cocktail antisense oligos both in vitro and in vivo (Saito et al., Unpublished).
3. For 2′OMePS, U (uracil) is used instead of T (thymidine).
4. Occasionally dystrophin-positive revertant fibers can be detected in dystrophic dog muscles (5, 12). Revertant fibers cannot be distinguished from antisense-mediated dystrophin expression by immunohistochemistry unless an epitope-specific antibody is used. Therefore, the expression level should be carefully compared with untreated controls.

Acknowledgments

Authors thank Drs. Terence Partridge, Stephanie Duguez (Children's National Medical Center, Washington DC), Masanori Kobayashi, Yoshitsugu Aoki, Takashi Saito, Katsutoshi Yuasa, Naoko Yugeta, Sachiko Ohshima, Jin-Hong Shin, Michiko Wada, Kazuhiro Fukushima, Satoru Masuda, Kazue Kinoshita, Hideki Kita, Shin-ichi Ichikawa, Yumiko Yahata, Takayuki Nakayama, Akinori Nakamura (National Institute of Neuroscience, Tokyo, Japan), Adam Rabinowitz, and Jonathan Beauchamp (Imperial College, London, UK), Qi-long Lu (Carolinas Medical Center) for discussions and technical assistance. This work was supported by the Foundation to Eradicate Duchenne, the Department of Defense CDMRP program, the Jain Foundation, the Crystal Ball of Virginia Beach (Muscular Dystrophy Association USA), the National Center for Medical Rehabilitation Research, the NIH Wellstone Muscular Dystrophy Research Centers, and the Ministry of Health, Labor, and Welfare of Japan (Research on Nervous and Mental Disorders, 16B-2, 19A-7; Health and Labor Sciences, Research Grants for Translation Research, H19-translational research-003, Health Sciences Research Grants Research on Psychiatry and Neurological Disease and Mental Health, H18-kokoro-019).

References

1. Hoffman, E. P., Brown, R. H., Jr., and Kunkel, L. M. (1987) Dystrophin: the protein product of the Duchenne muscular dystrophy locus. *Cell 51*, 919–928.
2. Yokota, T., Duddy, W., and Partridge, T. (2007) Optimizing exon skipping therapies for DMD. *Acta Myol 26*, 179–184.
3. Yokota, T., Takeda, S., Lu, Q. L., Partridge, T. A., Nakamura, A., and Hoffman, E. P. (2009) A renaissance for antisense oligonucleotide drugs in neurology: exon skipping breaks new ground. *Arch Neurol 66*, 32–38.
4. Yokota, T., Pistilli, E., Duddy, W., and Nagaraju, K. (2007) Potential of oligonucleotide-mediated exon-skipping therapy for Duchenne muscular dystrophy. *Expert Opin Biol Ther 7*, 831–842.
5. Yokota, T., Lu, Q. L., Partridge, T., Kobayashi, M., Nakamura, A., Takeda, S., and Hoffman, E. (2009) Efficacy of systemic morpholino exon-skipping in Duchenne dystrophy dogs. *Ann Neurol 65*, 667–676.
6. Lu, Q. L., Rabinowitz, A., Chen, Y. C., Yokota, T., Yin, H., Alter, J., Jadoon, A., Bou-Gharios, G., and Partridge, T. (2005) Systemic delivery of antisense oligoribonucleotide restores dystrophin expression in body-wide skeletal muscles. *Proc Natl Acad Sci USA 102*, 198–203.
7. Shimatsu, Y., Katagiri, K., Furuta, T., Nakura, M., Tanioka, Y., Yuasa, K., Tomohiro, M., Kornegay, J. N., Nonaka, I., and Takeda, S. (2003) Canine X-linked muscular dystrophy in Japan (CXMDJ). *Exp Anim 52*, 93-97.
8. Sharp, N. J., Kornegay, J. N., Van Camp, S. D., Herbstreith, M. H., Secore, S. L., Kettle, S., Hung, W. Y., Constantinou, C. D., Dykstra, M. J., Roses, A. D., and et al. (1992) An error in dystrophin mRNA processing in golden retriever muscular dystrophy, an animal homologue of Duchenne muscular dystrophy. *Genomics 13*, 115–121.
9. Ham, R. G. (1963) An improved nutrient solution for diploid Chinese hamster and human cell lines. *Exp Cell Res 29*, 515–526.
10. Jankowski, R. J., Haluszczak, C., Trucco, M., and Huard, J. (2001) Flow cytometric characterization of myogenic cell populations obtained via the preplate technique: potential for rapid isolation of muscle-derived stem cells. *Hum Gene Ther 12*, 619–628.
11. Yokota, T., Lu, Q. L., Morgan, J. E., Davies, K. E., Fisher, R., Takeda, S., and Partridge, T. A. (2006) Expansion of revertant fibers in dystrophic mdx muscles reflects activity of muscle precursor cells and serves as an index of muscle regeneration. *J Cell Sci 119*, 2679–2687.
12. Shimatsu, Y., Yoshimura, M., Yuasa, K., Urasawa, N., Tomohiro, M., Nakura, M., Tanigawa, M., Nakamura, A., and Takeda, S. (2005) Major clinical and histopathological characteristics of canine X-linked muscular dystrophy in Japan, CXMDJ. *Acta Myol 24*, 145–154.

Chapter 21

Whole Body Skeletal Muscle Transduction in Neonatal Dogs with AAV-9

Yongping Yue, Jin-Hong Shin, and Dongsheng Duan

Abstract

Gene therapy of muscular dystrophy requires systemic gene delivery to all muscles in the body. Adeno-associated viral (AAV) vectors have been shown to lead to body-wide muscle transduction after a single intravascular injection. Proof-of-principle has been demonstrated in mouse models of Duchenne muscular dystrophy and limb girdle muscular dystrophy. Before initiating clinical trials, it is important to validate these promising results in large animal models. More than a dozen canine muscular dystrophy models have been developed. Here, we outline a protocol for performing systemic AAV gene transfer in neonatal dogs. Implementing this technique in dystrophic dogs will accelerate translational muscular dystrophy research.

Key words: AAV: adeno-associated virus, Muscular dystrophy, Alkaline phosphatase, Gene therapy, Dog, Systemic gene transfer

1. Introduction

Gene therapy for muscle diseases faces a unique challenge. Since muscle is widely distributed throughout the body, a successful muscle gene therapy would require body-wide transduction. This barrier was surmounted in rodent models recently (1–4, 20). By exploring the unique capsid property of different adeno-associated virus (AAV) variants, investigators have achieved robust whole-body transduction in normal and dystrophic rodents with AAV-6, 8, and 9 (1–4, 20). The key issue now is whether a single intravenous AAV injection can transduce whole body muscle in a large animal model and ultimately in human patients.

Exploring systemic gene delivery in large animal models is extremely relevant in muscular dystrophy gene therapy research. The lack of the characteristic dystrophic phenotype has greatly

Dongsheng Duan (ed.), *Muscle Gene Therapy: Methods and Protocols*, Methods in Molecular Biology, vol. 709, DOI 10.1007/978-1-61737-982-6_21, © Springer Science+Business Media, LLC 2011

limited translational implication of rodent study results. On the other hand, the canine muscular dystrophy models display the typical clinical manifestations seen in human patients (5). It is expected that the results from studies performed in the dog models may yield a more accurate prediction on the potential outcomes of human trials. Duchenne muscular dystrophy (DMD) is by far the most common form of muscular dystrophies. Duchenne-like muscular dystrophy has been observed in more than a dozen canine breeds. Some of these canine models have been fully characterized at the genetic, biochemical, and clinical levels and experimental colonies have been established.

In this protocol, we outline the procedures for performing systemic AAV delivery in neonatal dogs. Traditionally, DMD patients are diagnosed between 2 and 5 years of age, when they cannot reach motor skill milestones (6). At this time, the progression of muscle diseases already results in clinically evident damage. Therapies initiated after this time may have already missed the best treatment window. Affordable chemical and molecular tests have been developed for neonatal DMD diagnosis (7). As a matter of fact, newborn DMD screening has been implemented in many places (8–10). Neonatal gene therapy would offer more to the parents of a newly diagnosed case than the medical, emotional, and reproductive consulting now contemplated. Since DMD is uniformly lethal, the FDA has indicated that it will apply special regulations (FDA regulation Title 21, Chapter I, Subchapter D, Part 312, Subpart E, Sections 312.80-84). In particular, the FDA "takes into consideration the severity of the disease and the absence of satisfactory alternative therapy" and the FDA will "exercise the broadest flexibility in applying the statutory standards." The lessons learned from neonatal canine gene therapy studies will be important stepping-stones to achieve systemic therapy in symptomatic patients in the future.

Besides detailing the neonatal systemic AAV delivery procedure, we also provide methods on newborn dog generation, muscle biopsy and necropsy, and transduction efficiency evaluation using the heat-resistant human placental alkaline phosphatase (AP) reporter gene.

2. Materials

2.1. Delivering an AAV-9 AP Vector Through the Jugular Vein in Newborn Dogs

1. Recombinant AAV-9 vector. In this protocol, we describe systemic AAV-9 delivery in newborn dogs using the RSV.AP vector. In this vector, AP gene expression is under the transcription regulation of the ubiquitous Rous sarcoma virus promoter (RSV) and the SV40 polyadenylation signal (see Note 1) (11).

2. Female dogs (see Note 2).
3. Semen from normal or affected dogs (see Note 3).
4. Ultramark 4 plus ultrasound machine (ATL-Philips, Andover, MA, USA) (see Note 4).
5. Xplorer X-ray machine (Imaging Dynamics Calgary, AB, Canada) (see Note 4).
6. Vicks digital rectal thermometer (Kaz, Hudson, NY, USA) (see Note 5).
7. Whelping facility. It includes a whelping box (The Jonart Whelping Box, Ada, MI, USA), white blankets, and an infrared heating lamp (Erin's Edge Dog Supply, Richfield, WI, USA) (Fig. 1).
8. Sharpie marker pen (Sanford, Oak Brook, IL, USA).
9. Braun electric shaver (Procter & Gamble, Cincinnati, OH, USA).
10. PetAg Esbilac puppy milk replacer (PETCO Animal Supplies, San Diego, CA, USA).
11. DNA extraction buffer: 50 mM Tris–HCl, pH 8.0, 100 mM EDTA, 100 mM NaCl, 1% SDS, 0.5 mg/mL proteinase K.

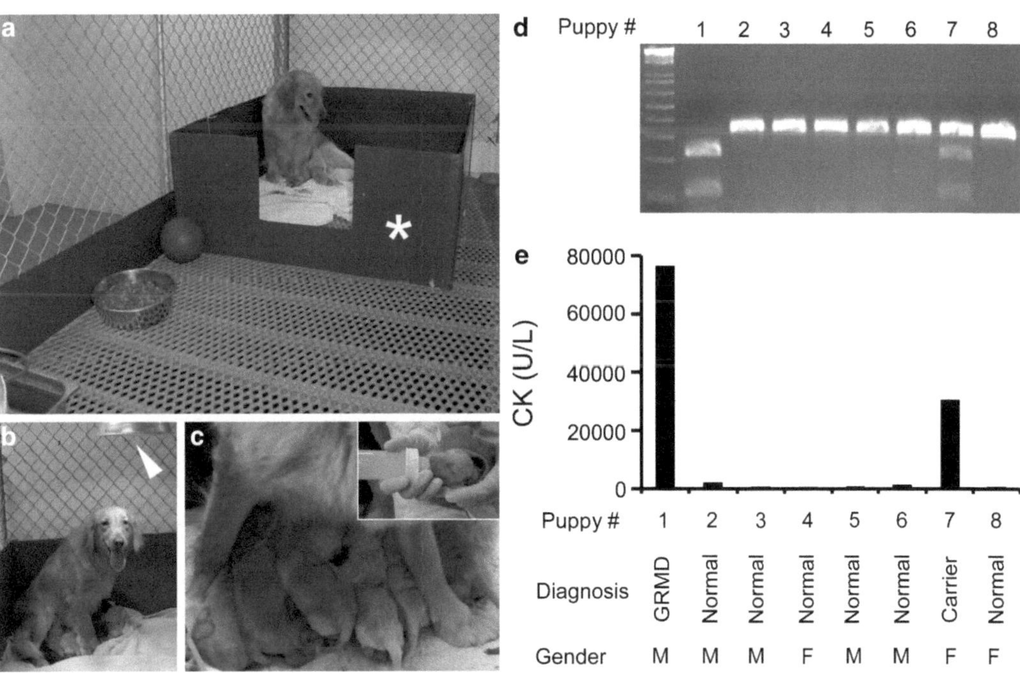

Fig. 1. Perinatal care and the GRMD genotyping. (**a**) Whelping room. *Asterisk*, whelping box. (**b**) A carrier dam and her puppies. *Arrowhead*, infrared heating lamp. (**c**) Breast feeding. *Insert*, bottle feeding for affected puppies. (**d**) Genotyping of a litter by PCR/Sau961 restriction fragment length polymorphism. Wild type allele, 328 bp; GRMD allele, 221 and 107 bp. (**e**) CK value from the same set of the litter. Puppy #1 is an affected dog. Puppy #7 is a carrier. Remaining puppies are normal dogs.

12. Other reagents for DNA extraction: 6 M NaCl, isopropanol, and 70% ethanol.
13. Model 5417C Eppendorf centrifuge (Eppendorf, Hamburg, Germany).
14. PCR primers for the GRMD genotyping. Forward primer, CTTAAGGAATGATGGGCATGGG; Reverse primer, ATGCATAGTTTCTCTTTCATGC (see Note 6) (Fig. 1) (12).
15. Green GoTaq Flexi PCR kit (Promega; Madison, WI, USA). The kit contains 5× Green GoTaq Flexi buffer and GoTaq DNA polymerase.
16. Eppendorf mastercycler personal PCR machine (Eppendorf).
17. *Sau*96I restriction enzyme (New England Biolab, Ipswich, MA, USA).
18. Betadine scrub (SmartPak Equine, Plymouth, MA, USA).
19. Heparin sodium (1,000 USP units/mL) (Abraxis Pharmaceutical Products, Schaumburg, IL, USA).
20. A 23G butterfly (Covidien, Mansfield, MA, USA).
21. A 10 mL syringe.
22. 0.9% Sodium chloride (Baxter Health Care Corporation, Deerfield, IL, USA).
23. Surgical supplies including sterilized gauze (Tyco Healthcare; Princeton, NJ, USA), 70% isopropyl rubbing alcohol (Medichoice, Mechanicsville, VA, USA), and sterilized scissors (World Precision Instruments, Sarasota, FL, USA).
24. Digital postal scale (Harbor Freight, Camarillo, CA, USA).
25. Thermophore heating pad (Medwing, Columbia, SC, USA).

2.2. Evaluating AP Expression

1. Preanesthetic cocktail: atropine sulfate (0.04 mg/kg), acepromazine maleate (0.02 mg/kg), and butorphanol tartrate (0.4 mg/kg) (IVX Animal Health, St. Joseph, MO, USA).
2. Anesthetic agents: propofol (up to 3 mg/kg) (IVX Animal Health), isoflurane (Halocarbon, River Edge, NJ, USA).
3. Veterinary anesthesia machine with ventilator and vaporizer (KEEBOSHOP, Chicago, USA).
4. 20–22G IV catheter (Becton-Dickinson Medical Supply, Franklin Lakes, NJ, USA).
5. Lactated Ringer's solution (Valley Vet Supply, Marysville, KS, USA).
6. 4% Chlorhexidine scrub (First Priority, Elgin, IL, USA).
7. Cephazolin, a generic antibiotic drug.
8. Muscle biopsy instruments: Sterile forceps and scissors (World Precision Instruments), sterile needle holders (Accurate

Surgical & Scientific Instruments Corp., Westbury, NY, USA). #10 Bard-Parker stainless steel surgical blades (Becton-Dickinson Medical Supply). 4-0 Vicryl suture and 4-0 Ethilon suture (Ethicon, Somerville, NJ, USA).

9. Carprofen (VÉTOQUINOL USA, Fort Worth, TX, USA).
10. Euthasol solution CIII (Virbac, Fort Worth, TX, USA): a veterinary euthanasia solution containing pentobarbital sodium and phenytoin sodium.
11. Stethoscope.
12. TBJ necropsy table (LABES of MA, Worcester, MA, USA).
13. Muscle necropsy instruments: Curved and straight scissors, tissue forceps, and hemostats (Biomedical Research Instruments, Silver Spring, MD, USA); tissue trimming knives and blades (Baxter Scientific, Columbia, MD, USA); 6 in. necropsy knife, rongeurs, bone cutting shears, scalpel blades, and handles (National Logistics Services, Westfield, MA, USA); heavy bone chisel, Virchow's skull breaker, bone cutting saw, polyethylene dissecting boards (Thermo Shandon, Pittsburgh, PA, USA).
14. 2-Methylbutane.
15. Liquid nitrogen.
16. Tissue-Tek OCT (Sakura Finetek USA, Torrance, CA, USA).
17. Microm HM505 EVP cryostat (Richard Allan Scientific, Kalamazoo, MI, USA).
18. 0.5% Glutaraldehyde in phosphate buffered saline (PBS).
19. PBS containing 1 mM $MgCl_2$.
20. AP prestaining buffer: 100 mM Tris–HCl, pH 9.5, 50 mM $MgCl_2$, 100 mM NaCl.
21. AP staining solution: 165 μg/mL 5 -bromo-4-chloro-3-indolylphosphate-p-toluidine (stock as 50 mg/mL in 100% dimethyformamide), 330 μg/mL nitroblue tetrazolium chloride (stock as 75 mg/mL in 70% dimethyformamide), 50 μM levamisol (13). Prepare at the time of use.
22. Monoclonal antibody for AP (1:8,000; Sigma, St. Louis, MO, USA).
23. Alex 594-conjugated rabbit antimouse antibody (Invitrogen, Grand Island, NY, USA).
24. Thermo Aqua-mount (Fisher Scientific, Waltham, MA, USA).
25. StemTAG AP activity assay kit (Cell Biolabs, San Diego, CA, USA). The kit includes the StemTAG AP activity assay substrate, cell lysis buffer, 10× stop solution, and the AP activity assay standard.

26. Porcelain mortar and pestle.
27. Model 5417C Eppendorf centrifuge (Eppendorf).
28. Bio-Rad protein concentration assay kit (Bio-Rad, Heracules, CA, USA).

3. Methods

3.1. Delivering an AAV-9 AP Vector Through the Jugular Vein in Newborn Dogs

1. Dog breeding. Deliver the semen to the uterus of a bitch twice at 48 and 72 h postovulation by artificial insemination (see Notes 2 and 3). Record the body weight weekly. One month postartificial insemination, confirm the pregnancy with ultrasound. Count the number of fetus by X-ray at the 60th day of gestation (see Note 4).
2. Puppy delivery. Record the body temperature of the bitch three times a day starting from the 58th day of pregnancy. Prepare for labor and delivery when temperature drops below 99°C and keeps dropping (see Note 5).
3. Record the birth time, body weight, hair color, and gender of each puppy (Fig. 2). Write the birth order on the back of the puppy with a marker pen. Shave ~1 cm^2 of the hair at the defined locations (such as the left shoulder, the right hip, etc.) as an additional marker (Fig. 2). For each puppy, crop a 0.5-in. long cord with a pair of sterile scissors. Place the puppy next to one of the nipples to allow it start nursing. Record the total number of puppies and placentas delivered. Make sure every puppy gets some colostrum (the milk from the dam in the first 24 h). Draw blood from the jugular vein and measure the serum creatine kinase level and liver function. Consult a veterinary doctor immediately in the case of difficult labor.

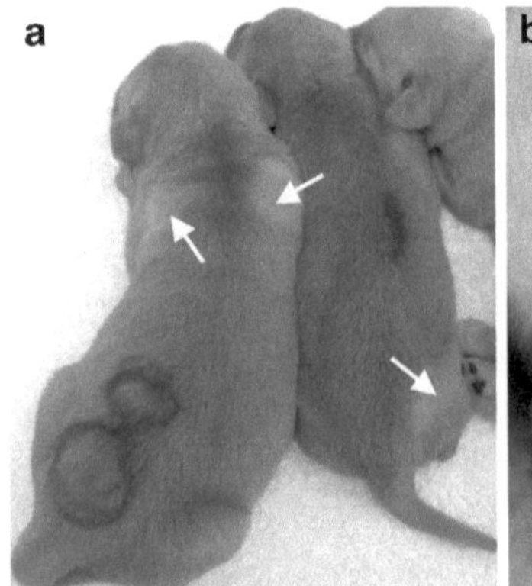

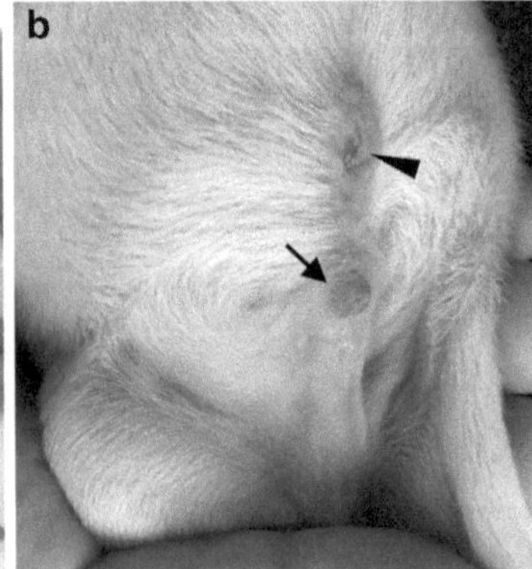

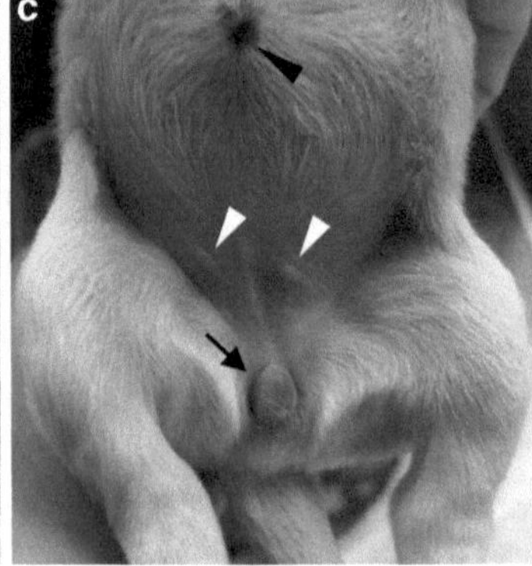

Fig. 2. Neonatal puppy images. (**a**) Newborn puppy identification. *Arrows*, areas in which body hair is shaved. (**b**) A newborn male puppy. *Arrow*, the male external genital; *arrowhead*, the umbilicus. (**c**) A newborn female puppy. *Arrow*, the female external genital; *black arrowhead*, the umbilicus; *white arrowheads*, nipple.

Carefully monitor the general health of the dam and puppies. Record the body weight of newborn puppies three times a day in the first 2 weeks and then twice a day in weeks 3 and 4. Start supplementing with the Esbilac puppy formula if the puppy is weak or does not nurse well (see Note 7).

4. Mince the cord with a razor blade. Add 750 μL DNA extraction buffer. Incubate at 55°C overnight on a spinning wheel. Add 250 μL 6 M NaCl. Incubate at room temperature for 1 h on a spinning wheel. Centrifuge at 16,000 × *g* for 30 min. Transfer the supernatant to a new tube. Add 600 μL isopropanol. Mix for 5 min by shaking. Centrifuge at 16,000 × *g* for 2 min. Wash the pellet with 70% ethanol. Centrifuge at 16,000 × *g* for 2 min. Air dry the pellet for 1 h. Resuspend the pellet in 50 μL PCR-quality water as the cord genomic DNA. Mix 0.5 μL cord DNA, 4 μL of 5× green GoTaq Flexi buffer, 1.6 μL of 25 mM $MgCl_2$, 0.4 μL of 10 mM dNTP, 0.5 μL of each primer (20 μM stock), 0.1 μL of GoTaq DNA polymerase, and 12.9 μL PCR-quality water. Start PCR with initial denaturation at 95°C for 2 min. Continue PCR with 35 cycles of 95°C 20 s denaturation, 58°C 20 s annealing, and 72°C 20 s extension. Digest 10 μL PCR product with 2 units of Sau96I at 37°C for 2 h. Examine in 2% agarose gel. Wild type allele yields a 328 bp band. Affected allele yields two bands at 221 and 107 bp (see Note 8) (Fig. 1).

5. Perform systemic AAV-9 delivery at 24–48 h after birth in conscious puppies. Calculate AAV-9 vector volume based on the body weight at the time of injection. A normal newborn puppy may tolerate up to 25 μL/g body weight (2.5×10^{11} vector genome particles/gram body weight) of AAV-9 vector. Prefill a 23G butterfly with heparinized saline (or prefill a 23G heparin-coated butterfly with saline) and then connect the butterfly to a 10 mL syringe containing the appropriate volume of AAV. One investigator should restrain the fore limbs and the head of the puppy with hands. Clean the neck skin with the Betadine scrub. Insertion of the butterfly to the jugular vein should be done by another investigator. Gently pull the plunger back and visualize the flashback of blood to confirm if the butterfly is correctly placed inside vein. Infuse AAV to the puppy at the speed of ~5 mL/min. Flush the butterfly with heparinized saline to deliver remaining AAV that is trapped in the connecting tubing. Press the needle entry site with a finger and at the same time retract the butterfly from the jugular vein. Continuously press the needle injection site to stop bleeding. Place the puppy on the heating pad to let it recover (see Note 9) (Fig. 3).

6. Carefully monitor the vital sign, general response, nursing, and body weight of AAV injected puppies. If they become sluggish, not gaining weight, dehydrated, or not nursing well, consult the veterinary doctor immediately (see Note 10).

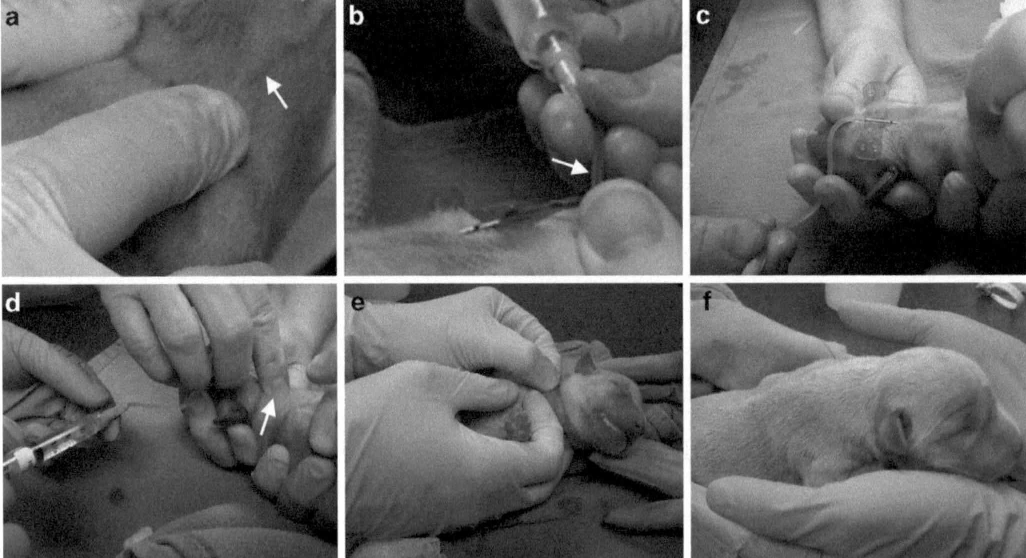

Fig. 3. The procedure for delivering AAV into a newborn puppy. (**a**) Reveal jugular vein by applying pressure on the clavicle bone. *Arrow*, jugular vein. (**b**) Confirm the correct position of the butterfly inside the jugular vein. *Arrow*, the flash back of blood. (**c**) AAV delivery. (**d**) Flush in AAV vectors inside the tubing and retrieve the butterfly while applying pressure on the needle penetration site. *Arrow*, needle penetration site. (**e**) Continuously apply pressure to the injection site to stop bleeding. (**f**) Puppy recovered after systemic AAV injection.

3.2. Evaluating AP Expression

1. Muscle biopsy. Withhold food for 12 h. Administer the preanesthetic drug cocktail via intramuscular injection. Ten minutes later, administer propofol via a bolus intravenous injection at the dose of 1.5 mg/kg. Administer additional propofol if needed (up to 3 mg/kg). Maintain anesthesia status with 2–3% of isoflurane. Place a 20–22G IV catheter to one of the major veins (such as the jugular, cephalic, femoral, or saphenous vein). Maintain on lactated Ringer's solution at a flow rate of 50 mL/kg/day for the duration of the procedure. Shave the body hair and scrub at least three times each with 4% chlorhexidine scrub and 70% ethanol before draping the incision site for aseptic surgery. Administer prophylactic antibiotic cephazolin at 22 mg/kg via IV catheter. Make a skin incision over the surface of the target muscle. Gently separate fat and connective tissues to reveal the target muscle. Remove a block of 1 × 0.5 × 0.5 cm muscle tissue with a surgical blade. Close the fascial layer with continuous stitch using absorbable 4-0 Vicryl suture. Close the skin with interrupted stitches using nonabsorbable 4-0 Ethilon suture. Administer carprofen twice a day at the dosage of 2.2 mg/kg for the first 24 h postsurgery to reduce postoperative pain. Provide dog with soft food after surgery. Monitor skin incision daily until it is completely healed (see Note 11) (Fig. 4).
2. Dog necropsy. Record experimental information (such as dog name, body weight, gender, project title, etc.). Collect the serum from the jugular vein. Euthanize the dog with intravenous

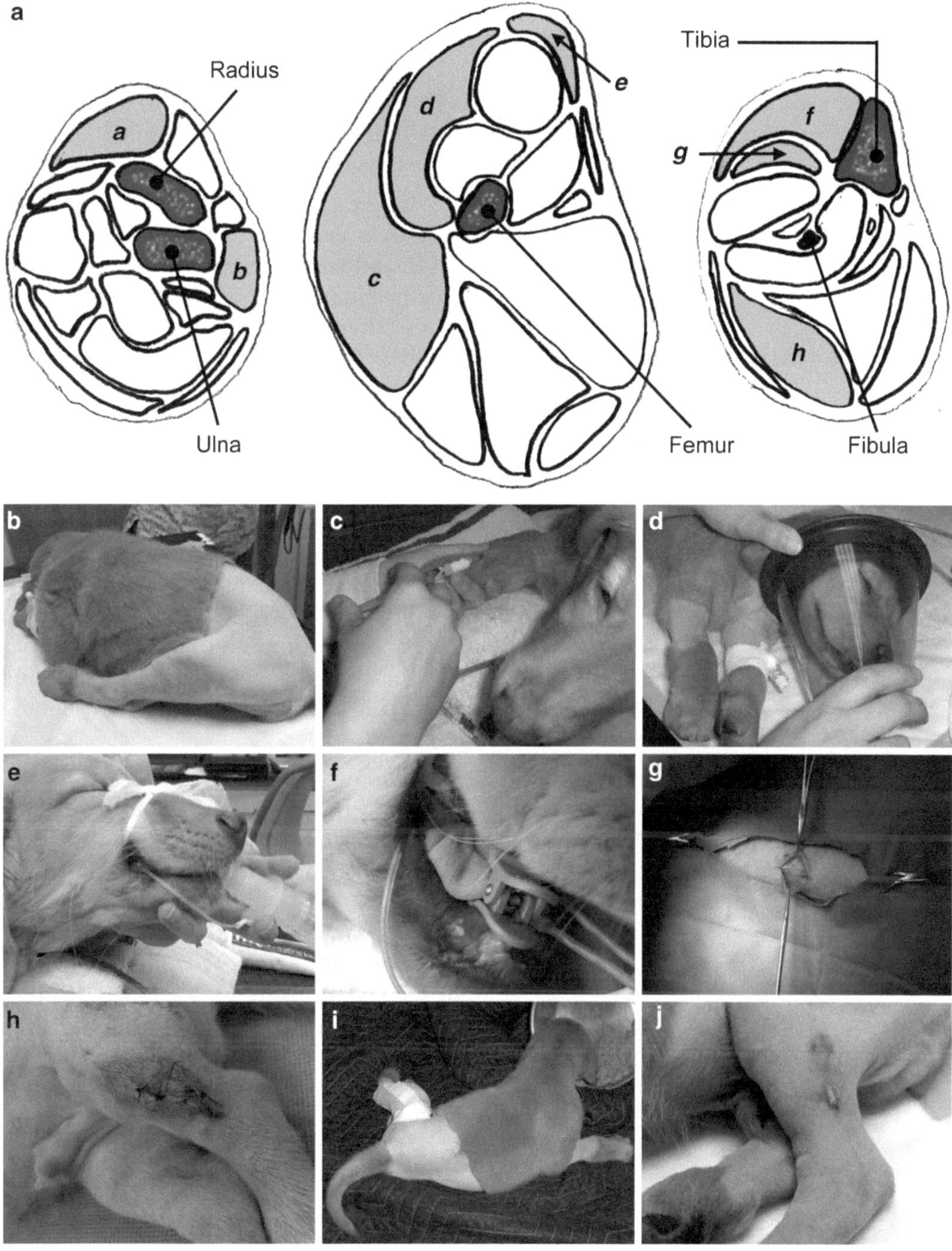

Fig. 4. Surgical muscle biopsy. (**a**) Schematic drawings of the lower forelimb (*left*), thigh (*middle*), and lower hind limb (*right*). Muscles commonly used for surgical biopsy are marked with small letters. *a* extensor carpi radialis; *b* extensor carpi ulnaris; *c* biceps femoris; *d* vastus laterialis; *e* cranial sartorius; *f* cranial tibialis; *g* long digital extensor; *h* the lateral head of gastrocnemius. (**b**) Prepare the skin for muscle biopsy and establish the intravenous route. (**c**) Inject preanesthetic drug cocktail. (**d**) Anesthetize the dog with isoflurane. (**e**) Intubate the trachea. (**f**) Clip the oximeter sensor to the tongue. (**g**) Make skin incision. (**h**) Suture the skin after removing the muscle. (**i**) Recover normal feeding after surgery. (**j**) Suture wound heal at 3 weeks postsurgery.

injection of Euthasol solution at the dose of 2 mL for the first 10 lb of the body weight and 1 mL per 10 lb for the remaining body weight. Confirm the death by cardiac auscultation with a stethoscope. Cut open the jugular vein to drain blood. Identify each skeletal muscle according to canine anatomy books (14–18). Harvest major body muscles and the internal organs (see Note 12).

3. Histochemical staining to evaluate AP expression. Embed the muscle sample in OCT and snap freeze the sample in liquid nitrogen-cooled 2-methylbutane (–155°C). Cut the muscle tissue block into 8 μm sections with a cryostat. Fix muscle sections in 0.5% glutaraldehyde for 10 min. Wash slides with 1 mM $MgCl_2$. Incubate at 65°C for 45 min. Wash with the AP prestaining buffer twice, 5 min each. Stain in the freshly prepared AP staining solution for 5–20 min (see Note 13) (Fig. 5).
4. Whole mount muscle AP histochemical staining. Cut the muscle to a thickness of ~0.2 in. and a size of ~1 × 1 in. Fix the muscle piece in 0.5% glutaraldehyde for 2 h at room temperature. Incubate with 1 mM $MgCl_2$ for 30 min at room temperature. Incubate at 65°C for 1.5 h to inactivate endogenous AP. Incubate with two exchanges of the AP prestaining buffer, 30 min each. Stain in freshly prepared AP staining solution for 60 min at 37°C (19).
5. Immunofluorescence staining for AP. Air dry 8 μm muscle cryosections. Wash with PBS for 5 min. Block with 20% rabbit serum at room temperature for 30 min. Wash with PBS twice, 5 min each. Incubate with the monoclonal AP antibody (1:8,000, diluted in 1% rabbit serum/PBS) at 4°C overnight. Wash with 1% rabbit serum/PBS three times, 5 min each. Incubate with the Alex 594-conjugated rabbit antimouse antibody (1:100, diluted in 1% rabbit serum/PBS). Wash with 1% rabbit serum/PBS three times, 5 min each. Coverslide with a drop of Aqua-mount (Fig. 5) (19).

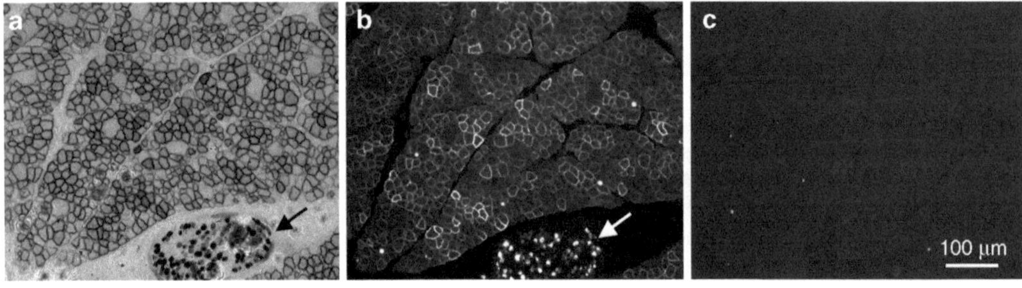

Fig. 5. Detect AP expression on muscle sections. A golden retriever dog was infected with AAV-9 AV.RSV.AP at 48 h of age and AP expression was examined at 6 months of the age. Representative images from serial sections of the cranial sartorius muscle. (**a**) AP histochemical staining. (**b**) AP immunofluorescence staining. (**c**) Immunofluorescence staining with the secondary antibody. *Arrow*, nerve.

6. Quantifying AP expression in muscle lysate. Pulverize ~100 mg muscle tissue in liquid nitrogen using a mortar and a pestle. Resuspend the muscle powder in 700 μL of cell lysis buffer. Incubate the crude lysate at 65°C for 60 min. Centrifuge at 16,000 × *g* for 1 min. Save the supernatant as muscle lysate. Measure the protein concentration with the Bio-Rad protein concentration assay kit. Add 5 μL muscle lysate to 50 μL StemTAG AP activity assay substrate. In another tube, add 5 μL cell lysis buffer to 50 μL StemTAG AP activity assay substrate. (This will be used as the blank for OD measurement.) Incubate at 37°C for 30 min. Add 50 μL 1× stop solution. Measure OD at 405 nm. Calculate raw AP activity using the formula of (OD 405 value × 20)/protein concentration (in μg/μL). Plot the raw AP activity value on the AP activity standard curve to get actual AP activity value (μM/μg) (see Note 14) (3, 19, 20).

4. Notes

1. AAV-9 vector used in our study is purified through three rounds of isopycnic CsCl ultracentrifugation and dialyzed through three changes of HEPES buffer. AAV-9 vectors can also be purchased from the Penn Vector at the University of Pennsylvania Gene Therapy Program (Philadelphia, PA) or the Powell Gene Therapy Center at the University of Florida (Gainesville, FL). It is crucial to check the endotoxin levels of the AAV stocks. We usually use the EndoSafe LAL gel clot test kit (Charles Rivers Laboratory, Wilmington, MA, USA). In general, the endotoxin level should be less than 5 EU/kg. We strongly recommend confirm AAV vector activity in murine models prior to the use for systemic gene delivery in dogs. We would like to point out that while the human placental AP protein is heat-resistant, the canine placental AP protein is heat-labile (21, 22). Besides the AP reporter gene, investigators may also use AAV vectors carrying other reporter genes such as the luciferase gene, the LacZ gene, the ntLacZ gene, or the GFP gene. The differences in the subcellular expression pattern and the assay sensitivity should be considered for data interpretation. The AP protein preferentially localizes at the cell membrane while the ntLacZ protein expresses in the nucleus. For reporter gene study, a ubiquitous promoter (such as the RSV, CMV, CB and CAG) is preferred since it provides information on AAV transduction in nonmuscle tissues. However, for gene therapy in a dystrophic canine model, a muscle specific promoter would be preferred.

2. A female dog usually starts her first heat cycle between the ages of 6–12 months. Most female dogs have their cycle every 6–8 months. When a bitch is in heat, one will notice an increase of urination, swelling of the vulva, and bloody discharge from the vagina. The heat period usually lasts ~20 days. The best time to breed is 48–72 h postovulation. Ovulation usually occurs at ~14 days after the bitch is in heat. A better way to determine the prime breeding time is by monitoring the progesterone level. Ovulation occurs at the progesterone level of 5 ng/mL.
3. We recommend artificial insemination rather than natural breeding. The fresh semen is preferred. However, the frozen semen can also be used. When the bitch is not ready, the semen may either be chilled or frozen. Chilled semen should be used within 24 h. It is important to check the quality of the semen before the use.
4. The gestation period for dogs is usually 62–64 days. The pregnancy can be confirmed at ~28 days into gestation with ultrasound detection of fetal heartbeats. X-rays can be taken around the 60th day of pregnancy to confirm a pregnancy and count the number of fetuses. The whelping date can also be predicted based on vaginal smears. It is usually 57 ± 1 days after the first day of the diestrus.
5. The normal dog body temperature is around 99–101°C. When it drops below 99°C and continues to drop, labor will start within 12–24 h. When it drops to 98°C, labor will occur within 2–12 h.
6. The reverse primer is located at the dystrophin gene intron 7. According to the NCBI sequence database (NC_006621), the underlined nucleotide should be a thymidine. According to Bartlett et al., the underlined nucleotide should be an adenine. In our primer, we used thymidine instead of adenine (12).
7. One can also use color nail polish to paint different claws to mark the puppy. When cutting the cord, make sure to get it from the baby side, not from the placenta side. We suggest use a different pair of sterilized scissors for each puppy. For each puppy, there should be a placenta. The placenta may not come out at the time the puppy is born. If it is retained, the dam will be at the risk of serious uterine infection. It is possible that the number of puppies delivered is different from that predicted by X-ray examination. The signs of difficult labor may include more than half hours of strong contractions without delivering a puppy, more than 4–6 h between two puppies, fail to deliver within 24 h after the temperature drops below 99°C. Colostrum is highly enriched with antibodies, vitamins electrolytes, and other nutrients. If you have

a small litter, you may want to freeze some colostrum for future use.

8. As an example, here we described the PCR protocol for the GRMD genotyping. Based on the dystrophic model you are working with, relevant genotyping assays should be performed. We have found that sometimes the serum CK level may yield an inaccurate diagnosis.
9. The systemic AAV-9 delivery procedure described here has been successfully used in dogs of different body size (such as the large size golden retriever and small size corgi) and in both normal and affected dogs. We have not seen a significant body weight difference between normal and affected puppies at birth. For this reason, we have applied the exactly same technique for all newborn puppies irrespective of the disease status. Although it is easy to monitor expression in nonmuscle tissues with a reporter gene vector, we recommend the use of a therapeutic vector in the case of affected puppies. To help visualize the jugular vein, one may need to place the thumb at the clavicle bone to stop the returning blood (Fig. 3a). The AAV injected puppy usually recovers within 2 min. If there is volume overload, it may lead to lung edema (coughing and increased lung sound on auscultation). If there is an anaphylactic reaction, the puppy will show hives or collapse. It is important to keep the puppy warm and hydrated.
10. We have experienced an adverse reaction in a puppy that received AAV-9 injection (25 μL/g body weight, 2.5×10^{11} vector genome particles/gram body weight) at 24 h after birth. The puppy became lethargic and less responsive after AAV injection. Laboratory test at 36 h post-AAV injection showed an increase of white blood cell count (2.22×10^4/μL; normal, $<1.7 \times 10^4$/μL), mean corpuscular volume (84 fl; normal, <77 fl), mean corpuscular hemoglobin (29.7 pg; normal <24.5 pg), banded neutrophil (1.11×10^4/μL; normal, $<0.3 \times 10^4$/μL), monocyte (6.22×10^4/μL; normal, $<1.35 \times 10^4$/μL), glucose (154 mg/dL; normal <128 mg/dL), potassium (5.1 mM; normal <4.9 mM), anion gap (23 mM; normal <15 mM), and phosphorus (9 mg/dL; normal <5.9 mg/dL). This puppy recovered after supplementary bottle feeding and subcutaneous injection of lactated Ringer's solution.
11. For preweaning puppies, we suggest withhold nursing for 4 h instead of 12 h. During biopsy, we constantly monitor the dog conditions with a veterinary pulse oximeter (Nonin Medical, Plymouth, MN, USA) and a veterinary ECG unit. The first muscle biopsy can be performed as early as 2 weeks post-AAV injection by experienced investigators. Additional biopsy can be performed at 1, 2, 3 months of age prior to necropsy at 6 months of age. If long-term study (>6 months) is planned,

additional biopsy can be performed per experimental needs. We have performed muscle biopsy in many different muscles including the extensor carpi radialis, extensor carpi ulnaris, biceps femoris, vastus laterialis, cranial sartorius, cranial tibialis, long digital extensor, and the lateral head of the gastrocnemius (Fig. 4). In the canine model of DMD, the cranial sartorius is one of the most severely affected muscles (23–25). This muscle is also efficiently transduced by AAV-9 (19). We suggest perform biopsy in the cranial sartorius muscle when testing therapeutic gene therapy in the canine DMD models. Mild swelling and temporary discomfort associated with the surgery may be anticipated. Needle biopsy is another commonly used method (Fig. 6). Although needle biopsy is less invasive, the amount of tissue obtained is smaller and the quality could be less ideal than what one gets from surgical biopsy. A special muscle biopsy needle is required (UCH Skeletal Muscle Biopsy Needle; Cadence Science, Lake Success, NY, USA) (Fig. 6a).

12. A comprehensive whole body necropsy is necessary to fully evaluate systemic gene transfer. We examine skeletal muscle transduction in selected limb, respiratory, and several other body muscles. For limb muscles, both sides are examined. For each limb muscle, collect three samples from the proximal end, middle belly, and the distal end. Following skeletal muscles are harvested including five muscles in the upper forelimb (the long, accessory, medial and lateral heads of the triceps brachii and biceps brachii muscles), four muscles in the lower forelimb (the extensor carpi radialis, extensor carpi ulnaris, flexor carpi ulnaris and superficial digital flexor muscles), nine

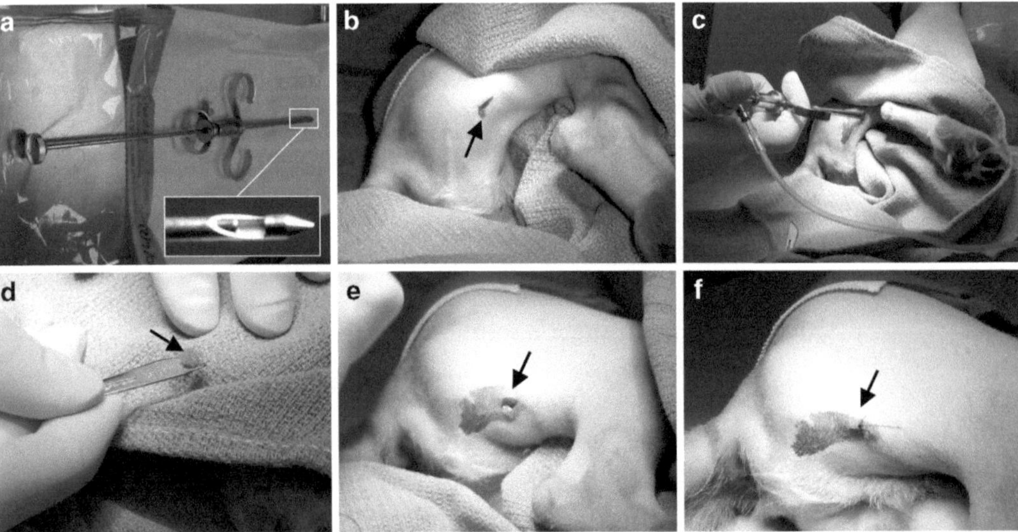

Fig. 6. Needle muscle biopsy. (**a**) Biopsy needle. The *insert* shows a close view of the tip of the needle. (**b**) Skin incision. (**c**) Needle inside the muscle. (**d**) Muscle sample obtained from needle biopsy (*arrow*). (**e**) Skin wound after biopsy. (**f**) Skin is sutured with a single stitch.

muscles in the thigh (the vastus laterialis, vastus intermediates, rectus femoris, biceps femoris, cranial and caudal sartorius, semimembranous, semitendinous and adductor muscles), four muscles in the lower hind limb (the cranial tibialis, long digital extensor and the medial and lateral heads of the gastrocnemius muscles), five head muscles (the occipitalis, frontoscutularis, long auricle levator and temporalis muscles and the tongue), four neck muscles (the sternohyoideus, sternocephalideus, cleidocervicalis and trapezius muscles), four chest muscles (the superficial and deep pectoral muscles, the external and internal intercostal muscles), the diaphragm (including eight samples from peripheral and central parts on the anterior, posterior, left and right sides, respectively), two back muscles (the supra and infra spinatus muscles), four abdominal muscles (the rectus, external oblique, internal oblique and transverses muscles), the gluteus muscle, and the latissimus dorsi muscle. These represent muscles at different anatomic positions in the body. They include both superficial (e.g., the biceps femoris muscle) and deep muscles (e.g., the adductor muscle). They also include muscles with different fiber type compositions. For example, the extensor carpi radialis muscle contains 0–45% type I fiber, the cranial sartorius muscle contains 46–75% type I fiber, and the flexor carpi ulnaris muscle contains 76–100% type I fiber (14). In the case of GRMD, these include muscles that are severely affected (e.g., the rectus femoris, cranial sartorius, cranial tibialis, temporalis, trapezius muscles, the tongue and the diaphragm), moderately affected (e.g., the triceps brachii, biceps brachii, extensor carpi radialis, gastrocnemius muscles), and lightly affected (e.g., the biceps femoris muscle) (23–25). Besides skeletal muscle, we also collect representative tissue samples from the heart (five locations including the left and right ventricles, left and right atriums and interventricular septum), smooth muscles (six locations including the aorta, vena cava, stomach, small and large intestine and bladder) and other internal organs including the brain, retina, lung, liver, spleen, pancreas, kidney, and ovary/testis.

13. Alkaline phosphatases (EC 3.1.3.1) (APs) are dimeric enzymes. Four different APs have been identified in humans including a nonspecific AP and three tissue-specific isozymes (intestinal AP, germ cell AP and placental AP). Only human placental AP is heat resistant. Crystal structure of human placental AP structure has been resolved at the 1.8 Å resolution (26). Comparative studies suggest that human placental AP gene may yield much stronger expression signal than that of LacZ (27, 28). AP positive cells are shaded in purple/blue color. In muscle, AP expression is usually first detected at the sarcolemma and staining at the sarcolemma usually is much stronger than that in the cytosol. When expression level is

low, it may only show up at the sarcolemma. The intensity of histochemical AP staining depends on expression levels. We usually determine the optimal staining time for each study by a pilot experiment. Always include muscle sections from an uninfected dog as negative controls. For muscle sections of the same project, all staining should be performed under the exactly same condition.

14. If the OD reading is higher than 2.0, dilute muscle lysate with water and repeat the assay. Ideally, the OD reading should be in the range of 0.1–2.0. For tissues from an internal organ (such as the lung, liver and kidney), the crude lysate can also be prepared using a tissue tearor. The AP activity standard curve can be obtained with twofold dilution of the AP activity assay standard provided in the kit.

Acknowledgments

The protocols were developed with the grant support from the National Institutes of Health (HL-91883, AR-49419 and AR-57209 to DD), the Muscular Dystrophy Association (DD) and the Parent Project Muscular Dystrophy (DD). We thank Drs. Dietrich Volkmann, Bruce Smith and Joe Kornegay for helpful discussion. We thank Robert J. McDonald, Jr., M.D. for the generous support to Duchenne muscular dystrophy research in the Duan lab.

References

1. Wang, Z., Zhu, T., Qiao, C., Zhou, L., Wang, B., Zhang, J., Chen, C., Li, J., and Xiao, X. (2005) Adeno-associated virus serotype 8 efficiently delivers genes to muscle and heart. *Nat Biotechnol* 23, 321–328.
2. Gregorevic, P., Blankinship, M. J., Allen, J. M., Crawford, R. W., Meuse, L., Miller, D. G., Russell, D. W., and Chamberlain, J. S. (2004) Systemic delivery of genes to striated muscles using adeno-associated viral vectors. *Nat Med* 10, 828–834.
3. Ghosh, A., Yue, Y., Long, C., Bostick, B., and Duan, D. (2007) Efficient whole-body transduction with trans-splicing adeno-associated viral vectors. *Mol Ther* 15, 750–755.
4. Pacak, C. A., Mah, C. S., Thattaliyath, B. D., Conlon, T. J., Lewis, M. A., Cloutier, D. E., Zolotukhin, I., Tarantal, A. F., and Byrne, B. J. (2006) Recombinant adeno-associated virus serotype 9 leads to preferential cardiac transduction in vivo. *Circ Res* 99, e3–e9.
5. Shelton, G. D., and Engvall, E. (2005) Canine and feline models of human inherited muscle diseases. *Neuromuscul Disord* 15, 127–138.
6. Emery, A. E. H., and Muntoni, F. (2003) Duchenne muscular dystrophy. Oxford University Press, Oxford.
7. Flanigan, K. M., von Niederhausern, A., Dunn, D. M., Alder, J., Mendell, J. R., and Weiss, R. B. (2003) Rapid direct sequence analysis of the dystrophin gene. *Am J Hum Genet* 72, 931–939.
8. Bradley, D. M., Parsons, E. P., and Clarke, A. J. (1993) Experience with screening newborns for Duchenne muscular dystrophy in Wales. *BMJ* 306, 357–360.
9. Drousiotou, A., Ioannou, P., Georgiou, T., Mavrikiou, E., Christopoulos, G., Kyriakides, T., Voyasianos, M., Argyriou, A., and Middleton, L. (1998) Neonatal screening for Duchenne muscular dystrophy: a novel semiquantitative application of the bioluminescence test for

creatine kinase in a pilot national program in Cyprus. *Genet Test* 2, 55–60.

10. Parsons, E. P., Clarke, A. J., Hood, K., Lycett, E., and Bradley, D. M. (2002) Newborn screening for Duchenne muscular dystrophy: a psychosocial study. *Arch Dis Child Fetal Neonatal Ed* 86, F91–F95.
11. Duan, D., Yue, Y., Yan, Z., Yang, J., and Engelhardt, J. F. (2000) Endosomal processing limits gene transfer to polarized airway epithelia by adeno-associated virus. *J Clin Invest* 105, 1573–1587.
12. Bartlett, R. J., Winand, N. J., Secore, S. L., Singer, J. T., Fletcher, S., Wilton, S., Bogan, D. J., Metcalf-Bogan, J. R., Bartlett, W. T., Howell, J. M., Cooper, B. J., and Kornegay, J. N. (1996) Mutation segregation and rapid carrier detection of X-linked muscular dystrophy in dogs. *Am J Vet Res* 57, 650–654.
13. Van Belle, H. (1976) Alkaline phosphatase. I. Kinetics and inhibition by levamisole of purified isoenzymes from humans. *Clin Chem* 22, 972–976.
14. Miller, M. E., and Evans, H. E. (1993) Miller's anatomy of the dog. Saunders, Philadelphia.
15. Goody, P. C. (1997) Dog anatomy : a pictorial approach to canine structure. J.A. Allen, London.
16. Budras, K.-D., McCarthy, P. H., Fricke, W., and Richter, R. (2002) Anatomy of the dog: an illustrated text. Schlèutersche, Hannover.
17. Kainer, R. A., and McCracken, T. (2003) Dog anatomy: a coloring atlas. Teton NewMedia, Jackson.
18. Evans, H. E., and DeLahunta, A. (2004) Guide to the dissection of the dog. Saunders, St. Louis.
19. Yue, Y., Ghosh, A., Long, C., Bostick, B., Smith, B. F., Kornegay, J. N., and Duan, D. (2008) A single intravenous injection of adeno-associated virus serotype-9 leads to whole body skeletal muscle transduction in dogs. *Mol Ther* 16, 1944–1952.
20. Bostick, B., Ghosh, A., Yue, Y., Long, C., and Duan, D. (2007) Systemic AAV-9 transduction in mice is influenced by animal age but not by the route of administration. *Gene Ther* 14, 1605–1609.
21. Goldstein, D. J., Rogers, C., and Harris, H. (1982) Evolution of alkaline phosphatases in primates. *Proc Natl Acad Sci U S A* 79, 879–883.
22. Moak, G., and Harris, H. (1979) Lack of homology between dog and human placental alkaline phosphatases. *Proc Natl Acad Sci U S A* 76, 1948–1951.
23. Kornegay, J. N., Cundiff, D. D., Bogan, D. J., Bogan, J. R., and Okamura, C. S. (2003) The cranial sartorius muscle undergoes true hypertrophy in dogs with golden retriever muscular dystrophy. *Neuromuscul Disord* 13, 493–500.
24. Valentine, B. A., and Cooper, B. J. (1991) Canine X-linked muscular dystrophy: selective involvement of muscles in neonatal dogs. *Neuromuscul Disord* 1, 31–38.
25. Nguyen, F., Cherel, Y., Guigand, L., Goubault-Leroux, I., and Wyers, M. (2002) Muscle lesions associated with dystrophin deficiency in neonatal golden retriever puppies. *J Comp Pathol* 126, 100–108.
26. Le Du, M. H., Stigbrand, T., Taussig, M. J., Menez, A., and Stura, E. A. (2001) Crystal structure of alkaline phosphatase from human placenta at 1.8 A resolution. Implication for a substrate specificity. *J Biol Chem* 276, 9158–9165.
27. Bell, P., Limberis, M., Gao, G., Wu, D., Bove, M. S., Sanmiguel, J. C., and Wilson, J. M. (2005) An optimized protocol for detection of E. coli beta-galactosidase in lung tissue following gene transfer. *Histochem Cell Biol* 124, 77–85.
28. Xu, Z., Yue, Y., Lai, Y., Ye, C., Qiu, J., Pintel, D. J., and Duan, D. (2004) Trans-splicing adeno associated viral vector-mediated gene therapy is limited by the accumulation of spliced mRNA but not by dual vector coinfection efficiency. *Hum Gene Ther* 15, 896–905.

Chapter 22

A Translatable, Closed Recirculation System for AAV6 Vector-Mediated Myocardial Gene Delivery in the Large Animal

JaBaris D. Swain, Michael G. Katz, Jennifer D. White, Danielle M. Thesier, Armen Henderson, Hansell H. Stedman, and Charles R. Bridges

Abstract

Current strategies for managing congestive heart failure are limited, validating the search for an alternative treatment modality. Gene therapy holds tremendous promise as both a practical and translatable technology platform. Its effectiveness is evidenced by the improvements in cardiac function observed in vector-mediated therapeutic transgene delivery to the murine myocardium. A large animal model validating these results is the likely segue into clinical application. However, controversy still exists regarding a suitable method of vector-mediated cardiac gene delivery that provides for efficient, global gene transfer to the large animal myocardium that is also clinically translatable and practical. Intramyocardial injection and catheter-based coronary delivery techniques are attractive alternatives with respect to their clinical applicability; yet, they are fraught with numerous challenges, including concerns regarding collateral gene expression in other organs, low efficiency of vector delivery to the myocardium, inhomogeneous expression, and untoward immune response secondary to gene delivery. Cardiopulmonary bypass (CPB) delivery with dual systemic and isolated cardiac circuitry precludes these drawbacks and has the added advantage of allowing for control of the pharmacological milieu, multiple pass recirculation through the coronary circulation, the selective addition of endothelial permeabilizing agents, and an increase in vector residence time. Collectively, these mechanics significantly improve the efficiency of global, vector-mediated cardiac gene delivery to the large animal myocardium, highlighting a potential therapeutic strategy to be extended to some heart failure patients.

Key words: Vector-mediated transgene delivery, Adeno-associated virus, Intramyocardial injection, catheter-based coronary delivery, Cardiopulmonary bypass gene delivery, Gene therapy, Cardiac gene delivery, Large animal myocardium, molecular cardiac surgery, MCARD™

Dongsheng Duan (ed.), *Muscle Gene Therapy: Methods and Protocols*, Methods in Molecular Biology, vol. 709, DOI 10.1007/978-1-61737-982-6_22, © Springer Science+Business Media, LLC 2011

1. Introduction

Clinically relevant, vector-mediated myocardial gene therapy requires efficient transduction of a significant percentage of diseased cardiac myocytes and long-term transgene expression. These requirements may be satisfied through consideration of three essential factors: (a) the vector system, which must demonstrate tropism to the myocardium and the details of the specific vector construct; (b) the therapeutic gene of interest, which ideally must target the underlying biological signaling process implicated in the etiology of heart failure; and (c) the technique of gene delivery (1). Over the past decade, there have been substantial advances in all these areas but achieving high-efficiency, global transgene delivery to the large animal myocardium and subsequently developing a translatable method for use in humans has still proven to be technically challenging.

A number of vector-mediated cardiac gene delivery methods have been explored. The most attractive of them, to date, involves catheter-based, percutaneous infusion of vector into the coronary arteries. The benefits of this technique include its minimal invasiveness, the possibility of transgene delivery to the whole myocardial territory, and the delivery of vector genomes using endovascular coronary catheterization – a procedure for which there is established clinical experience. These advantages notwithstanding, efficiency of intracoronary delivery is highly variable among studies, and the short residence time of vector within the coronary circulation is a major deficiency. Attempts to resolve this shortcoming have employed a number of modifications, including concomitant coronary vein occlusion (2), coronary artery and sinus occlusion (3–5), pressure-regulated retro-infusion of coronary veins during coronary artery occlusion (6), and sequential cross-clamping of the aorta and pulmonary artery while executing catheter-based infusion via the aortic root on the beating heart (7). Still, none of these attempts have demonstrated satisfactory levels of clinical relevance nor globally efficient, transgene expression in the large animal myocardium.

Surgical methods for vector-mediated myocardial gene transfer, therefore, have emerged as a possible solution to the global myocardial gene delivery problem. With subtle technical variations, the basic fundamental of each surgical technique involves intracoronary vector infusion, following cardioplegic arrest during cardiopulmonary bypass (CPB). Collectively, studies have revealed that it is feasible to use CPB for global myocardial gene delivery (8–10), and that mild to moderate body hypothermia and cold cardioplegic solution do not impair gene transfer. In fact, relative myocardial ischemia may increase endothelial permeability and enhance the efficiency of gene transfer (9–12).

Further, the CPB technique can be configured such that it significantly reduces extracardiac transgene expression (13).

Bridges et al. (10) and Davidson et al. (8) first hypothesized that CPB may facilitate cardiac-selective gene transfer using recombinant replication-deficient adenovirus. However, unlike Davidson et al. (8) who utilized a single-pass perfusion technique, Bridges et al. created an isolated "closed loop" recirculating model of vector-mediated cardiac gene delivery in the large animal heart using CPB with an antegrade delivery approach, allowing for vector recirculation for 20 min (10). Although transfection efficiency was improved with this technique, it was later demonstrated that complete surgical isolation of the heart in situ, using CPB with high-pressure retrograde coronary sinus infusion with multiple-pass recirculation of vector through the heart results in an increase of several orders of magnitude in cardiac marker gene activities compared with controls, receiving retrograde infusion of adenovirus without CPB and without cardiac isolation (13). These results validate this surgical technique as a potentially translatable approach for cardiac gene therapy appropriate for use in clinical practice.

Success in global, efficient and cardiac-specific myocardial transduction via retrograde infusion notwithstanding, the intrusiveness and attendant morbidity of CPB remains as relative disadvantages of this surgical gene delivery method. These limitations, however, would presumably be mitigated by the fact that a significant subset of patients for whom the technique may be ultimately utilized would likely be undergoing CPB for other reasons (such as valve replacement, coronary artery bypass surgery or left ventricular assist device placement). Additionally, modern advances in robotic cardiac surgery and other minimally invasive approaches add great promise to this surgical cardiac gene therapy delivery platform as it is further developed.

Here, we detail a novel surgical technique which integrates a closed retrograde transcoronary sinus transgene delivery system utilizing CPB with the creation of a dual perfusion circuitry that isolates the cardiac circuit from that of the systemic circuit and allows for a multiple recirculation sequence. This technique, succinctly referred to as molecular cardiovascular surgery with recirculation delivery (MCARD™), holds the advantages of (1) increased vector residence time in the cardiac circuit, (2) decreased biodistribution of vector genomes in the systemic circulation, and (3) increased perfusion pressure and microvascular permeability in the transcoronary venous system. Cumulatively, these advantages effectively maximize the biodistribution of vector genomes in the isolated cardiac circuit available for vector-mediated cardiac-specific gene therapy, resulting in global and efficient gene expression in the large animal myocardium with minimal untoward collateral expression. Here, we have used the sheep model as an example. Nevertheless, the materials and methods outlined here are also applicable to other large animal models.

2. Materials

2.1. Vector

1. rAAV-6.CMV.EGFP (see Note 1).

2.2. Preoperative Care and Preparation

1. Ovine subject of dorsett pedigree (body weight: 35 ± 3.7 kg). Both male and female may be used.
2. Rumen tube (Webster Veterinary, Sterling, MA, USA).
3. Laryngoscope (Welch Allyn Diagnostics, Skaneateles Falls, NY, USA).
4. 20G intravenous catheters (Columbus Serum, Columbus, OH, USA).
5. 20 Injection caps (Columbus Serum).
6. Heating pad on transport table (Owens and Minor, Paulsboro, NJ, USA).
7. Pulse oximeter (Medical Monitors, Parker, CO, USA).
8. Grooming clippers (Columbus Serum).
9. ECG pads (Owens and Minor).
10. Transduction gel (Owens and Minor).
11. Heparin/Saline intravenous solution (1:1,000).
12. Assorted syringes and needles (Columbus Serum).
13. Tourniquet (Columbus Serum).
14. Silk tape-assorted sizes (Columbus Serum).
15. Lacri-lube eye lubricant (Owens and Minor).
16. Anesthesia kiosk (DRE Veterinary, Louisville, KY, USA).
17. Induction mask (DRE Veterinary).
18. Thermometer (Webster Veterinary).
19. Chlorhexadine scrub, solution and alcohol (Columbus Serum).
20. Lidocaine gel/lubrication (Columbus Serum).
21. Sterile gauze (4 × 4) (Owens and Minor).
22. Sterile gloves (Owens and Minor).
23. Sterile drape (Owens and Minor).
24. Scalpel blade (Fisher Scientific, Pittsburgh, PA, USA).
25. 5.5 and 7 Fr triple lumen intravenous catheters (Columbus Serum).
26. Ketamine (100 mg/mL IV).

27. Diazepam (5 mg/mL IV).
28. Buprenorphine (0.3 mg/mL IM).
29. Flumeglumine (50 mg/mL IM).
30. Propofol (10 mg/mL IV).
31. Isoflurane (USP 100 mL).
32. Glycopyrrolate (0.2 mg/mL IM).

2.3. Cardiopulmonary Bypass with Isolated Cardiac Circuit

1. Heparin (1,000 U/mL IV).
2. Fentanyl (0.05 mg/mL IV).
3. Amicar (250 mg/mL IV).
4. Plegisol.
5. Normosol-R intravenous solution.
6. Phenylephrine (10 mg/mL).
7. 25% Mannitol (20 mL vial).
8. Epinephrine (1 mg/mL).
9. Cimetidine (150 mg/mL).
10. Solumedrol (62.5 mg/mL).
11. Lidocaine (20 mg/mL).
12. Bupivacaine (5 mg/mL).
13. Naxcel (1–2 mg/kg IV).
14. K-Penicillin G (1.6–2.3 million units IV).
15. Gentamycin.
16. Esmolol.
17. Benadryl.
18. Hetastarch (500 mL IV).
19. Suture (Covidien, Mansfield, MA, USA). pledgets, umbilical tape, #4-0 pledgeted Prolene on taper needle, #4-0 nonpledgeted prolene on taper needle, #0 Silk ties, #2-0 Silk pop-off on taper needle, #1 Vicryl on taper needle, #2-0 Vicryl on taper needle, #2-0 Nylon on cutting needle, #5 steel sternal wires.
20. Surgical instruments (Aesculap USA, Center Valley, PA) – #10-blade, #15-blade, #11-blade, #3 knife handle, weitlaner, army/navy retractor, Debakey forceps, scissors, mosquito hemostat, needle holder, tubing clamps, rib spreader, suture forceps, Kelly clamp, Schnidt, large right angle, wire cutters.
21. Cannulas/catheters (Medtronic, Minneapolis, MN, USA).
22. Perfusion equipment (DRE Veterinary) – 13 Fr Terumo retrograde catheter, 26 Fr right angle cannula, 26 Fr straight cannula, 12 and 14 Fr arterial catheters, 9 Fr DLP aortic root cannula, centrifugal pump, ¼′siliconized tubing, Bentley bubble oxygenator, siliconized polycarbonate connectors.

23. 10^{14} Genome copies of rAAV-6.CMV.EGFP.
24. Echocardiogram machine (Fisher Scientific).
25. Arterial line monitor, connection apparatus and pressure tubing (Owens and Minor).
26. 16G angio-catheter (Owens and Minor).
27. Miscellaneous equipment (Fisher Scientific) – Two 28 Fr right angle chest tubes, pleurevac, Heimlich valves.
28. 60 μg Vascular endothelial growth factor (VEGF) (Pepro Tech, Rocky Hill, NJ).

2.4. Intramyocardial Injection

1. Ketamine (4 mg/kg IV).
2. Diazepam (0.5 mg/kg IV).
3. Buprenorphine (3 mcg/kg IM).
4. Flumeglumine (1–2 mg/kg IM).
5. Propofol (4–8 mg/kg IV).
6. Isofluorane, USP 100 mL.
7. Glycopyrrolate (0.02 mg/kg IM).
8. Fentanyl (50 mcg patch).
9. Bupivacaine (2 mg/kg).
10. Naxcel (1–2 mg/kg IV).
11. Dexamethasone (0.5 mg/kg IV).
12. Guaifenesin (15 mL PO).
13. I-stat cartridges (CG4+, CG8+ and ACT) (Heska Corporation, Loveland, CO, USA).
14. I-stat machine (Heska Corporation).
15. Suture (Covidien). #2-0 Silk pop-off on a taper needle, #1 Vicryl on a taper needle, #2-0 Vicryl on a taper needle, #2 prolene on taper needle, #2-0 Nylon on cutting needle.
16. Surgical instruments (Aesculap USA) and preoperative assist equipment (Columbus Serum). rib spreader, #11-blade, #10-blade, #15-blade, #3 knife handle, army navy, Debakey forceps, scissors, mosquito hemostats, hemostats, needle holder, (6.5–7.5) Fr endotracheal tube laryngoscope (Welch Allyn Diagnostics), 24–28 Fr chest tube, pleurevac, Rumen tube, triple lumen catheter, 25–27G needles, 1 cc syringes, rectal temperature probe (Webster Veterinary).
17. 10^{13} Genome copies of rAAV-6.CMV.EGFP.

2.5. Postoperative Care

1. Flumeglumine (50 mg/mL IV).
2. Hetastarch (500 mL IV).
3. Normosol-R intravenous solution.

4. Potassium (40 mEq IV).
5. K-Penicillin G (1.6–2.3 million units IV).
6. Gentamicin (360 mg IV).
7. Furosemide (50–100 mg IV).
8. Miscellaneous equipment. Supplemental oxygen via nasal cannula, EKG machine, I-stat machine, CG8+/CG4+ I-stat cartridges, lactic acid monitor, CVP manometer (DRE Veterinary), urine dip stick, rectal temperature probe (Webster Veterinary).

2.6. Euthanasia, Necropsy, and Tissue Harvest

1. Pentobarbitol (100–200 mg/kg).
2. Rib cutters (Aesculap USA).
3. #10-blade, #22-blade (Fisher Scientific).
4. #3 knife handle, #4 knife handle (Aesculap USA).
5. Forceps (Aesculap USA).
6. Dry ice.
7. Mayo and Metzenbaum scissors (Aesculap USA).
8. RNAlater reagent (Qiagen Sciences, Germantown, MD, USA).
9. Tissue-Tek O.C.T. Compound (Sakura-Finetek) (Torrance, CA, USA).
10. Peel-A-Way® disposable tissue embedding molds.
11. 10% Formalin.
12. PBS, 1×.
13. 2-Methylbutane.
14. 70% Ethanol.

2.7. PCR: Tissue Isolation and Polymerase Chain Reaction Sequence

1. DNeasy Blood and Tissue Kit (Qiagen Sciences). Store at room temperature (RT).
2. SYBR Green Supermix (Bio-Rad, Hercules, CA, USA). Store at –20°C.
3. Sheared salmon sperm DNA.
4. dH_2O.
5. GFP-specific polymerase primer (5′-TATATCATGGCCGA CAAGCA(left); 5′GAACTCCAGCAGGACCATGT (right)) (Sigma, St Louis, MO, USA).

2.8. Western Blotting

1. Passive lysis buffer (Promega, Madison, WI, USA). Store at –20°C.
2. 1% protease inhibitor (Pepstatin/Aprotinin) (Sigma). Store at –20°C.
3. dH_2O.

4. 70% Ethanol.
5. BCA protein assay kit (Pierce, Rockford, IL, USA). Store at RT.
6. NuPAGE LDS sample buffer (Invitrogen, Carlsbad, CA, USA). Store at RT.
7. NuPAGE Sample Reducing Agent (Invitrogen). Store at 4°C.
8. NuPAGE MOPS SDS Running Buffer (Invitrogen). Store at RT.
9. 500 μL NuPAGE antioxidant (Invitrogen). Store at 4°C.
10. 10% NuPAGE Bis-Tris gel (Invitrogen). Store at 4°C.
11. Novex sharp prestained protein standard (Invitrogen). Store at −20°C.
12. MagicMark XP Western standard (Invitrogen). Store at −20°C.
13. NuPAGE transfer buffer (Invitrogen). Store at RT.
14. Invitrolon PVDF filter paper sandwich (Invitrogen). Store at RT.
15. Methanol.
16. Western breeze chromogenic western blot immunodetection kit (Invitrogen). Store at 4°C.

3. Methods

3.1. Vector Design and Production

Single stranded and self complementary recombinant AAV-6 vectors, encoding enhanced green fluorescent protein gene (EGFP) under the control of the human cytomegalovirus immediate early enhancer/promoter with a splice donor/acceptor sequence and polyadenylation signal from the human β-globin gene are produced (see Note 1).

1. Seed human embryonic kidney 293 cells at a density of 1.5×10^7 cells/plate in 150 mm Corning tissue culture dishes the day prior to transfection.
2. For each plate, use 12 μg of pXX680 plasmid (the adenoviral helper plasmid), 10 μg of pXR6 plasmid (the plasmid containing AAV2-rep and AAV6-cap), and 6 μg of the ITR containing plasmid pTRUFR for production of single stranded vectors or pTR-CMV-GFP for production of self complementary vectors and for polyethylenimine mediated transfection.
3. Use between 200 and 400 plates for the production of 1×10^{14} DNase resistant viral particles (drp).

4. Three days posttransfection, rinse cells from the plates and pellet by centrifuge at 1,300 × *g*. Then freeze at –80°C.
5. For ten plates of cells, resuspend the pellet in 5 mL of 25 mM Tris-HCl pH8.5, on ice.
6. Sonicate the cells at 26% amplitude for 25 × 1 s bursts with a Brandson digital sonifier on ice.
7. Treat the lysate with 50 μL of leupeptin (10 mg/mL) and 25 μL of DNase I (10 U/μL) for 30–60 min at 37°C.
8. Centrifuge the lysate at 4,200 × *g* for 10 min in a Sorvall legend RT centrifuge.
9. Layer the clarified lysate onto an iodixanol gradient (14), and centrifuge for 1 h at 822,000 × *g* in an SW-70.1 rotor.
10. Remove the recombinant virus from the gradient by needle puncture between the 60 and 40% iodixanol interface.
11. Purify the virus further by FPLC using HiTrap QHP columns.
12. Apply AAV to the column, wash with 10 column volumes of 25 mM Tris-HCl pH 8.5 and elute in 25 mM Tris-HCl pH 8.5 with a linear 25 mM to 1 M NaCl gradient.
13. Set flow speed of the pump to 1 mL/min for all column steps.
14. Collect and dialyze elute against 1 × PBS supplemented with 5% sorbitol.
15. Determine the drp or viral genomes (vg) by performing slot blot hybridization of purified vector genomes and by SYBR Green real-time PCR using the forward primer 5′-ACGTAAACGGCCACAAGTTC-3′and reverse primer 5′-AAGTCGTGCTGCTTCATGTG-3′.
16. Freeze all AAV stocks at –80°C in single aliquots of 1×10^{14} vg particles and thaw during the surgical procedure.

3.2. Preoperative Care and Preparation

All animals are properly handled and receive humane care in compliance with the National Institutes of Health and the institutional guidelines established by the Institutional Animal Care and Use Committee.

1. Transport the ~1-year-old ovine subject of dorsett pedigree from the colony to the preparation room.
2. Clip the forelimbs and prepare for cannulation of the right cephalic vein with a 16G catheter.
3. Secure the catheter with a silk tape.
4. Anesthetize with 4 mg/kg ketamine and 0.5 mg/kg diazepam, delivered IV.
5. Position the sheep in sternal recumbency on transport table.

6. Administer 6 mg/kg propofol intravenously via the cephalic vein catheter.
7. Using a straight-edge laryngoscope and endotracheal suction (for clearance of secretions where necessary), visualize the vocal cords.
8. Once vocal cords are in adequate view, insert the deflated, regular-cuffed 7.5 Fr endotracheal tube (or alternate size depending upon sheep's laryngeal space and vocal cord aperture), lubricated with viscous lidocaine, into endotracheal passage between vocal cords.
9. Inflate the balloon cuff with 20 ml of air.
10. Confirm placement via auscultation and connect endotracheal tube to the mechanical ventilation apparatus, letting 1–2.5% isoflurane.
11. Administer 0.02 mg/kg of glycopyrrolate intramuscularly.
12. Insert the rumen tube, lubricated with lidocaine gel/KY jelly, by feeding the tube gently, but firmly, over the base of the tongue into the esophagus and down into the rumen tract.
13. Confirm placement either via the presence or smell of rumen fluid.
14. Clip the remaining forelimb and hind limbs, inserting 16G catheters in the cephalic and saphenous veins.
15. Cleanse the areas with chlorhexidine scrub followed by chlorhexidine solution. Convert the forelimb catheters to triple lumen catheters using the Seldinger technique.
16. Using Seldinger technique as well, place 7 Fr triple lumen catheter in the left jugular vein, which can be used intraoperatively for medication delivery and preserved postoperatively for CVP monitor.
17. Cannulate the posterior auricular artery of either ear with a 21G catheter to provide arterial line access for postoperative monitor.
18. Place EKG pads on clipped forelimbs and secure in place with silk tape.
19. Administer K-penicillin G 22,000 IU/kg intravenously, slowly-followed by gentamicin 6.6 mg/kg IV 1 h prior to incision.
20. Clip the sheep from neck to groin and prepare to enter the operative suite.

3.3. Cardiopulmonary Bypass with Isolated Cardiac Circuit

1. Once in operative suite, position the sheep in supine position and secure all four limbs.
2. Prep sheep in sterile fashion from neck to knees, using chlorohexidine scrub and solution.

3. Drape the entire sheep in sterile fashion, subsequently exposing the thorax, and right groin into the surgical field.
4. Surgically expose the right femoral artery via cut-down procedure in standard fashion.
5. Insert a 16G angiocatheter into the artery and connect to arterial line apparatus and monitor via pressure tubing.
6. Once arterial line is placed and appropriate waveform and BP are confirmed, secure line using 2-0 silk pop-off suture.
7. Shift to the right carotid artery. Expose the artery via cut-down procedure as it was done with the right femoral artery in standard fashion; in preparation for introduction of the carotid catheterization to facilitate CPB (see Notes 2–5).
8. After securing the delineated right coronary artery from adjacent vascular anatomy, secure with 2-0 Silk tie suture, and cover until bypass configuration begin.
9. Begin preparation of the thorax for exposing the cardiac window via median sternotomy. Exposure is initially gained via electrocautery, beginning the incision at the level of the sternal notch and advancing inferiorly, two finger breadths above the xiphoid process.
10. Advance the bone saw onto the field and proceed with creating the median sternotomy, starting from the superiorly at the level of the sternal notch and proceeding distally to at least two finger breadths above the inferior aspect of the sternal plate, taking care not to divide the most inferior portion of the xiphoid (see Note 6).
11. Reflect the lung parenchyma and enter the pericardium, exposing the midline positioned cardiac window (see Notes 7 and 8).
12. Perform a transepicardial echocardiographic examination to assess initial function of the heart (prebypass period).
13. Place a pledgeted 4-0 Prolene purse string using a rosette of pledgets around the right atrial appendage for superior vena cava (SVC) cannulation (see Notes 9 and 10).
14. Place a 0-silk, heavy suture is doubly around the SVC and connect to a tourniquet.
15. Using pericardial pledgets and 4-0 Prolene suture, a horizontal mattress, place a pledgeted suture on the ascending aorta, approximately 1 cm distal to the aortic root.
16. Ensnare the aorta and pulmonary artery by exclusion using umbilical tape.
17. Place a purse string on the right atrium, adjacent to the atrioventricular groove. This will become the cannulation site for the coronary sinus catheter. Place an additional purse string,

also a rosette of Teflon pledgets, the right atrium near its junction with the inferior vena cava (IVC) to accommodate IVC cannulation.

18. Give heparin (130 U/kg) and check active coagulation time (ACT). Ensure ACT is greater than 400 before proceeding.
19. Bring the systemic lines onto the surgical field.
20. Cannulate the right carotid artery using a 14 Fr cannula.
21. Place the aortic cannula, using a DLP cardioplegic cannula (containing a vent limb).
22. Cannulate the SVC is cannulated using a 26 Fr right angle cannula. These cannulae are then connected to a Y connector and connected to the venous limb of the pump circuit.
23. Place the coronary sinus cannula (retrograde catheter) (see Notes 11 and 12).
24. After placing the SVC cannula, initiate partial CPB.
25. Once the IVC cannula is placed, initial full CPB (see Note 13).
26. Ligate the right azygous and left (hemiazygous) veins with a 4-0 prolene suture (see Note 14).
27. Snare the IVC with a double loop of silk suture.
28. Place a temporary tourniquet around the posterior-descending vein, using 4-0 prolene.
29. Bring the in-flow lines from the coronary reservoir circuit and out-flow lines onto the surgical field.
30. Recirculate the coronary reservoir solution through a stopcock with care to avoid entry of the viral solution into the systemic circulation or onto the field.
31. Place a stopcock on the DLP aortic root cannula. Connect the parallel arm of the stopcock to the cardioplegia lines.
32. Place a purse string suture onto the tip of the apex of the left ventricle. Make a stab in the middle of the purse string. Place a left ventricular vent cannula into the left ventricular cavity and clamp.
33. Place a pledgeted purse string in the right ventricular outflow tract. Then place a right ventricular vent cannula into the right ventricle and clamp. Snare the purse string. Leave the volume in the heart. Construct the cardiac circuit.
34. Connect the left ventricle, right ventricle, and aortic root vents via a Y connector to the venous limb of the cardiac circuit.
35. Connect the arterial limb of the cardiac circuit to the coronary sinus catheter.
36. Cool the heart to 30°C with bypass heat exchanger and the assisted application of iced saline.

37. Reduce aortic flow.
38. Apply the aorta cross-clamp and administer cardioplegia via the aortic root cannula in antegrade fashion.
39. Isolate the coronary circuit by tightening the SVC snare, the IVC snare; cross-clamping the pulmonary artery and tightening the temporary snare around the posterior descending vein (see Notes 15 and 16) (Fig. 1).
40. Remove all excess volume and air from the circuit by flowing forward and discarding any volume in the pop-off reservoir once the heart is decompressed.
41. Flow should resume until the coronary sinus driving pressure equals 70–80 mmHg.
42. Flow averages should be 100–150 ml/min. (The calculated cardiac blood volume is less than 30 cc and the dead space in the catheter is only about 5 ml; so, the intravascular space is saturated with vector solution.)
43. Deliver the virus solution (approximately 10^{14} genome copies of rAAV-6.CMV.EGFP and 60 μg of VEGF165 in total of 30 mL PBS). Allow solution to circulate for 20 min (see Note 17).

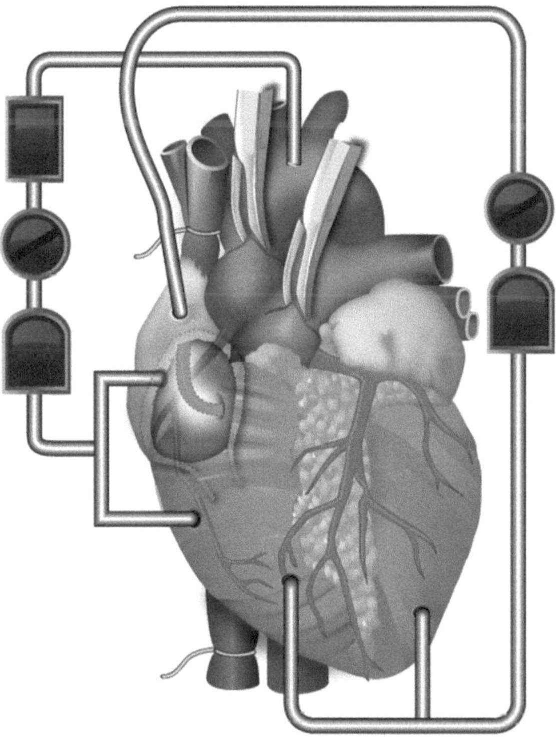

Fig. 1. Cardiopulmonary bypass with isolated cardiac circuit for cardiac-specific gene delivery.

44. Restore flow over 1 min to approximately 100–150 mL/min, with coronary sinus pressure equaling 70–80 mmHg.
45. Remove the coronary sinus catheter.
46. Flush the coronary circuit antegrade with 1 L of Hetastarch.
47. After the flow has been reduced, remove the aortic and pulmonary artery cross-clamps and the IVC, SVC, and pulmonary vein snares.
48. Restore flow and initiate rewarming (see Note 18).
49. Convert the RV and LV cannulae to systemic vents.
50. Administer epinephrine at 1–2 mcg/min.
51. Administer lidocaine 50 mg bolus and then infuse at 1 mg/min.
52. Give 5 mg of amicar IV.
53. Remove the aortic root cannula.
54. Remove RV cannula.
55. Once the heart is contracting well, remove the LV cannula.
56. Wean the animal from bypass.
57. Once off bypass, give protamine, and check ACT.
58. Remove carotid access cannula, and repair the artery using 7-0 prolene suture.
59. Irrigate the pericardial space copiously with warm normal saline.
60. Place bilateral chest thoracostomy tubes, and close the sternal incision with seven to eight 2-0 steel wires in standard fashion (The thoracostomy tubes should subsequently be placed to wall suction through a multiliter pleur-evac.).
61. Repair subcutaneous and skin layers with 1 Vicryl and 2-0 Nylon suture, respectively.
62. Remove the femoral arterial line and repair the artery, primarily using 7-0 Prolene.
63. Repair the overlying fascia and skin with 2-0 Vicryl and 2-0 Nylon, respectively.
64. Disconnect the thoracostomy tubes from the pleur-evac and secure Heimlich valves at end of each, covered with surgical gloves on the end to catch residual output.
65. Transfer the sheep to the postoperative care area after it recovers from anesthesia (see Notes 19–22).

3.4. Intramyocardial Injection

1. Proceed with Preoperative Care and Preparation (see Subheading 3.2).
2. Transport the sheep to the operative suite and place the animal in a right lateral decubitus position.

3. Connect the sheep to the anesthesia kiosk, letting anesthesia with 1–2.5% isoflurane.
4. Connect ECG leads and supportive lines.
5. Prep and drape the operative field in sterile fashion, using chlorhexadine scrub and solution.
6. Enter the chest via a standard left thoracotomy intercostal incision made with electrocautery, approximately 7 cm in length at the fourth intercostal space.
7. Retract the lung parenchyma, bringing the cardiac window into view.
8. Excise the pericardium, exposing the beating heart.
9. Stabilize the operative pericardial window by securing the reflected pericardium with 2-0 Nylon stitch.
10. Evenly distribute approximately 40 injections of rAAV-6.CMV. EGFP into the myocardium of the left ventricle. Each injection should be carefully spaced 1 cm apart and less than 1 cm deep, making sure that the injection is not into the endocardial space. A tuberculin syringe tends to work best (Fig. 2).
11. Inspect the injection field for residual hemorrhage.
12. Loosely approximate the pericardium using 2-0 Nylon.
13. Place a 28 Fr thoracostomy tube and connect to low, continuous wall suction via a multiliter pleur-evac.
14. Repair the thoracotomy incision in usual fashion, using 2-0 Vicryl suture.
15. Extubate in sequential fashion (see Notes 19–24).

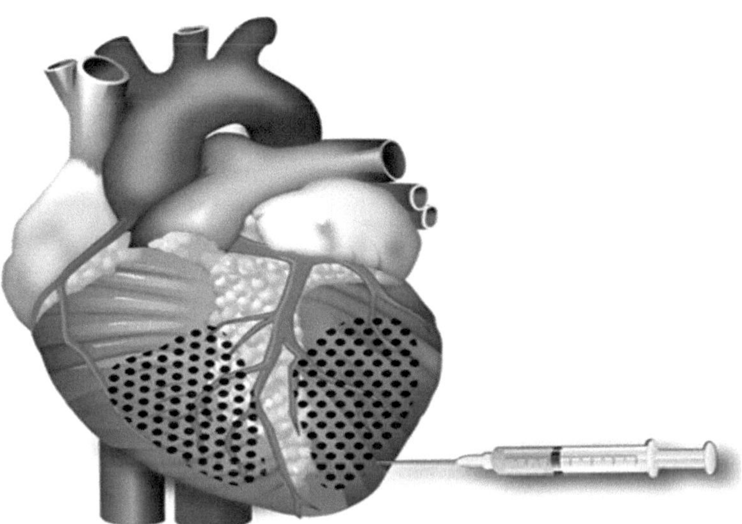

Fig. 2. Intramyocardial injection gene delivery.

16. Discontinue the thoracostomy tube. Repair the defect with a 2-0 Nylon purse-string stitch.
17. Transport the sheep to the postoperative care area and proceed with routine recovery and postoperative management.

3.5. Postoperative Care

The immediate postoperative care is perhaps the most critical component of the investigation, secondary to the surgical procedure itself. It is the vigilant attention paid to the physiologic parameters and aggressive intervention which typically begets favorable outcomes. The primary components of the acute postoperative care include an array of monitoring which provides clues on the animals' viability. Supplemental or supportive medications are used to augment the animals own support mechanisms; including intravenous hydration, antibiotics, analgesics, and nourishment (see Notes 19–26).

1. After the animal recovers from anesthesia, extubate while in sternal recumbency.
2. Transport the animal to the postoperative care area, which should be separate from the animal colony and designated as the Intensive Care Unit.
3. Affix a nasal cannula to the animal's rostrum with 2-0 Nylon suture and connect the cannula to humidified oxygen.
4. Initiate oxygen supplementation at 5 L/min until the sheep is active and able to maintain a pulse oximetry value of 94% or greater.
5. Monitor the animal's physiologic parameters, beginning with measurement of its temperature, pulse, respirations (TPR), and rumen contractions – obtained every 4 h and documented in the animal record.
6. Monitor SaO_2 every 2 h until stable at 94% or greater. Once appropriate saturation is observed, monitor of this parameter can transition to every 4 h until the animal is successfully weaned to room air.
7. Using the jugular central venous catheter, which can be placed during the preoperative preparation (see Subheading 3.2) or intraoperatively, connect a central venous pressure (CVP) manometer to obtain CVP values every 4 h. This will assist in assessing the animal's hydration status. The target CVP value is 4–10 cmH_2O. Normosol-R or hetastarch is used to bolus the animal at 500 cc intervals to maintain a CVP greater than 4. No more than 1,500 mL of Hetastarch should be delivered within the course of 24 h.
8. Use electrocardiography to evaluate for arrhythmias every 4–8 h or more frequently if deemed necessary in cases of concern.
9. If arterial access remains available, obtain an arterial blood gas every 2–4 h to evaluate the animal's acid-base level and to

elucidate electrolytes which may need repletion. In the absence of an arterial access, obtain venous samples from the jugular CVP catheter.

10. Collect urine samples every 12 h for the purpose of evaluating the urine specific gravity and urine ketone levels.
11. Monitor urine output every 2 h.
12. Examine packed cell volume (PCV), hemoglobin, liver, and kidney function parameters and lactate every 4 h until baseline values are obtained.
13. Observe and record excrement production and chest thoracostomy tube output.
14. Administer the following medication. K-Penicillin 1.6 million units IV every 6 h-slow infusion, Gentamicin 360 mg (3.6 mL) IV every 24 h, Flumeglumine (Banamine) 50 mg (1 mL) IV every12 h, and Furosemide 50–100 mg IV prn if CVP is greater than 10 cm H_2O or crackles are auscultated.
15. Supplement with intravenous administration of Hetastarch 500 mL IV, slow infusion (100 mL/h) and Norm-R + 2.5% dextrose + KCl 40 mEq/L, infused until the animal is drinks well or if CVP exceeds 10.
16. Begin nourishment with water when sheep is alert and hay may be given when the animal is standing but neither is offered prior to 6 h, postoperatively.
17. Add 5% dextrose to fluids if the animal remains *nil per os* (NPO) the next morning. Maintain the blood glucose to less than 200 mg/dL.
18. If the animal remains NPO beyond the acute postoperative period, evaluate rumen contents for protozoa activity and transfaunation should be considered with veterinarian guidance.

3.6. Euthanasia, Necropsy, and Tissue Harvest

1. Several weeks after surgical intervention (mean 4–6 weeks), euthanize the sheep by intravenous injection of a lethal dose pentobarbital (100–200 mg/kg).
2. Position the animal in supine fashion.
3. Remove the sternal plate with a pair of rib cutter. Make sure not to damage the underlying organs.
4. Excise the heart and lungs en bloc as one contiguous unit.
5. Keep dissection in a rostral–caudal fashion. Take samples from the following tissues including the left ventricular (LV) apex, LV lateral wall, LV anterior wall, LV posterior wall (take two pieces from two different position), ventricular septum (take two pieces from two different position), right ventricular (RV) free wall, RV outflow tract, left atrium, right atrium, left lung, right lung, left lobe of the liver, right lobe of the liver, spleen, left diaphragm, right diaphragm, left testis, right testis,

left triceps, left intercostals muscle, left quadriceps, and left kidney (right samples of each can be taken as well – sidedness of sample is arbitrary.

6. Extract total RNA. Cut a cube of freshly dissected tissue in the size of 7–8 mm long and less than 5 mm thick. Place the tissue in an RNAlater tube. Incubate at 4°C overnight and transfer to –20°C the next day (see Note 27).
7. Embed tissue for cryosections (see Notes 28 and 29).
8. Extract DNA and protein (see Note 30).
9. Prepare the tissue for whole mount analysis (see Note 31).

3.7. PCR: Tissue Isolation and Polymerase Chain Reaction Sequence

1. Isolate the DNA *from* blood was isolated by *the* High Pure Nucleic Acid Kit (Roche Applied Science) and the DNA from tissue was purified with Qiagen Blood & Tissue DNeasy kit following the protocols. Absorbance at 260 nm (A_{260}) was measured for each DNA sample using the NanoDrop (ND-1000) spectrophotometer. A total amount of 100 ng template DNA per reaction was then taken for q-*PCR* assay.
2. Perform q-PCR analysis in optical 96-well plates using iQ SYBR Green Supermix to monitor double-stranded DNA (dsDNA) synthesis. All samples were analyzed in duplicate or triplicate in volume of 25 μL. Each reaction consisted of 12.5 μL 2× SYBR Green Supermix, 1 μL of specific primer mix, and 100 ng of DNA template.
3. Use a tenfold dilution series of pTRUFR GFP (5 μL containing 1.45×10^8 copies to 1.45×10^3 copies) as the external quantitation standard.
4. Calculate the copy number based on the molecular weight of pTRUFR GFP.
5. Perform PCR for standard curve and unknown samples simultaneously in the same run.
6. Use the MyiQ software to extrapolate the concentration from the standard curve.
7. Perform reactions using the default two-step amplification plus melting curve protocol with GFP specific forward 5′-TATATCATGGCCGACAAGCA-3′ and reverse 5′-GAAC-TCCAGCAGGACCATGT-3′ primers.
8. The reaction conditions should be optimized at: 95°C for 3 minutes; 40 cycles of 95°C for 1 mm and 60°C for 45 s.
9. Begin the melt curve protocol immediately after amplification and have it to consist of 1 min at 55°C followed by 80–10 s steps with a 0.5°C increase in temperature at each step.
10. Generate threshold values for threshold cycle (C_t) determination using the MyiQ software.

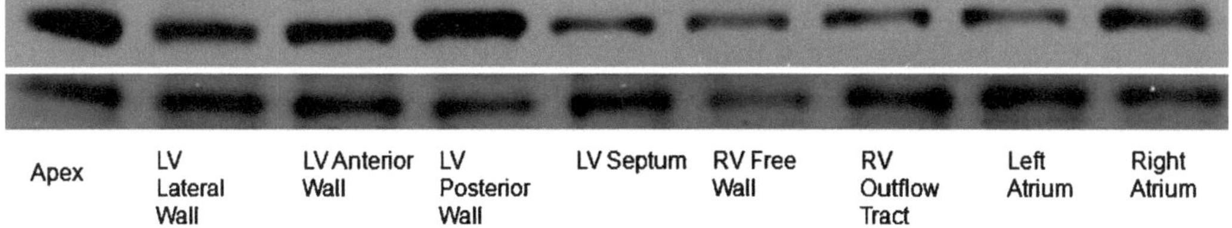

Fig. 3. Representative western blots of GFP from cardiopulmonary bypass with isolated cardiac circuit and intramyocardial injection gene delivery.

3.8. Western Blotting

1. Disrupt frozen tissues (heart, liver, kidney, lung, diaphragm, muscles and testis) using the Omni TH Tissue Homogenizer in RIPA lysis buffer supplemented with PMSF, protease inhibitor cocktail, and sodium orthovanadate.
2. Remove tissue debris by centrifugation at 10,000×*g* for 10 min.
3. Quantify resulting protein extract concentrations by the Bradford method.
4. Load a total of 65 μg of each protein sample in one well and resolve in a NuPAGE 12%-BT gel.
5. Run the NuPAGE gel under reducing conditions in NuPAGE MOPS SDS Running Buffer according to manufacturer's instructions for the XCell SureLock Electrophoresis Cell. Molecular weight standards are 10 μL Novex Sharp Prestained Protein Standard for chromogenic detection or 5 μL MagicMark XP Western standard for chemiluminescent detection.
6. For Western-blot analysis proteins, electrotransfer to PVDF membranes at 40 V and 4°C overnight in XCell II™ Blot Module.
7. Reveal immunodetection by using a polyclonal anti-EGFP and anti-GAPDH antibody with the WesternBreeze™ Chromogenic Kit.
8. Expression level approximations should be taken by comparing the band intensity of the GFP protein to a housekeeping GAPDH gene product by quantitative densitometric analysis (Fig. 3).

4. Notes

1. Though this specific outline utilizes enhanced green fluorescence protein (EGFP) as its reporter gene to resolve the proof of concept, this technique is well-suited and applicable for use with therapeutic vector constructs for genetic modulation of ventricular function in the large animal model.

2. It is absolutely critical that the anesthesiologist closely monitors the animal's cardiovascular stability during prebypass period to avoid systemic vasodilatation with afterload reduction, tachycardia, and attempt to preserve sinus rhythm.
3. Intraoperative monitoring of all essential physiologic parameters is imperative for a successful outcome. These measuring devices include both invasive and noninvasive measuring parameters. Essential monitors should include: pulse oximetry, invasive blood pressure, echocardiography, temperature, and ECG.
4. The operative field is prepped to include the neck (right carotid artery), chest (with places for drainage), and right groin (right femoral artery). Only abdomen and genital region remains completely undisclosed throughout the entirely of the procedure.
5. Intravenous access is established through two peripheral venous lines and two central venous accesses prior to the induction of general anesthesia. Cannulation of the central venous circulation permits rapid central administration of drugs and fluids and measure of CVP. Cannulation of the right femoral artery for blood pressure monitoring and preparation of the carotid artery is essential for ease of initiation of bypass.
6. A midline sternotomy is performed using the bone saw, taking care not to divide the xiphoid process, as it is very easy to damage the stomach and diaphragm. Attention should be devoted to identifying the middle of the sternum, realizing that the sternal anatomy is narrowed in comparison to most other large animals. Deviation from the sternal outline will result in rib splicing or cutting.
7. When accessing the midline exposure, it usually is not necessary to remove the thymic tissue and pericardial fat.
8. Once midline bony structures are exposed, surgical gelfoam is preferred over bone wax because experience has demonstrated that it decreases the incidence of sternal wound infection.
9. The purse string for cannulation should be wide and reliable enough to accommodate the cannula with ease, but not too lax as to cause additional strain on the surrounding cardiac tissue when tightened. Special attention should be devoted to the aortic root purse string because sheep have a very short ascending aorta. For the same reason, care must be taken to observe adequate space for placement of the aortic cross-clamp. In the sheep model, the descending aorta begins with an acute angle only after a few centimeters of the ascending aorta. Care not taken in this placement will result in concomitant cross-clamp of both the ascending and descending aorta.
10. The described sequence of cannulation is preferable because it minimizes bleeding before bypass initiation and, if necessary, allows for quick initiation of artificial circulation.

11. Standard retrograde cannula is used for cannulation of the coronary sinus. Care must be taken to inspect and confirm that the catheter in positioned properly in the coronary sinus. The best results have been received when the catheter tip was placed in midway position after deviation of the posterior descending vein. Placement is confirmed with injection of methylene blue and evaluated by its distribution throughout the major cardiac veins in the right and left ventricles.
12. To avoid coronary sinus rupture, no more than 3 ml of saline is injected into the retrograde catheter balloon.
13. After starting CPB, close attention is paid to the adequacy of venous drainage, the arterial line pressure, the arterial blood oxygenation, and decompression of the heart.
14. A prominent left hemiazygos vein enters on the left side of the sheep heart and drains into the coronary sinus. This configuration is a main difference between sheep and human anatomy of coronary venous system. Ligation of the left hemiazygos vein confirms complete cardiac isolation and prevents systemic loss of viral particles.
15. This technique has been modified to place two snares around each of the IVC and SVC, allowing for more secure control of these major vessels and enforcement of the cardiac isolation. Once the snares are tightened, better results are seen when the heart decompression occurs.
16. Preference has dictated the tendency to snare the vein that runs in the posterior intraventricular groove prior to the vector delivery. This vein is usually just proximal to the coronary sinus balloon. Ligation of this vein is preferred since it drains proximal to the coronary sinus balloon and thus prevents decompression of the coronary venous system into the right atrium during coronary sinus perfusion. The snare is removed after the isolation interval.
17. Vector infusion should proceed in correct port of the retro grade catheter under constant pressure not to exceed 80–100 mmHg over approximately 10 s.
18. Prior to stopping CPB, the patient is rewarmed to 36°C, the heart is defibrillated and the lungs are ventilated, ensuring that intraoperative atelectasis has resolved. Cardiac rhythm is monitored; acid–base status and plasma electrolytes are reviewed. Inotropic drugs are started at low flow rates. After weaning, cardiac filling and contractility is monitored by echocardiography. When cardiac performance is stable, all catheters and cannulas are removed and cardiac inotropic support is ceased. Protamine can be administered in half-doses until ACT is appropriate.
19. The immediate postoperative care phase is critical to animal survival, as is a flawless surgical procedure.

20. When extubating the animal, care is taken to observe jaw motion, eye movement, and axial motion. The animal is positioned in sternal recumbency to prevent aspiration and to encourage recovery.
21. To prevent the thick mucous from building up in the sheep's throat, we take following precautions before the sheep is extubated and rumen tube is removed. These include: (1) administering glycopyrrolate 0.01 mg/kg IM 20 min before coming off of isoflurane; (2) administering dexamethasone 0.5 mg/kg IV at the end of procedure and the total of both doses is not to exceed 50 mg; (3) administering guaifenesin 15 mL PO into rumen through rumen tube, 20 min before coming off of isoflurane.
22. Continuous monitoring of physiologic functions is important because of the possibility of hemodynamic instability. A basic initial hemodynamic assessment includes heart rate and rhythm, CVP, and electrocardiogram to exclude ischemia and conduction abnormalities.
23. Fluid management includes colloid infusions or crystalloid solutions. Pharmacological support (inotropic agents) is administered when necessary based on hemodynamic parameters deemed.
24. Postoperative blood loss from the chest thoracostomy tubes is greater than 100 cc/h; ACT, hemoglobin, and hematocrit levels are checked immediately. One must then consider a surgical source of bleeding.
25. It is not recommended to routinely transfuse blood that has been obtained from another donor sheep routinely due to the possible risk of complications like renal dysfunction and febrile reactions secondary to immune reactivity. In practice, however, the majority of sheep receive 1–2 U of blood from the donor sheep – not to exceed 2 U, to support the animal, particularly in cases of low hematocrit and hypovolemia after CPB.
26. Postbypass, the sheep myocardium has demonstrated increased arrhythmogenic sensitivity (atrial and ventricular fibrillation) Care must be taken when handling or contacting the postbypass heart. Lidocaine and other antiarrhythmic drugs can be used to counter the effects if encountered.
27. As quickly as possible to avoid RNA degradation, cut a cube of tissue around 7–8 mm edge length in half, so that it is no more than 5 mm thick. Place the tissue in RNAlater tube. It can be kept at RT until completion of procedure. It should stay in the 4°C overnight, and then transferred to –20°C the next day.
28. Place a 7–8 mm thick piece of tissue in 5 mL formalin. The samples are kept at RT overnight. The next day, remove the formalin and wash each sample with PBS. After the wash, replace the PBS with 70% ethanol.

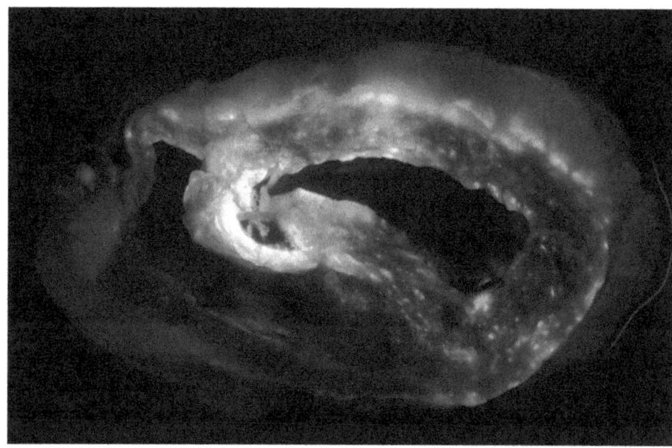

Fig. 4. GFP fluorescence of whole mount cross section of the ovine myocardium at the midpapillary level, including both the left ventricle and the right ventricle.

29. For OCT embedding, prepare a methylbutane/dry ice bath. Fill a mold with OCT, taking care to avoid bubbles. It helps to run the OCT down the side, and to keep the bottle on its side when not in use. Orient a section of tissue in the center of the gel, making sure that it is completely surrounded. If necessary, more OCT is added. Place the mold in the bath such that it cools slowly from the bottom. The OCT will start to turn white. Once it is completely solid, it can be stored in the dry ice container, then the sample is placed in the –20°C freezer.
30. For DNA/protein extraction, fill cryotubes with the remaining tissue, and store in the dry ice container until the completion of the tissue procurement. Store the samples in –80°C freezer until needed for DNA/protein analysis.
31. Whole mount slices from the heart are created by incising a 1.5 cm thick cross section approximately 2.5–3 cm from the base of the apex at the level of the midpapillary muscles, using a nonserrated, straight-edged dissection knife – taking care to include the RV, LV, and intraventricular septum (Fig. 4). The specimen is placed in PBS on ice and later photographed under blue light, viewed through a yellow filter.

Acknowledgments

Acknowledgements are extended to Dr. Marina Sumaroka and Catherine Tomasulo of the University of Pennsylvania, Dr. Joseph Rabinowitz of Thomas Jefferson University, Department of Translational Medicine and Dr. Rose Nolen-Walston and Dr. JanLee Jansen of the University of Pennsylvania School of Veterinarian Medicine.

References

1. Bernecker, O. Y., del Monte, F., Hajjar, R. (2003) Gene therapy for the treatment of heart failure-calcium signaling. *Semin Thorac Cardiovasc Surg* 15, 268–276.
2. Hayase, M., del Monte, F., Kawase, Y., et al. (2005) Catheter-based antegrade intracoronary viral gene delivery with coronary venous blockade. *Am J Physiol Heart Circ Physiol* 288, H2995–H2100.
3. Emani, S. M., Shah, A. S., Bowman, M. K., et al. (2003) Catheter-based intracoronary myocardial adenoviral gene delivery: importance of intraluminal seal and infusion flow rate. *Mol Ther* 8, 306–313.
4. Logeart, D., Hatem, S. N., Heimburger, M., et al. (2001) How to optimize in vivo gene transfer to cardiac myocytes: mechanical or pharmacological procedures? *Hum Gene Ther* 12, 1601–1610.
5. Wright, M. J., Wightman, L. M. L., Latchman, D. S., et al. (2001) In vivo myocardial gene transfer: optimization and evaluation of intracoronary gene delivery in vivo. *Gene Ther* 8, 1833–1839.
6. Boekstegers, P., von Degenfeld, G., Giehrl, W., et al. (2000) Myocardial gene transfer by selective pressure-regulated retroinfusion of coronary veins. *Gene Ther* 7, 232–240.
7. Parsa, S. J., Reed, R. C., Walton, G. B., et al. (2005) Catheter-mediated subselective intracoronary gene delivery to the rabbit heart: introduction of a novel method. *J Gene Med* 7, 595–603.
8. Davidson, M. J., Jones, J. M., Emani, S. M., et al. (2001) Cardiac gene delivery with cardiopulmonary bypass. *Circulation* 104, 131–133.
9. Jones, J. M., Wilson, K. H., Koch, W. J., et al. (2002) Adenoviral gene transfer to the heart during cardiopulmonary bypass: effect of myocardial protection technique on transgene expression. *Eur J Cardiothorac Surg* 21, 847–852.
10. Bridges, C. R., Burkman, J. M., Malekan R., et al. (2002) Global cardiac-specific transgene expression using cardiopulmonary bypass with cardiac isolation. *Ann Thorac Surg* 73, 1939–1946.
11. Jones, J. M., Koch, W. J. (2005) Gene therapy approaches to cardiovascular disease. *Meth Mol Med* 112, 15–35.
12. Ikeda, Y., Gu, Y., Iwanada, Y., et al. (2002) Restoration of deficient membrane proteins in the cardiomyopathic hamster by in vivo cardiac gene transfer. *Circulation* 105, 502–508.
13. Bridges, C. R., Gopal, K., Holt, D. E., et al. (2005) Efficient myocyte gene delivery with complete cardiac surgical isolation in situ. *J Thorac Cardiovasc Surg* 130, 1364–1370.
14. Zolotuchin, S., Byrne, B.J., Mason, E., Zolotuchin, I., Potter, M., Chestnut, K. et al. (1999) Recombinant adeno-associated virus purification using novel methods improves infectious titer and yield. *Gene Ther* 6, 973–985.

Chapter 23

Method of Gene Delivery in Large Animal Models of Cardiovascular Diseases

Yoshiaki Kawase, Dennis Ladage, and Roger J. Hajjar

Abstract

Cardiovascular disease is a major cause of morbidity and mortality in contemporary societies. While progress in conventional treatment modalities is making steady and incremental gains to reduce this disease burden, there remains a need to explore new and potentially therapeutic approaches. Gene therapy, which was initially envisioned as a treatment strategy for inherited monogenic disorders, has been found to hold broader potential that also includes acquired polygenic diseases, such as atherosclerosis, arrhythmias, and heart failure. Advances in the understanding of the molecular basis of conditions such as these, together with the evolution of increasingly efficient gene transfer technology, have placed some cardiovascular pathophysiologies within the reach of gene-based therapy. In fact, gene therapy holds great promise as a targeted treatment for cardiovascular diseases. One of the major hurdles for effective cardiovascular gene therapy is the delivery of the viral vectors to the heart. In this chapter, we will present the various types of delivery techniques in preclinical, large animal models of cardiovascular diseases.

Key words: Cardiovascular diseases, Gene therapy, Minimally invasive, Clinical application, Viral vectors

1. Introduction

Cardiovascular disease is a major cause of morbidity and mortality in contemporary societies. While progress in conventional treatment modalities is making steady and incremental gains to reduce this disease burden, there remains a need to explore new and potentially therapeutic approaches. Gene therapy, for example, was initially envisioned as a treatment strategy for inherited monogenic disorders. It is now apparent that gene therapy has broader potential that also includes acquired polygenic diseases, such as cardiovascular diseases. Advances in the understanding of the molecular basis of conditions such as these, together with the evolution of increasingly efficient gene transfer technology, have

Dongsheng Duan (ed.), *Muscle Gene Therapy: Methods and Protocols*, Methods in Molecular Biology, vol. 709, DOI 10.1007/978-1-61737-982-6_23,

placed some cardiovascular pathophysiologies within the reach of gene-based therapy.

This chapter focuses on gene therapy strategies that have targeted diseases of the cardiovascular system. Gene and vector delivery systems are presented from the perspective of designing a successful cardiovascular gene therapy protocol in large animal models. These strategies are discussed mindful of the constraints of contemporary gene transfer technology and the demands imposed by the pathophysiology of interest. This is done with the aim of instilling in the mind of the readers the likelihood of successful therapeutic intervention with any given strategy.

1.1. Antegrade Injection

Basically, this method mimics the physiological coronary flow pattern. According to the data from the injection of fluorescent microspheres, this method can achieve the most homogeneous distribution of injected material of all the methods (Figs. 1 and 2). Considering the extensive experience in coronary catheterization procedures, this method can be the most ideal for the delivery of vectors. However, in a number of studies, simple intracoronary injection with in vivo models only achieved a very low transduction (1).

Several methods have been used to increase the transduction efficacy of the antegrade injection approach. These include coadministration of drugs, simultaneous occlusion of coronary venous supply, and the use of a circulatory device (2–5). A number of agents that can increase the permeability of the vessel have been used to increase the efficacy of transduction (2, 3).

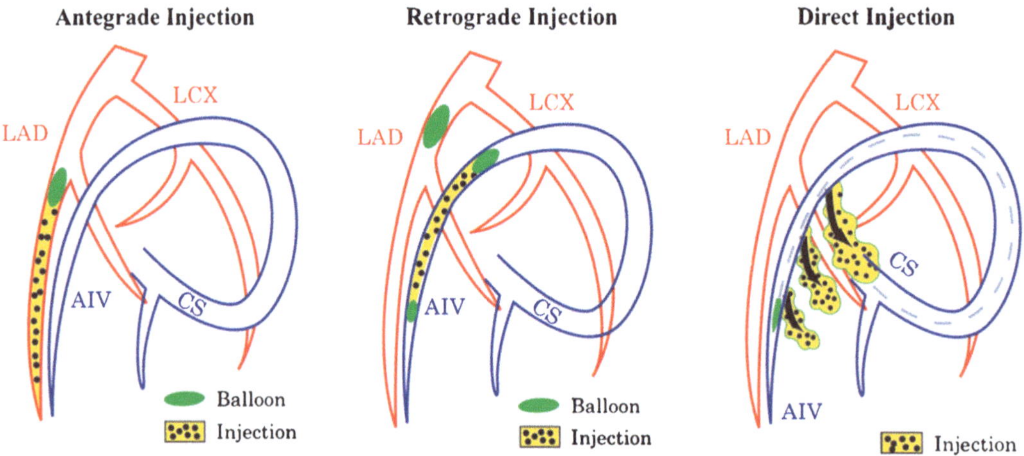

Fig. 1. The depiction of catheter-based three delivery strategies (modified from Hoshino et al. (12)). *Left*: Antegrade injection via coronary arteries. *Center*: Retrograde injection via coronary veins. *Right*: Direct injection via the coronary sinus. *LAD* left anterior descending coronary artery; *AIV* anterior interventricular coronary vein; *LCX* left circumflex coronary artery; *CS* coronary sinus; *IVUS* intravascular ultrasound imaging system.

Hayase et al. (4) showed that the concomitant blockage of the coronary artery and venous system is superior to the single occlusion of coronary artery for the gene transfer.

The use of the circulatory device is another valid approach (5). Similar to the other organs, cardiac circulation is supplied through the coronary arteries and returned to the venous system through the coronary sinus. The V-Focus delivery system allows for a more invasive, percutaneous procedure to establish isolated regional perfusion between the coronary arteries and the coronary sinus (6). The concept of this circulation device is to minimize the exposure of the therapeutic agent into the systemic circulation. If closed circulation is firmly established with this device, theoretically the only unavoidable loss from the circuit occurs via the Thebesian veins, which communicate directly with the cardiac chambers and not with the coronary sinus. In most patients, the blood flow to these vessels is <10% of the total coronary circulation. Careful avoidance of the backflow of infused material from the coronary artery during slow administration and the confirmation of the nonexistence of leakage of blood from the coronary sinus are mandatory.

1.2. Retrograde Injection

This method is using the same conduct of vessels as antegrade injection with retrograde fashion (Fig. 1). Using the coronary vein, which is disease free in most cases, for the route to deliver the therapeutic material is attractive especially in the clinical setting where the patient has a diseased coronary artery. The increase of the efficacy has been achieved by pressure-regulated retroperfusion using a specialized device in this approach (7). However, the distribution of virus from simple occlusion of coronary venous system followed by retrograde injection is limited due to the pressure from antegrade flow. To achieve broad distribution of virus, the concomitant occlusion of the coronary artery to eliminate the effect of the pressure from antegrade flow is prerequisite (8). This physiological limitation somehow diminishes the advantage of this method to use the coronary vein for the safe administration of virus compared to the artegrade approach.

1.3. Direct Injection

In order to improve cardiac gene transfer the existing barriers in biodistribution, cellular uptake and intracellular trafficking have to be taken into account (9). When a vector is applied by a remote or intravenous injection, the first-pass effect of the liver and spleen limits the efficacy of injection. Neutralizing antibodies and T-cell responses further decreases the available virus. Injection into a vessel also requires the virus to cross the endothelial barrier. All of these obstacles can be reduced if the vectors are injected directly into the targeted area.

Direct injection of virus into the myocardium is performed either by surgical approach or percutaneous delivery. The surgical

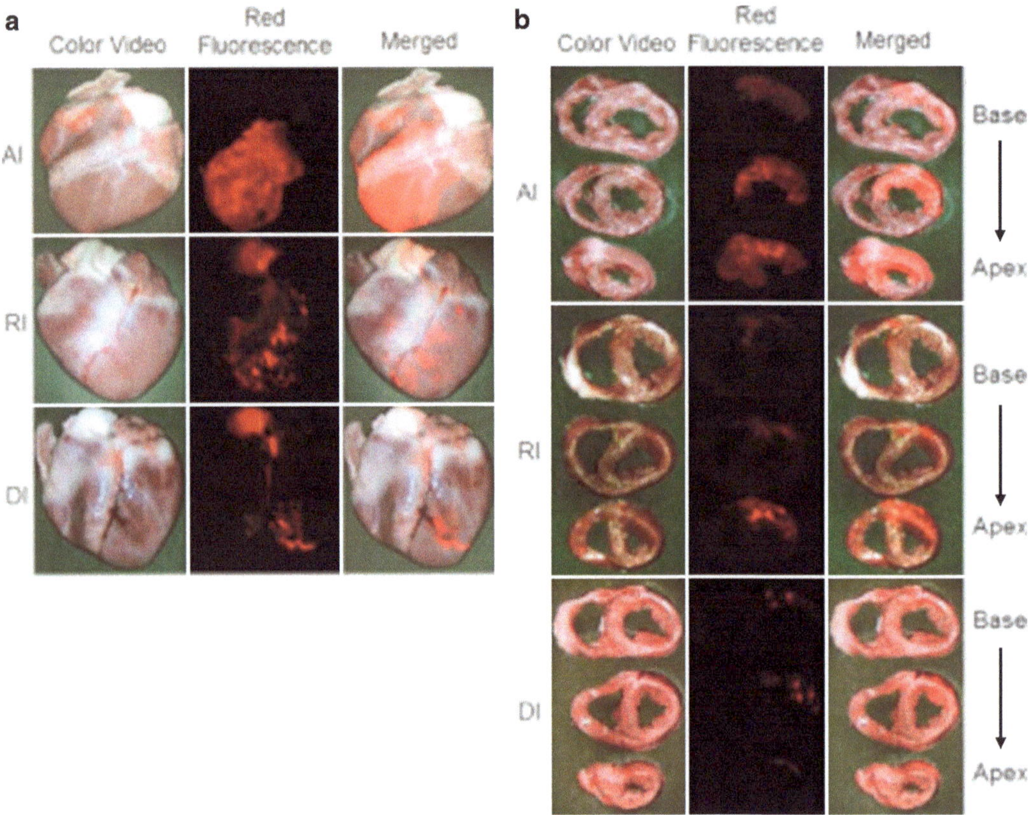

Fig. 2. Distributions of 10-μm red-fluorescent microspheres using three different delivery strategies (modified from Hoshino et al. (12)). (**a**) Surface images of whole hearts. (**b**) Slice images of hearts from (**a**) *Color video* color video images; *Red fluorescence* red fluorescent images; *Merged* merged images; *AI* antegrade injection via coronary arteries; *RI* retrograde injection via coronary veins; *DI* direct injection via the coronary sinus.

approach is considered to be the most invasive approach because of the significant morbidity associated with the thoracotomy procedures to access the heart. There are several catheters that can be used for percutaneous direct injection (10–12). Regardless of the route of access, the leakage of the virus from the injection site and limited distribution of the virus (Fig. 2) have been reported and thought to be the main limitation of this method especially with large animals (13). To avoid this leakage, the amount of solution in each injection and the endomyocardial approach seem to increase the retention of injected therapeutic material in the myocardium (14).

1.4. Pericardial Delivery

Pericardial space has been investigated for the potential delivery pathway for drug, virus, and ablation catheters (15–17). The rationale underlying this approach is the anatomical proximity between the pericardium and the myocardium and the nature of the closed space. The access to the pericardial sac can be achieved either by surgical or percutaneous method (18). Simple administration of the virus into the pericardial sac can only cause superficial

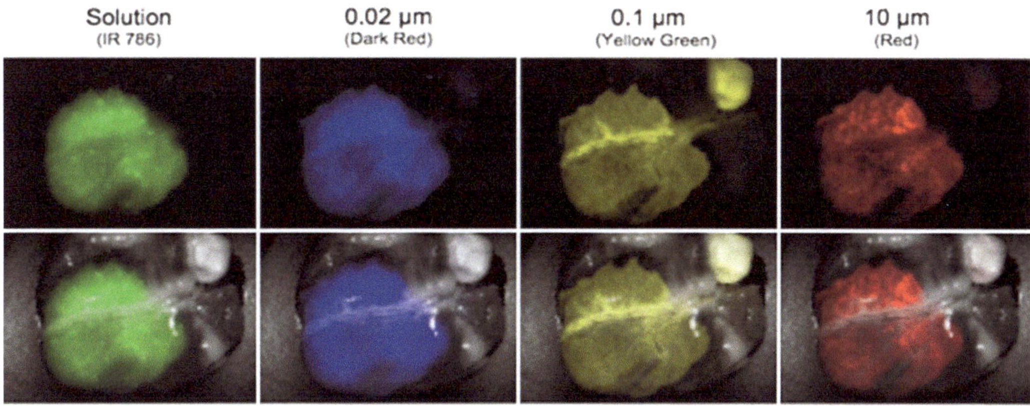

Fig. 3. The difference in homogeneity of the distribution of injected different sizes of microspheres using antegrade injection. Solution, images of the distribution of IR 786 dye; 0.02 µm, images of the distribution of 0.02 µm dark red fluorescent microspheres; 0.1 µm, images of the distribution of 0.1 µm yellow green fluorescent microspheres; 10 µm, images of the distribution of red fluorescent microspheres.

infection. Coadministration of collagenase and hyaluronidase, that disrupt the pericardial cellular and extracellular barriers, can increase the penetration of virus at the cost of increasing the systemic contamination and unfavorable damage to the heart structure (16).

The conceptual figures for the percutaneous antegrade, retrograde, and direct injection are depicted in Fig. 1. The distribution of 20 µm fluorescent microspheres injected using each method is depicted in Fig. 2. The most important information from Fig. 2 is that the distribution of microspheres in retrograde and direct injection is limited compared to that of antegrade injection. The distribution of microspheres in pericardial sac injection is limited to the surface of the heart.

In addition to the administration procedure, the size of injected therapeutic material does matter for the homogeneity of distribution. Using antegrade injection, different sizes of microspheres result in different distribution patterns (Fig. 3). The smaller the size of injected microspheres, the better the homogeneity of distribution of injected microspheres. Of note, the distribution pattern of 0.02 µm of microspheres is almost identical to that of liquid.

2. Materials

2.1. Antegrade Injection

1. Analgesia, anesthesia, and antibacterial drugs that are approved by the animal committee at your facility for all procedures. Heparin Sodium 100 U/kg IV for percutaneous vascular access.

2. A heat therapy pump (T/PUMP) for all procedures (Gaymar Industries, Orchard Park, NY, USA).
3. A mechanical ventilator for large animal for all procedures.
4. A CATH LAB I catheter pack (Medline Industries, Mundelein, IL, USA).
5. A custom manifold kit with high pressure tubing and a syringe (Merit Medical Systems, South Jordan, UT, USA).
6. A guide wire accessory kit with COPILOT (Abbott Vascular, Santa Clara, CA, USA).
7. Contrast agent for angiogram.
8. An introducer sheath for vascular access.
9. An inflation device for a balloon catheter.
10. A guiding catheter for coronary artery access (see Note 1).
11. A 0.014 in. coronary guide wire with soft tip for coronary artery access.
12. An angioplasty balloon for the occlusion of coronary artery. The balloon size should be selected depending on the size of the pig (see Note 2).
13. A guiding catheter for coronary sinus access (see Note 3).
14. A 0.014–0.035 in. hydrophilic wire for the coronary venous access.
15. A balloon catheter for the concomitant occlusion of coronary vein. This is usually larger than the balloon catheter used to occlude the coronary artery. In addition, the lumen should be appropriate in size to allow the passage of the wire selected to engage the coronary vein.
16. An extracorporeal VFocus circuit with oxygenator (VKardia, Minneapolis, MN, USA). A special balloon catheter for coronary sinus occlusion comes with this system.

2.2. Retrograde Injection

1. The material for the coronary artery and vein access is the same as Subheading 2.1.
2. A specialized instrument set for the pressure-regulated retroperfusion (custom made of researcher group in Germany) (7).

2.3. Surgical Direct Injection

1. A thoracotomy kit (4 towel clamps, 2 small rib spreaders, 4 forceps, 2 Langenbeck retractors, 1 Iris Scissors, 1 Mayo scissors, 1 Metzenbaum scissors, 1 needleholder, 1 lung grasping forceps, 1 large right angle forceps, 1 small right angle forceps, 14 hemostats (Miltex, York, PA, USA)).
2. Disposable scalpel (Miltex).
3. CONMED Cautery pen (CONMED Corporation, Utica, NY, USA).

4. 2-0 VICRYL, 2-0 ETHILON sutures (Ethicon, Somerville, NJ, USA).
5. A 24 Fr silicone thoracic drain tube.
6. A 1 mL syringe and a 27G needle with modified L type shape to control the amount and place of injection (see Note 4).
7. Catheters for percutaneous direct injections (10, 11). The same as Subheading 2.1.

2.4. Percutaneous Direct Injection

The same as Subheading 2.3.

2.5. Surgical Pericardial Delivery

1. A thoracotomy kit and surgical instruments (the same kit as Subheading 2.3).
2. An introducer sheath for both surgical and percutaneous approach (the same sheath as Subheading 2.1).
3. An 18G Tuohy needle for percutaneous approach (19).
4. StarClose SE for percutaneous approach (Abbott Vascular).

2.6. Percutaneous Pericardial Delivery

The same as Subheading 2.5.

3. Methods

3.1. Antegrade Injection

1. Access to the femoral or carotid artery and femoral or jugular vein with the Seldinger method (see Note 5).
2. Administer heparin sodium at the dose of 100 U/kg IV.
3. Advance the guide wire with the guiding catheter to artery system. This will prevent accidental damage to the aorta system. Special care should be taken when the guiding catheter passes the aortic arch.
4. After engaging the coronary artery, place a coronary guide wire to increase the stability of the catheter position (see Note 6). For concomitant coronary artery occlusion, advance the angioplasty balloon along with the coronary guide wire.
5. When concomitant venous occlusion or circulatory system is used, advance the guide wire with the guiding catheter to venous system (see Note 7). Artiodactyla has an azygous vein connected to the coronary sinus. For concomitant venous occlusion, select the great cardiac vein with a hydrophilic wire and advance occlusion balloon to great cardiac vein (see Note 8). For circulatory system, the occlusion of the azygous vein is ideal to prevent the contamination of blood flow from an azygous vein using additional balloon or surgical approach. After placing the artery and venous catheters, connect the

catheters to the circulatory device. Occlude the coronary sinus and start the circulatory device. Optimization of pump flow is determined by progressively increasing the roller pump speed to achieve a pump head pressure of –80 to –100 mmHg on the venous side (5).

6. Administer drugs and viral vector from the side port of the Y-connector. Mixture of heparinized blood into vehicle can prevent the ischemic damage to the heart.

   In case of concomitant venous occlusion, occlude the artery and the vein with an occlusion balloon at the same time. Then inject therapeutic material and wait for several minutes. In case of the circulatory device, administrate the therapeutic material from the administration port of the circulatory system.

7. Flush drugs and viral vector by the mixture of heparinized blood and vehicle.

   In case of concomitant venous occlusion, aspirate injected therapeutic material from the tip of coronary venous balloon. In case of the circulatory device, collect blood into a disposal back to minimize the systemic contamination of therapeutic material.

8. Withdraw the guiding catheters, the balloons, and the wires.
9. Withdraw the sheaths from the artery and the vein and achieve hemostasis by applying direct pressure to the site for several minutes (see Note 9).

3.2. Retrograde Injection

The method for the coronary artery and vein access is the same as Subheading 3.1.

1. After placing the artery and venous catheters, occlude the coronary artery first.
2. Occlude the coronary vein.
3. Administer drugs and viral vector from the lumen hole of the venous occlusion catheter.
4. Wait several minutes and retrieve injected drugs or viral vector from the lumen hole of the venous occlusion catheter.
5. Deflate both balloons and withdraw catheters.
6. Withdraw sheaths and achieve hemostatis.
7. For the pressure-regulated retroperfusion, please refer to the original paper (7).

3.3. Surgical Direct Injection

1. The animal is anesthetized, intubated, and placed on the ventilator. After placing the animal in right lateral recumbency, the thoracic area is clipped and scrubbed in preparation for surgery.

2. Perform intercostal nerve block with an approved local anesthesia drug on left side of the chest.
3. Create ~10 cm incision on the skin. Achieve hemostatis using a cautery pen.
4. Locate the fourth intercostal space (see Note 10).
5. Enter the fourth intercostal space and expand the surgical field by using the small rib spreaders. Pay special attention not to damage the lung and pericardial sac (see Note 11).
6. Open up the pericardial sac and expose the heart.
7. Inject drugs or viral vector using 1 mL syringe and a 27G L-shaped needle. Maximum 100 uL/1 injection site is ideal to minimize the leakage from injection site (see Note 12). Continuously monitor ECG throughout the procedure since multiple direct injections sometimes cause ventricular fibrillation.
8. Create ~3 cm skin incision over the sixth intercostal space area and use 2-0 ETHILON to place a suture to close the opening when the drain is pulled.
9. Insert the silicon thoracic drain tube through this incision into the thoracic space.
10. Close the intercostal space and the muscle layer using 2-0 VICRYL.
11. Close the skin layer using 2-0 ETHILON.
12. Aspirate air from the chest cavity using the silicon thoracic drain tube.
13. Withdraw the silicon thoracic drain tube while keeping a negative pressure and close the incision using 2-0 ETHILON.

3.4. Percutaneous Direct Injection

1. For the percutaneous approach, the method for the access of coronary artery is the same as Subheading 3.1.
2. Advance a specialized catheter for this procedure through the arterial sheath (10, 11).
3. Identify damaged or ischemic area using a corresponding device for each injection catheter.
4. Navigate the injection catheter to the target site and perform injection. Maximum 100 uL/single injection site is ideal.
5. Withdraw catheters and sheaths.
6. Achieve hemostatis.

3.5. Surgical Pericardial Delivery (Surgically and Percutaneously)

1. The surgical procedure is the same as Subheading 3.3 up to exposure of the pericardial sac.
2. Create a purse string suture with 2-0 silk in the top layer of the pericardial sac.

3. Make a small incision in the pericardial sac within the area of the purse string suture.
4. Insert a sheath dilator into the pericardial space (see Note 13).
5. Inject drugs and viral vectors through this dilator.
6. While tightening the purse string suture, withdraw the dilator. Completely close and tie the purse string.
7. The rest of the surgical procedure is the same as Subheading 3.3.

3.6. Percutaneous Pericardial Delivery

1. Perform a skin puncture with a 18G Tuohy needle using a subxiphoid approach (19).
2. Confirm the direction of the needle with both A-P and LAO views using fluoroscopy (see Note 14).
3. Use contrast agent to check the position of the tip of the needle and advance the needle under fluoroscopy guidance.
4. Puncture the pericardial sac and confirm it with the injection of contrast agent or a wire (see Note 15).
5. Place a sheath using the Seldinger method.
6. Inject drugs and viral vectors through this sheath.
7. Close the hole on the pericardial sac using StarClose SE (see Note 16).

4. Notes

1. A Hockey Stick catheter is suitable for the femoral approach and an Amplatz Right catheter is suitable for the carotid approach.
2. The shorter balloon is preferred to minimize the damage of the endothelium from the expansion of the balloon. Usually 1–2 atm is enough to occlude the coronary artery.
3. An Amplatz Left catheter is suitable for the femoral approach to engage the coronary sinus and Judkins Right is suitable for the jugular vein approach.
4. The L-type shape can be made with forceps by bending the needle around 5 mm from the tip. This shape prevents the needle from being inserted into the left ventricular cavity yet delivers the virus far from the puncture hole by advancing the needle tip parallel to the ventricular wall.
5. The femoral and the carotid arteries can also be accessed by a cut down method.

6. The administration of viral vector into coronary artery looks simple, but it is tricky especially in some animals. In pig and sheep, the left coronary artery system usually has a shorter left main trunk, so the selective engagement of the guiding catheter is necessary in each coronary artery. In most cases (78%), pig has a dominant RCA, supplying the posterior septum and the atrioventricular node via the posterior descending coronary artery (20). In pig coronary artery, the position of the inserted catheter is sometimes difficult to keep because of the anatomical problem. The stabilization of the catheter with a coronary guide wire and holding of the catheter itself is sometimes mandatory. Deep engagement of the catheter sometimes causes an obstruction of the coronary artery and formation of thrombus. A large amount of heparin administration, the continuous monitoring of the catheter position under fluoroscopy, and checking ECG for signs of ischemic change are important.
7. The coronary sinus is located between the atrium septum and the right ventricle. An injection of contrast from the coronary artery will help to detect the location of the coronary sinus.
8. To prevent the perforation of the coronary vein, forming a loop with the tip of the wire is recommended.
9. You can use 1–2 mL of Protamine Sulfate IV over 5 min to accelerate hemostasis. Although the frequency is low, Protamine Sulfate can cause shock or hypotension. Continuously monitor the signal of the pulse oximeter to catch this phenomenon. If you lose the signal of the pulse oximeter, check the color of the nipples or mucous membranes to identify if the pig is really in a shock. If the color is white, inject 1–2 mL of atropine sulfate IV or, in severe case, inject 0.1–2 mL of epinephrine IV. With timely administration of these drugs, the pig will recover without any adverse effect.
10. Depending on the position or the size of the heart, the fifth intercostal space may be more suitable for this approach. If fluoroscopy is available in the operation room, visualization of the silhouette of the heart helps to determine an appropriate access site.
11. When dissecting the intercostal tissue, be careful not to damage the internal thoracic artery.
12. The parallel insertion of the needle tip is ideal to prevent the perforation of the left ventricular wall, yet it allows the viral vector to be delivered a long distance from the injection site.
13. Use a wire that comes with the sheath set. With this as a guide, you can deliver the dilator to the place where you want to place the catheter.

14. It is recommended to stop mechanical ventilation during this procedure to avoid the damage of the lung by the needle.
15. You usually feel that you puncture through the membrane twice. (The first one is the pleura and the second one is the pericardial membrane.) When you inject contrast to confirm the position of the needle tip, make sure that the tip of the needle is not inside the left ventricular muscle layer. It can cause ventricular fibrillation.
16. Make sure that the tension on the pericardial membrane is not too much or too little using fluoroscopy. Too much of tension can rupture the pericardial sac while too little tension leads to the failure of the placement of the closure device in an appropriate position. Leave pig on its back for a while to prevent dislodgement of the closure device.

References

1. Logeart, D., Hatem, S.N., Heimburger, M., Le Roux, A., Michel, J.B., Mercadier, J.J. (2001) How to optimize in vivo gene transfer to cardiac myocytes: mechanical or pharmacological procedures? *Hum Gene Ther* 12, 1601–1610.
2. Suzuki, G., Lee, T.C., Fallavollita, J.A., Canty, J.M, Jr. (2005) Adenoviral gene transfer of FGF-5 to hibernating myocardium improves function and stimulates myocytes to hypertrophy and reenter the cell cycle. *Circ Res* 96, 767–775.
3. Lai, N.C., Roth, D.M., Gao, M.H., et al. (2000) Intracoronary delivery of adenovirus encoding adenylyl cyclase VI increases left ventricular function and cAMP-generating capacity. *Circulation* 102, 2396–2401.
4. Hayase, M., Del Monte, F., Kawase, Y., et al. (2005) Catheter-based antegrade intracoronary viral gene delivery with coronary venous blockade. *Am J Physiol Heart Circ Physiol* 288, H2995–H3000.
5. Kaye, D.M., Preovolos, A., Marshall, T., et al. (2007) Percutaneous cardiac recirculation-mediated gene transfer of an inhibitory phospholamban peptide reverses advanced heart failure in large animals. *J Am Coll Cardiol* 50, 253–260.
6. Preovolos, A.C., Mennen, M.T., Bilney, A., Mariani, J., Kaye, D.M., Power, J.M. (2006) Development of a novel perfusion technique to allow targeted delivery of gene therapy – the V-Focus system. *J Extra Corpor Technol* 38, 51–52.
7. Raake, P.W., Hinkel, R., Muller, S., et al. (2008) Cardio-specific long-term gene expression in a porcine model after selective pressure-regulated retroinfusion of adeno-associated viral (AAV) vectors. *Gene Ther* 15, 12–17.
8. Raake, P., von Degenfeld, G., Hinkel, R., et al. (2004) Myocardial gene transfer by selective pressure-regulated retroinfusion of coronary veins: comparison with surgical and percutaneous intramyocardial gene delivery. *J Am Coll Cardiol* 44, 1124–1129.
9. Muller, O.J., Katus, H.A., Bekeredjian, R. (2007) Targeting the heart with gene therapy-optimized gene delivery methods. *Cardiovasc Res* 73, 453–462.
10. Smits, P.C., van Langenhove, G., Schaar, M., et al. (2002) Efficacy of percutaneous intramyocardial injections using a nonfluoroscopic 3-D mapping based catheter system. *Cardiovasc Drugs Ther* 16, 527–533.
11. Kornowski, R., Leon, M.B., Fuchs, S., et al. (2000) Electromagnetic guidance for catheter-based transendocardial injection: a platform for intramyocardial angiogenesis therapy. Results in normal and ischemic porcine models. *J Am Coll Cardiol* 35, 1031–1039.
12. Hoshino, K., Kimura, T., De Grand, A.M., Yoneyama, R., Kawase, Y., Houser, S., Ly, Q.H., Kushibiki, T., Kurukawa, Y., Ono, K., Tabata, Y., Frangioni, J.V., Kita, T., Hajjar, R.J., Hayase, M. (2006) Three catheter-based strategies for cardiac delivery of therapeutic gelatin microspheres. *Gene Ther* 13, 1320–1327.
13. Anderl, J.N., Robey, T.E., Stayton, P.S., Murry, C.E. (2009) Retention and biodistribution of microspheres injected into ischemic myocardium. *J Biomed Mater Res A* 88, 704–710.

14. Grossman, P.M., Han, Z., Palasis, M., Barry, J.J., Lederman, R.J. (2002) Incomplete retention after direct myocardial injection. *Catheter Cardiovasc Interv* 55, 392–397.
15. Xiao, Y.F., Sigg, D.C., Ujhelyi, M.R., Wilhelm, J.J., Richardson, E.S., Iaizzo, P.A. (2008) Pericardial delivery of omega-3 fatty acid: a novel approach to reducing myocardial infarct sizes and arrhythmias. *Am J Physiol Heart Circ Physiol* 294, H2212–H2218.
16. Vassalli, G., Bueler, H., Dudler, J., von Segesser, L.K., Kappenberger, L. (2003) Adeno-associated virus (AAV) vectors achieve prolonged transgene expression in mouse myocardium and arteries in vivo: a comparative study with adenovirus vectors. *Int J Cardiol* 90, 229–238.
17. Sosa, E., Scanavacca, M., d'Avila, A., Oliveira, F., Ramires, J.A. (2000) Nonsurgical transthoracic epicardial catheter ablation to treat recurrent ventricular tachycardia occurring late after myocardial infarction. *J Am Coll Cardiol* 35, 1442–1449.
18. Hou, D., March, K.L. (2003) A novel percutaneous technique for accessing the normal pericardium: a single-center successful experience of 53 porcine procedures. *J Invasive Cardiol* 15, 13–17.
19. d'Avila, A., Neuzil, P., Thiagalingam, A., et al. (2007) Experimental efficacy of pericardial instillation of anti-inflammatory agents during percutaneous epicardial catheter ablation to prevent postprocedure pericarditis. *J Cardiovasc Electrophysiol* 18, 1178–1783.
20. Weaver, M.E., Pantely, G.A., Bristow, J.D., Ladley, H.D. (1986) A quantitative study of the anatomy and distribution of coronary arteries in swine in comparison with other animals and man. *Cardiovasc Res* 20, 907–917.

Chapter 24

Percutaneous Transendocardial Delivery of Self-Complementary Adeno-Associated Virus 6 in the Canine

Lawrence T. Bish, Meg M. Sleeper, and H. Lee Sweeney

Abstract

Achieving efficient cardiac gene transfer in a large animal model has proven to be technically challenging. Prior strategies have employed cardio-pulmonary bypass or dual catheterization with the aid of vasodilators to deliver vectors, such as adenovirus, adeno-associated virus (AAV) or plasmid DNA. While single-stranded AAV vectors have shown the greatest promise, they suffer from delayed expression, which might be circumvented by using self-complementary vectors. We have recently optimized a cardiac gene transfer protocol in the canine using a percutaneous transendocardial injection catheter to deliver an AAV vector under fluoroscopic guidance. Percutaneous transendocardial injection of self-complementary AAV (scAAV)-6 is a safe, effective method for achieving efficient cardiac gene transfer to approximately 60% of the myocardium.

Key words: Gene therapy, Cardiac, Canine model, AAV, Vector delivery, Adeno-associated virus

1. Introduction

Gene therapy has great therapeutic potential. Adeno-associated virus (AAV) mediated gene therapy has been described extensively in the literature resulting in efficient cardiac gene transfer in small animal models, including mice (1), rats (1), and hamsters (2). In addition to being efficient, AAV-mediated transgene expression is also stable, an advantage over other commonly used vectors such as adenovirus or plasmid DNA (3). However, AAV vectors have the significant disadvantage of delayed expression, taking nearly 1 month to reach full expression (4), which would limit their usefulness in a rapidly progressing cardiomyopathy. This limitation might be overcome by using the recently developed self-complementary AAV (scAAV) vectors, which package a double-stranded genome and thus bypass the need

Dongsheng Duan (ed.), *Muscle Gene Therapy: Methods and Protocols*, Methods in Molecular Biology, vol. 709,
DOI 10.1007/978-1-61737-982-6_24, © Springer Science+Business Media, LLC 2011

for complementary strand synthesis. This approach offers the advantage of faster onset of expression that may also be more efficient than traditional single-stranded AAV (ssAAV) vectors, albeit at the cost of halving the possible size of the expression cassette (5).

While cardiac gene transfer has been extremely successful in small animal models, delivery to the heart in large animal models and humans has proven to be technically challenging. Several delivery methods have been investigated with varying degrees of success using AAV, adenovirus, or plasmid DNA as vectors. Pericardial instillation of vector results in gene transfer that is restricted to the epicardium (6). Direct, transepicardial injection of vector following left thoracotomy allows delivery throughout the left ventricular free wall (LVFW) but is highly invasive and cannot target the interventricular septum (IVS) (7). Infusion of vector into the coronary arteries can lead to efficient gene transfer, but optimal transfer often requires highly invasive cardio-pulmonary bypass (8) or dual catheterization of a coronary artery and vein (9) with the use of potentially dangerous pharmacological vasodilators. Promising preclinical and Phase I/II results have been obtained using NOGA left ventricular electromechanical mapping to guide transendocardial injections of either plasmid DNA or adenovirus via a percutaneously inserted injection catheter (10). Although this method is relatively noninvasive, it requires creation of a 3D map of the heart using very specific and expensive equipment prior to injection. In addition, gene transfer vectors used in these studies have been associated with inflammation and unstable expression in the case of adenovirus and low-efficiency, unstable expression in the case of plasmid DNA (11).

In this chapter, we describe a simple, relatively noninvasive protocol in the canine which results in highly efficient, global cardiac gene transfer by scAAV6 following delivery via a percutaneous injection catheter. This procedure is superior to previously reported large animal cardiac gene transfer techniques because it combines the advantages of direct injection with those of a noninvasive percutaneous delivery system. Because it results in high-level transgene expression throughout LVFW and IVS, this technique could be useful to treat diseases, which affect the heart globally, such as cardiomyopathy. Furthermore, localized delivery is also possible, and importantly can be achieved in poorly perfused areas. This offers a tremendous advantage over vascular approaches in treating regional ischemia, which is important considering that ischemic cardiomyopathy accounts for approximately 40% of all heart failure cases (12). (This Introduction has been adapted from Bish et al. (13)).

2. Materials

2.1. Cardiac Catheterization and Vector Delivery

1. Anesthetic reagents: Preanesthetic (diazepam 0.5 mg/kg IV, hydromorphone 0.05 mg/kg IV, atropine 0.01 mg/kg IV, buprenorphine 0.01 mg/kg IM), induction (propofol 1–2 mg/kg IV boluses to effect, etomidate 0.5 mg/kg IV boluses to effect), maintenance (Isoflurane 2–3% inhaled), intraoperative analgesia (fentanyl citrate 0.005 mg/kg/min), postoperative analgesia (fentanyl citrate 50 μg/h transdermal patch placed prior to surgery), replacement fluid (normal saline 2–4 mL/kg/h IV).
2. Emergency medication: epinephrine (0.002–0.01 mg/kg IV), phenylephrine (0.1–1.0 mg/kg IV), dopamine (10 mcg/kg/min IV), lidocaine (2 mg/kg IV bolus, then 50 μg/kg/min) (see Note 1).
3. Suture (3-0 Vicryl, 4-0 Vicryl, 3-0 PDS, 3-0 silk) (Ethicon, Somerville, NJ, USA).
4. Standard surgical instrument pack, sterile drapes, sterile gowns, sterile gloves.
5. 2% Chlorhexidine surgical scrub.
6. Avanti Introducer, 7 Fr, 5.5 cm (Cordis, Miami, FL, USA).
7. Iohexol (Omnipaque) contrast solution (240 mg Iodine/mL) (GE Healthcare, Princeton, NJ, USA).
8. SR200 MyoCath percutaneous injection catheter (115 cm, 25G needle, 8 Fr, medium curve) (Bioheart, Sunrise, FL, USA).
9. scAAV6 vector (5×10^{11} gc/kg) (see Note 2).
10. Procedure room equipped for fluoroscopy.

2.2. Assessment of Cardiac Gene Transfer Efficiency

1. Sodium barbiturate: 80 mg/kg IV.
2. 4% Paraformaldehyde solution in PBS. Stored at 4°C.
3. 20% Sucrose solution in PBS. Stored at 4°C.
4. Tissue-Tek OCT compound (Sakura Finetek, Torrance, CA, USA).
5. Vectashield 4′-6-diamidino-2-phenylindole (DAPI) mounting media (Vector Laboratories, Burlingame, CA, USA).
6. Cardiac troponin T antibody, goat polyclonal (Santa Cruz Biotechnology, Santa Cruz, CA, USA).
7. Alexa Fluor 555 donkey anti-goat antibody (Invitrogen Molecular Probes, Eugene, OR, USA).
8. Leitz DMRBE fluorescent microscope (Leica, Bannockburn, IL, USA).
9. Micro Max digital camera (Princeton Instruments, Trenton, NJ, USA).

10. Image Pro Plus software (Media Cybernetics, Bethesda, MD, USA).
11. Open Lab software (Improvison, Waltham, MA, USA).
12. 2-Methylbutane (Sigma, St. Louis, MO, USA).
13. 5% Bovine serum albumin (BSA) in PBS.
14. Peel-away embedding mold (22×40 mm) (Electron Microscopy Sciences, Hatfield, PA, USA).

3. Methods

3.1. Cardiac Catheterization and Vector Delivery

All canine studies must follow National Institute of Health guidelines and be approved by the appropriate Institutional Animal Care and Use Committee. Animals should undergo induction of anesthesia with the above listed medications and be intubated, mechanically ventilated, and placed in the left lateral recumbency position. The surgery site should be shaved and scrubbed three times with 2% chlorhexidine solution, and the animal should be draped in a sterile fashion.

1. Make a 5 cm vertical incision in the right mid-cervical region in the location of the jugular groove to expose the right common carotid artery (see Note 3). Mobilize the artery circumferentially, and secure it proximally and distally with 3-0 silk sutures.
2. Perform a right carotid arteriotomy in the portion of the artery between the two silk sutures, and insert the 7 Fr introducer into the artery.
3. Flush approximately 0.5 mL of heparinized blood from the dog through the injection port of the SR200 Myocath injection catheter (Fig. 1) (see Note 4).

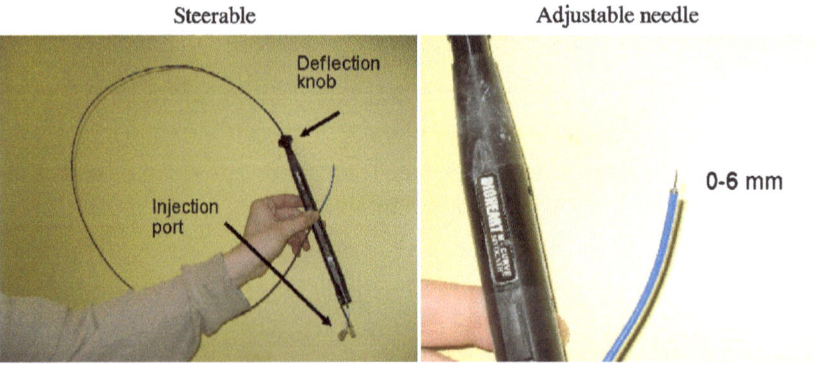

Fig. 1. SR200 Myocath M-Curve Injection Catheter. This figure demonstrates the various features of the Myocath injection catheter including the deflection knob used for steering the catheter, the injection port, and the adjustable length core needle.

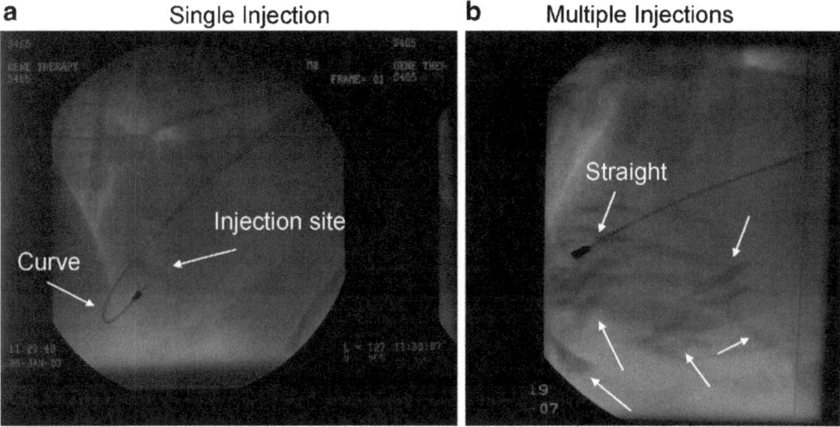

Fig. 2. Delivery of scAAV6 vector via deformable injection catheter under fluoroscopic guidance. (**a**) Note the catheter in the curved position with the needle deployed, and note a single opacity resulting from a single injection. (**b**) Note the catheter in the straight position with the needle retracted, and note multiple opacities following multiple injections. Adapted from Bish et al. (13).

4. Turn on the fluoroscope. Advance the catheter through the introducer into the right common carotid artery, and continue to advance the catheter in a retrograde manner under fluoroscopic guidance until the catheter is visualized in the left ventricle (see Note 5).
5. Prepare the vector solution by mixing 5×10^{11} gc/kg AAV vector and 2 mL Iohexol in a final volume of 15 mL sterile normal saline (see Note 6). Load the virus into three 5 mL syringes for injection. Push 0.5 mL of vector solution into the catheter (injection port) prior to the first injection to account for dead space. Perform 60 injections of 250 μL each under fluoroscopic guidance to target the IVS and LVFW from the base to the apex (see Notes 7 and 8) (Fig. 2). Steer the catheter to these regions of the heart by using the deflection knob (see Note 9) (Fig. 1). Adjust the core needle length (0–6 mm) during injection to target endocardium to epicardium (see Note 10) (Fig. 1). After the last syringe is empty, perform the final two injections by injecting 2× 250 μL saline through the injection port to utilize the vector solution remaining in the catheter dead space.
6. Remove the catheter and introducer. Tie the proximal and distal silk sutures to ligate the right common carotid artery (see Note 11). Close the muscle layer with 3-0 Vicryl sutures. Close the subcutaneous tissue with 4-0 Vicryl sutures and close the skin with a running subcuticular 3-0 PDS suture.

3.2. Assessment of Cardiac Gene Transfer Efficiency

The protocol below will describe assessment of cardiac gene transfer efficiency of the EGFP reporter gene (see Note 12). Our cardiac gene transfer protocol results in transduction of approximately 60% of the myocardium (13).

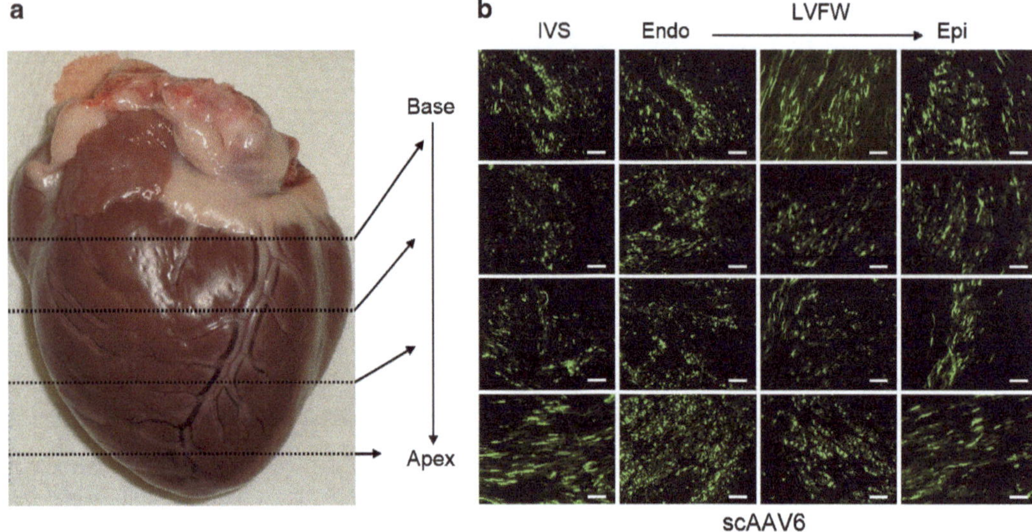

Fig. 3. Analysis of gene transfer efficiency following injection of scAAV6 vector. (**a**) For analysis, the heart was divided into four equal sections through the short axis from base to apex. (**b**) GFP expression in representative cryosections of the heart 7–10 days following injection of 5×10^{11} gc/kg of scAAV6-CB-EGFP. *IVS* interventricular septum; *LVFW* left ventricular free wall; *Endo* endocardium; *Epi* epicardium. *Scale bar*, 200 μm. Adapted from Bish et al. (13).

1. Euthanize dogs 7–10 days following reporter gene delivery by injecting sodium barbiturate (see Note 13). Perform a left thoracotomy or median sternotomy, and explant the heart.
2. Divide the heart into four sections along the short axis from the apex to base (see Fig. 3a), and fix it in 4% paraformaldehyde overnight at 4°C.
3. Wash the tissue 3× 10 min in PBS, and dehydrate in 20% sucrose overnight at 4°C.
4. Place the tissue in a peel-away embedding mold, and fill the mold with OCT embedding compound. Freeze the tissue by submerging it in liquid nitrogen-cooled 2-methylbutane. Prepare 10 μm cryosections.
5. Perform immunofluorescence staining for cardiac troponin T (see Note 14). Wash slides in PBS for 10 min. Block in 5% BSA for 15 min. Incubate 1 h at room temperature in cardiac troponin T antibody at 1:50 dilution (in 5% BSA). Wash 3× 10 min in PBS. Incubate 1 h at room temperature in dark with Alexa Fluor 555 donkey anti-goat antibody at 1:200 dilution (in 5% BSA). Wash 3× 10 min in PBS. Mount slides with Vectashield DAPI media.
6. Visualize GFP and cardiac troponin T fluorescence using, for example, a Leitz DMRBE fluorescent microscope equipped with a Micro Max digital camera interfaced with Image Pro Plus software. For each animal and each wavelength (488 for GFP, 555 for cardiac troponin T), record four representative photographs under constant exposure conditions using the

10× objective lens. Record images from the IVS and LVFW from the endocardium to the epicardium in each of the four heart sections. Quantify the area positive for GFP and cardiac troponin T using the Open Lab software. Set the threshold for detection above background levels that can be measured in an untreated control dog. Calculate the ratio of the area positive for GFP vs. cardiac troponin T, and report this value as the percentage of cardiomyocyte area transduced by the vector. See Fig. 3b for representative results.

4. Notes

1. The most common adverse event associated with the injection procedure is cardiac arrhythmia. In our experience, the arrhythmias have been amenable to lidocaine bolus and infusion and have always resolved by the time the animal has recovered from anesthesia.
2. We have previously identified scAAV6 to be a superior cardiac gene transfer vector via the percutaneous transendocardial delivery route when compared to traditional ssAAV vectors and serotypes 8 and 9 (13). We achieved high levels of cardiac expression of the EGFP reporter gene using the constitutive chicken β-actin promoter with CMV enhancer (CB). Extracardiac expression was relatively minimal compared to the other vectors examined. If tissue specificity is absolutely necessary, a cardiac-specific promoter could be used, although expression levels achieved using the tissue-specific promoters have traditionally been lower than those achieved with the constitutive promoters. It should also be noted that the packaging capacity of scAAV (~2.5 kb) is half that of ssAAV (~5 kb), so larger transgenes may require the use of an ssAAV vector, albeit at the cost of less efficient expression (13).
3. The pulsatile nature of the carotid artery can be used to distinguish it from the internal jugular vein. In this protocol, we describe arterial catheterization to access the left heart. However, if delivery to the right heart is desired, venous catheterization could be performed. In addition, in larger animals (>10 kg), the femoral artery can be used for arterial access.
4. A previous study has shown that flushing the catheter used for gene transfer with heparinized blood can prevent vector inactivation (14). In addition, be sure that all injections are pushed through the injection port, not the sheath port.
5. It can sometimes be difficult to advance the catheter past the aortic valve. If this occurs, repeatedly withdraw and advance

the catheter in order to pass it through the valve during systole when the valve is open.

6. Adding contrast to the vector solution allows the investigator to visualize injection sites in the myocardium under fluoroscopy and to determine where to target future injections.
7. This protocol has been optimized in dogs ranging in size from 5 to 10 kg. As the vector does not spread far from the injection site, a greater number of injections are associated with increased efficiency of global cardiac gene transfer. We have determined 60 injections to be adequate in this size range. A greater number of injections may be necessary in larger animals, although we have not optimized delivery in dogs >10 kg at this time.
8. In this protocol, we describe global cardiac gene transfer. However, the protocol can be modified to target delivery to a specific region of the heart, and the contrast in the vector solution can be used to confirm regional delivery under fluoroscopy. In this case, the total number of injections may likely be reduced based on the proportion of the heart being targeted.
9. When steering the catheter, be sure that the needle is retracted into the sheath to avoid unnecessary damage to the cardiac endothelium.
10. In 5–10 kg dogs, we generally adjust the needle in the 2–4 mm range. Extending the needle too far will lead to a transmural injection into the pericardium, which can be visualized with fluoroscopy. If the needle is not extended adequately, an absence of contrast dye visualization in the myocardium will be evident on fluoroscopy.
11. Canines have extensive collateral circulation. Both carotid arteries and both femoral arteries can be ligated without adverse effects.
12. We have optimized the delivery protocol using the EGFP reporter gene vector. Investigators may choose to perform preliminary experiments with an EGFP reporter vector to assess their proficiency with the protocol before delivering the therapeutic vector.
13. Canines are euthanized at 7–10 days to assess EGFP expression before immune-mediated destruction of transduced cells occurs.
14. We costained the hearts for cardiac troponin T expression so that we could determine the percentage of cardiomyocyte area that was transduced. Any cardiac-specific marker would be suitable for this purpose.

Acknowledgments

This work was funded by a grant from the National Heart, Lung, and Blood Institute (P01-HL059407) (to H.L.S.), the Parent Project Muscular Dystrophy (to H.L.S.), and by T32-HL-007748 (to L.T.B).

References

1. Bish, L. T., Morine, K., Sleeper, M. M., Sanmiguel, J., Wu, D., Gao, G., Wilson, J. M., and Sweeney, L. (2008) AAV9 provides global cardiac gene transfer superior to AAV1, AAV6, AAV7, and AAV8 in the mouse and rat. *Hum Gene Ther* 19, 1359–1368.
2. Hoshijima, M., Ikeda, Y., Iwanaga, Y., Minamisawa, S., Date, M. O., Gu, Y., Iwatate, M., Li, M., Wang, L., Wilson, J. M., Wang, Y., Ross, J., Jr., and Chien, K. R. (2002) Chronic suppression of heart-failure progression by a pseudophosphorylated mutant of phospholamban via in vivo cardiac rAAV gene delivery. *Nat Med* 8, 864–871.
3. Woo, Y. J., Zhang, J. C., Taylor, M. D., Cohen, J. E., Hsu, V. M., and Sweeney, H. L. (2005) One year transgene expression with adeno-associated virus cardiac gene transfer. *Int J Cardiol* 100, 421–426.
4. Gao, G. P., Alvira, M. R., Wang, L., Calcedo, R., Johnston, J., and Wilson, J. M. (2002) Novel adeno-associated viruses from rhesus monkeys as vectors for human gene therapy. *Proc Natl Acad Sci USA* 99, 11854–11859.
5. McCarty, D. M., Fu, H., Monahan, P. E., Toulson, C. E., Naik, P., and Samulski, R. J. (2003) Adeno-associated virus terminal repeat (TR) mutant generates self-complementary vectors to overcome the rate-limiting step to transduction in vivo. *Gene Ther* 10, 2112–2118.
6. Lazarous, D. F., Shou, M., Stiber, J. A., Hodge, E., Thirumurti, V., Goncalves, L., and Unger, E. F. (1999) Adenoviral-mediated gene transfer induces sustained pericardial VEGF expression in dogs: effect on myocardial angiogenesis. *Cardiovasc Res* 44, 294–302.
7. Ferrarini, M., Arsic, N., Recchia, F. A., Zentilin, L., Zacchigna, S., Xu, X., Linke, A., Giacca, M., and Hintze, T. H. (2006) Adeno-associated virus-mediated transduction of VEGF165 improves cardiac tissue viability and functional recovery after permanent coronary occlusion in conscious dogs. *Circ Res* 98, 954–961.
8. Bridges, C. R., Gopal, K., Holt, D. E., Yarnall, C., Cole, S., Anderson, R. B., Yin, X., Nelson, A., Kozyak, B. W., Wang, Z., Lesniewski, J., Su, L. T., Thesier, D. M., Sundar, H., and Stedman, H. H. (2005) Efficient myocyte gene delivery with complete cardiac surgical isolation in situ. *J Thorac Cardiovasc Surg* 130, 1364–1370.
9. Sasano, T., Kikuchi, K., McDonald, A. D., Lai, S., and Donahue, J. K. (2007) Targeted high-efficiency, homogeneous myocardial gene transfer. *J Mol Cell Cardiol* 42, 954–961.
10. Fuchs, S., Dib, N., Cohen, B. M., Okubagzi, P., Diethrich, E. B., Campbell, A., Macko, J., Kessler, P. D., Rasmussen, H. S., Epstein, S. E., and Kornowski, R. (2006) A randomized, double-blind, placebo-controlled, multicenter, pilot study of the safety and feasibility of catheter-based intramyocardial injection of AdVEGF121 in patients with refractory advanced coronary artery disease. *Catheter Cardiovasc Interv* 68, 372–378.
11. Li, J. J., Ueno, H., Pan, Y., Tomita, H., Yamamoto, H., Kanegae, Y., Saito, I., and Takeshita, A. (1995) Percutaneous transluminal gene transfer into canine myocardium in vivo by replication-defective adenovirus. *Cardiovasc Res* 30, 97–105.
12. Baldasseroni, S., Opasich, C., Gorini, M., Lucci, D., Marchionni, N., Marini, M., Campana, C., Perini, G., Deorsola, A., Masotti, G., Tavazzi, L., and Maggioni, A. P. (2002) Left bundle-branch block is associated with increased 1-year sudden and total mortality rate in 5517 outpatients with congestive heart failure: a report from the Italian network on congestive heart failure. *Am Heart J* 143, 398–405.
13. Bish, L. T., Sleeper, M. M., Brainard, B., Cole, S., Russell, N., Withnall, E., Arndt, J.,

Reynolds, C., Davison, E., Sanmiguel, J., Wu, D., Gao, G., Wilson, J. M., and Sweeney, H. L. (2008) Percutaneous transendocardial delivery of self-complementary adeno-associated virus 6 achieves global cardiac gene transfer in canines. *Mol Ther* 16, 1953–1959.

14. Marshall, D. J., Palasis, M., Lepore, J. J., and Leiden, J. M. (2000) Biocompatibility of cardiovascular gene delivery catheters with adenovirus vectors: an important determinant of the efficiency of cardiovascular gene transfer. *Mol Ther* 1, 423–429.

INDEX

Dongsheng Duan (ed.), *Muscle Gene Therapy: Methods and Protocols*, Methods in Molecular Biology, vol. 709, DOI 10.1007/978-1-61737-982-6, © Springer Science+Business Media, LLC 2011

D

E

F

G

H

I

J

L

M

N

O

P

R

S

MIX
Papier aus verantwortungsvollen Quellen
Paper from responsible sources
FSC® C105338

If you have any concerns about our products,
you can contact us on
ProductSafety@springernature.com

In case Publisher is established outside the EU,
the EU authorized representative is:
Springer Nature Customer Service Center GmbH
Europaplatz 3, 69115 Heidelberg, Germany

Printed by Libri Plureos GmbH
in Hamburg, Germany